CONTROL OF PERMAN
SYNCHRONOUS M

Control of Permanent Magnet Synchronous Motors

Sadegh Vaez-Zadeh

School of Electrical and Computer Engineering
University of Tehran

OXFORD
UNIVERSITY PRESS

OXFORD
UNIVERSITY PRESS

Great Clarendon Street, Oxford, OX2 6DP,
United Kingdom

Oxford University Press is a department of the University of Oxford.
It furthers the University's objective of excellence in research, scholarship,
and education by publishing worldwide. Oxford is a registered trade mark of
Oxford University Press in the UK and in certain other countries

First Edition published in 2018

Published in the United States of America by Oxford University Press
198 Madison Avenue, New York, NY 10016, United States of America

British Library Cataloguing in Publication Data
Data available

Library of Congress Control Number: 2017943624

ISBN 978–0–19–874296–8

DOI 10.1093/oso/9780198742968.001.0001

Dedicated to my father, distinguished Professor and renowned scholar
Mohammad Vaez-Zadeh Khorasani (1925–2016)
who inspired me the most

Foreword

It is my pleasure to write the Foreword for this book, *Control of Permanent Magnet Synchronous Motors* by Professor Sadegh Vaez-Zadeh, an Electrical Engineering Professor at the University of Tehran. Professor Vaez-Zadeh has excellent credentials as an author for this book.

It is well known that PM machine drives are becoming increasingly popular over induction motor drives in industrial applications. A PM machine with a high energy magnet is more expensive than an induction motor, but it provides the advantages of higher efficiency, reduced size, and lower inertia that makes its lifetime cost lower. With rising energy costs, this is becoming ever more important.

The book begins with a concise review of power electronic inverters, permanent magnet materials, and PM synchronous motors that includes an informative look at the related market. Then, it discusses PMSM mathematical models as needed for different control methods of the drives. Major drive control methods are then covered in three consecutive chapters. The book not only reviews the traditional vector and scalar control techniques (including direct torque control) as given in standard textbooks, but also covers the recent predictive and dead-beat control techniques. Sensorless control and parameter estimation as salient features of modern motor drives are covered well in the last two chapters of the book. There are problems and references at the end of each chapter. In addition, the book includes practical simulation, design, and performance examples. The book will be very popular among graduate students, researchers, and application engineers in industry, and I recommend it as a text for university students. I look forward to the book's success.

Bimal K. Bose

Emeritus Endowed Chair Professor in Power Electronics

The University of Tennessee, Knoxville, USA

Preface

This book is a timely response to the emergence of controlled permanent magnet synchronous motors as the fastest growing electric drive system in both academic and market arenas. This is to be prepared for the much-anticipated mainstream drive in the near future, despite the fluctuating price of permanent magnet materials. Electric AC drives, as a broad multidimensional system, are an essential part of man's life today in industry, transportation, home appliances, office equipment, aerospace, etc. Electric motor control as the brain of drive systems manages all parts of the system to meet the needs of the drive application. Therefore, motor control has evolved as a field being contributed to by various disciplines, including electric machines, power electronics, microelectronics, control theory, software engineering, new materials, etc. The advancement in the contributing disciplines by unprecedented research and development has led largely to the design of high performance AC motor control systems. The progress in power electronics, mainly the production of high frequency, high power, and low loss devices, and their integration into compact and intelligent power modules, has provided a determining impact on the advancement of motor control systems as the device-switching pattern is a major part of a motor control system. The field has also been considerably influenced by state of the art microelectronic means. Dedicated fast and self-sufficient digital signal processors (DSPs) and field-programmable gate arrays (FPGAs) are the main advancing processing chips now that realize the implementation of motor control systems for manufacturing and research purposes. Professional adaptation of control theories and techniques, including variable and parameter estimation schemes to electric machines, improves the performance of controlled machines and reduces the total cost. Software engineering has played a definite role in developing motor control routines and algorithms, and simulation programs and packages.

High-energy rare earth permanent magnet materials, in connection with innovative motor topologies have widened the opportunities for constructing many useful motor types, accommodating a wider range of performance characteristics. Permanent magnet synchronous (PMS) motors, with their indispensable control system, in particular, have emerged as the fastest growing electric machines in both research and manufacturing due to their inherent high efficiency, compactness, and control durability. The growth comes in an era when the environmental crisis and high energy prices work hand in hand as the strong driving force behind severe measures for energy and material savings which can be achieved by PMS motors. PMS motors, traditionally built in low power ratings, are increasingly being built and used in medium power ratings now, and even are predicted to replace induction motors as the workhorse of industry in the future. In addition to the previously mentioned applications, emerging applications, such as unmanned aircraft systems known as drones, autonomous electric vehicles, and smart buildings built in connection with contactless power transfer and the Internet, are opening an unforeseeable era in the use of controlled PMS motors. Because of the current and emerging applications, PMS motor control has been the focus of increasing research on subjects like modeling with saturation, loss and harmonics included, optimal design, and parameter selection for achieving desirable performances, specific operating modes and control performance enhancements, parameter and position estimation and sensorless control schemes, etc. These, in turn, are demanding more credible, profound, and up-to-date sources of information devoted to controlled PMS motors for academic and industry uses.

A number of excellent books on electric motor control have been published in the past 10 years, each covering the control of different AC motors. However, none of these books focus solely on PMS motors, although some books are devoted to brushless machines in general, including brushless DC (BLDC), synchronous reluctance, and PMS motors. In addition, the books cover plenty of useful topics on design aspects of the machines, but do not allow enough space for control aspects of PMS motors. As a result, many features of PMS motor control systems, including major recent advances, have not been addressed sufficiently. Therefore, a growing need for the treatment of the PMS motor control as a whole has emerged as more researchers and engineers become involved in the field. This book is an attempt to respond to this need by studying PMS motor control in a broader, more detailed, and still deeper sense. The main features of the book can be highlighted as follow:

1. A thorough and unified modeling of PMS motors for control applications in all major reference frames, taking into account iron loss and magnetic saturation, which provides a solid common basis for all other chapters, as frequently referred to in the rest of the book;
2. Systematic presentation and analysis of major control methods for PMS motors including vector control, direct torque control, predictive control, deadbeat control, and combined control in various appropriate reference frames;
3. Analysis and design of control systems to achieve various operating modes of motor operation including maximum torque per ampere, maximum torque per voltage, unity power factor, and minimum loss, under major control methods in various reference frames considering motor and inverter constraints;
4. Treatment of a wide variety of position and speed estimation schemes and sensorless control systems, emphasizing their features and limitations;
5. Identification and estimation of major motor parameters by several offline and online schemes, and the means for incorporating them into motor control systems;
6. About two-hundred figures, each fully described in the text;
7. Many end-chapter problems to serve as complementary resources for comprehending various aspects of the presented materials and for facilitating the book to be used as a text;
8. Lists of important classical works, in addition to selected major recent developments, on PMS motor control in the form of end-chapter bibliographies.

This book coherently investigates about 70 control and estimation systems, thus providing a rather comprehensive account of PMS motor control. Nevertheless, the author has intentionally disregarded much material found in the literature, although they are useful in themselves to prevent this book from losing focus and to avoid a handbook type of publication.

Graduate students, academics, and engineers with a basic background in electric machines, power electronics, and control can use this book. The author's experience is that AC motor control can be taught by teaching the control of PMS motors first. Control of induction and other types of AC motors can then be offered with less effort and more effectiveness. Pursuing this approach, this book can be used as a text for a major part of a graduate course on the control of AC motors. Alternatively, it can be used for a course fully devoted to the control of PMS motors. New researchers in the field, including graduate students, may find this book useful in systematically reaching a level to pursue advanced topics in many branches of PMS motor control. Experienced researchers and faculty members with a background in the field will hopefully find this book an essential resource on their bookshelf. This book has been authored by eyes open to the needs of developers of motor control systems, without going through very technical materials, which are found in motor drive manufacturing resources such as application notes. The author greatly welcomes suggestions for a possible new edition of the book. Readers may interact with the author through the book's companion website, which contains additional material about this book.

S. Vaez-Zadeh
February 2017

Website

Please see the link to the book's companion website at "http://www.oup.co.uk/companion/PMSControl2018" for additional materials, including:

- Accompanying slides for lecturers
- Sample material from book
- Chapter summaries

Acknowledgments

This book, like any other single-authored book, means that the author has worked on every sentence, formula, and figure. However, I would like to sincerely acknowledge my profound debt to the many people who helped pave the way for the book. I must start by naming Professor V. I. John, Queen's University, Kingston, Ontario, Canada; with his great compassion I pursued my PhD thesis on the control of permanent magnet synchronous motors during 1993–1997, after becoming familiar for the first time with motor drives in a graduate course offered by Professor P. C. Sen in 1991 at the same university. I am greatly indebted to Professor M. A. Rahman from the University of Newfoundland, Canada, for his superb advice and generosity in implementing a complete PMS motor control system in his laboratory during the long days of June to October 1996. I am also truly grateful to Professor H. Toliyat from Texas A&M University for his very kind words of encouragement and for suggesting expert reviewers for the book proposal. I appreciate a lot Professor B. K Bose from University of Tennessee for kindly reviewing the proposal, and providing very encouraging feedback and useful suggestions. Professor E. Levi from Liverpool John Moores University is appreciated for kindly accepting to review the proposal. Also, I am thankful to anonymous reviewers selected by the publisher for their constructive suggestions. I owe much to the authors of many publications that are referred to in the book. This book is an outcome of 20 years of reading, teaching, and researching on PMS motor control with my graduate students. I also gratefully acknowledge the assistance of my PhD student, E. Sarani, for arranging the list of references according to the publisher's format. The agreement of IEEE to allow me to reuse figures from my papers published in its transactions and conference proceedings is acknowledged.

I am in much debt to the School of Electrical and Computer Engineering, University of Tehran, for giving me the opportunity to spend as much time as needed in preparing this book. The support of Iran National Science Foundation is also appreciated. The staff at Oxford University Press have made an excellent effort to professionally deliver the book. Of these, I would especially like to thank the Senior Commissioning Editor, Ms. S. Adlung and the Assistant Commissioning Editor, Ms H. Konishi, for their kind support.

Finally, I appreciate the patience and understanding of my wife Fereshteh Aghati, my son Mohammad, and my lovely daughters Maryam and Mahya during the long period of authoring the book.

Contents

Glossary of Symbols

$\boldsymbol{A}(\boldsymbol{F})$	system matrix
$\boldsymbol{B}$	input matrix
B	viscous coefficien
B_{max}	maximum flux density
$\boldsymbol{C}(\boldsymbol{H})$	output matrix
$\boldsymbol{D}$	disturbance matrix
e_a, e_b, e_c	total back EMFs (total induced voltages) in stator phases
e_d, e_q	total d- and q-axis EMFs; d- and q-axis induced (speed) voltages
$\bar{e}_i$	induced (internal) voltage space vector
E_m	magnitude of sinusoidal permanent magnet back EMFs at steady-state
e_{ma}, e_{mb}, e_{mc}	phase winding permanent magnet back EMFs
e_{mD}, e_{mQ}	D- and Q-axis magnet EMFs
e_{md}, e_{mq}	d- and q-axis magnet EMFs
E_{rms}	rms value of total induced voltage in a phase at steady-state
f	inverter output frequency
f_c	carrier wave frequency
$\boldsymbol{G}$	gain matrix
$G_d(s), G_q(s)$	linearized transfer functions of motors along the d- and q-axes
$I(I_s)$	magnitude of phase current at steady-state
i_a, i_b, i_c	three-phase winding currents
i_D, i_Q	direct- and quadrature-axis components of stator current in two-axis stationary reference frame
i_d, i_q	direct- and quadrature-axis components of stator current in rotor reference frame
i_{dc}, i_{qc}	iron loss current components of i_d and i_q
i_{dL}, i_{qL}	d- and q-axis current limits
i_{dT}, i_{qT}	torque-producing components of i_d and i_q
I_{rms}	rms value of phase current at steady-state
i_s	magnitude of stator current space vector
$\bar{i}_s$	stator current space vector
i_{sL}	limit of stator current space vector
i_x, i_y	direct- and quadrature-axis components of stator current in stator flux linkage reference frame
J	moment of inertia
$\boldsymbol{K}$	Kalman gain matrix
K_e	eddy current loss coefficient
K_h	hysteresis loss coefficient
$\boldsymbol{L}(\boldsymbol{L}_s)$	impedance matrix

L_1	inductance component due to the space fundamental air gap flux linkage
L_2	inductance component due to rotor position dependent flux linkage
L_{aa}, L_{bb}, L_{cc}	phase-winding self-inductances
L_d	direct axis inductance
L_q	quadrature axis inductance
$M_{ab}, N_{bc}, \ldots M_{ca}$	mutual inductances
p	derivative operator
P	number of pole pairs
$\boldsymbol{P}, \boldsymbol{Q}, \boldsymbol{R}$	covariance matrices
P_{ag}	motor air gap power
P_{Cu}	motor windings (copper) loss
P_{DC}	inverter input power
P_e	eddy current loss
P_{el}	electromagnetic power
PF	power factor
P_h	hysteresis loss
P_{in}	instantaneous electrical input power
P_L	motor electrical loss
P_m	motor mechanical power
P_{out}	motor output power
R_c	iron loss resistance
R_s	stator phase winding resistance
S_a, S_b, S_c	inverter switching functions corresponding to phases a, b, and c
T_c	carrier wave period
T_e	electromagnetic torque
T_{eL}	torque limit
T_f	friction torque
T_L	load torque
T_m	magnet torque
T_Q	reactive torque
T_r	reluctance torque
T_s	sampling time
$\boldsymbol{u}(k)$	system input vector
V	magnitude of phase voltage at steady-state
v_a, v_b, v_c	phase winding voltages
v_d, v_q	direct- and quadrature-axis voltage components in rotor reference frame
v_D, v_Q	direct- and quadrature-axis voltage components in two-axis stationary reference frame
V_{DC}	inverter DC link voltage
V_{rms}	rms value of phase voltage at steady-state
v_s	magnitude of stator voltage space vector
$\bar{v}_k, k = 0 - 7$	inverter space vector voltages
$\bar{v}_s$	stator voltage space vector

v_{sL}	stator voltage limit
v_x, v_y	x- and y-axis voltage components in stator flux linkage reference frame
$\boldsymbol{x}(k)$, $\boldsymbol{x}(k+1)$	system state vector at the current and the next instants
$\boldsymbol{x}(t_o)$	initial state vector
$\boldsymbol{y}(k)$	system output vector
α	angle of stator current space vector in rotor reference frame (with respect to the d-axis)
α_s	angle of stator current space vector in stationary reference frame (with respect to the phase a axis)
γ	angle of stator voltage space vector in rotor reference frame (with respect to the d-axis)
δ	angle of the stator flux linkage space vector in rotor reference frame (with respect to the d-axis); load angle
δ_L	load angle limit
δ_s	angle of the stator flux linkage space vector in stationary reference frame (with respect to the phase a axis)
ΔT	torque hysteresis band
$\Delta\lambda$	flux linkage hysteresis band
η	efficiency
θ_m	mechanical rotor position
θ_r	rotor position defined as the electrical angle of rotor pole axis or rotor d-axis, from the axis of stator phase winding a
λ_a, λ_b, λ_c	phase winding flux linkages corresponding to phases a, b, and c
λ_D, λ_Q	direct- and quadrature-axis flux linkage components in two-axis stationary reference frame
λ_d, λ_q	direct- and quadrature-axis flux linkage components in rotor reference frame
λ_m	maximum flux linkage of a phase, produced by magnet poles
λ_s	magnitude of the stator flux linkage space vector
$\overline{\lambda}_s$	stator flux linkage space vector
λ_{sL}	stator flux linkage limit
λ_x, λ_y	direct- and quadrature-axis flux linkage components in stator flux linkage reference frame
ξ	demagnetizing coefficient
ξ_L	limit of demagnetizing coefficient
ρ	saliency ratio
$\boldsymbol{\sigma}(t)$	noise matrix
τ	torque flag
φ	flux linkage flag
χ	angle of stator current space vector in stator flux linkage reference frame (with respect to the x-axis)
ω_e	angular frequency of supply
ω_m	mechanical angular speed of motor
ω_r	electrical angular speed of motor
ω_s	synchronous speed

Subscripts

h	harmonic
i	injected signal
$k/k-1$	value at the kth instance based on the measurements up to the $(k-1)$th instance

Superscripts

–	space vector
*	reference or commanded
^	estimated
·	derivative
p	predicted
T	matrix transpose

Abbreviations

ASDSP	application-specific DSP
BLDC	brushless direct current
CC	combined control
CSI	current source inverter
DBC	deadbeat control
DSP	digital signal processor
DTC	direct torque control
DY	dysprosium
EKF	extended Kalman filter
EMF	electromagnetic force
IGBT	insulated-gate bipolar transistor
IPM	interior permanent magnet
LMC	loss minimization control
LPF	low-pass filter
mmf	magnetomotive force
MOSFET	metal-oxide semiconductor field effect transistor
MPC	model predictive control
MRAS	model reference adaptive system
MTPA	maximum torque per ampere
MTPV	maximum torque per voltage
MW	megawatt
NdFeB	neodymium iron boron
PC	predictive control
PF	power factor

PI	proportional integral
PM	permanent magnet
PMS	permanent magnet synchronous
PWM	pulse width modulation
RF	reference frame
RLS	recursive least-squares
RTD	resolver to digital
SC	scalar control
SVM	space vector modulation
VC	vector control
VSI	voltage source inverter
VVVF	variable voltage variable frequency

Introduction

1

The focus of this book is on the control of permanent magnet synchronous (PMS) motors. It is anticipated that readers have enough background regarding the related aspects of a controlled motor system. Nevertheless, an overview of the system is presented to clarify the interconnections of the control system to other parts in the whole controlled motor system. In addition, major parts of the total system are overviewed. The most common power converter used in PMS motor drives, i.e., pulse width modulated voltage source inverter, is briefly discussed. PMS motor configurations and operating principles are also presented after considering characteristics of permanent magnet materials. The control system, including different control methods, and position and parameter estimation schemes are then reviewed.

1.1 Overview of a controlled motor system

It is said that the universe does not have any functions except processing energy and information. A motor system, as a whole, performs both functions in its own scale. It converts electrical energy to mechanical energy, as required by the application. It may also have limited capability in the opposite direction, i.e., converting mechanical energy to electrical energy on occasions, such as regenerating braking. The system consists of motor supply and control. The former processes electrical energy and delivers it to the motor for further processing, i.e., conversion to mechanical energy; the latter processes information to control the former. Therefore, the supply, control, and motor are highly interconnected. In this section, an overview of a controlled PMS motor system is presented, as seen in Fig. 1.1, before considering its parts in more detail in the following sections.

The energy process of a motor system begins with the power supply. It provides electrical energy to the power converter. It is known that the most convenient way to supply a motor, including a PMS motor, is to convert a fixed DC power to a variable voltage and variable frequency power supply. Therefore, the power supply

Control of Permanent Magnet Synchronous Motors. Sadegh Vaez-Zadeh. © Sadegh Vaez-Zadeh 2018.
Published in 2018 by Oxford University Press. DOI 10.1093/oso/9780198742968.001.0001

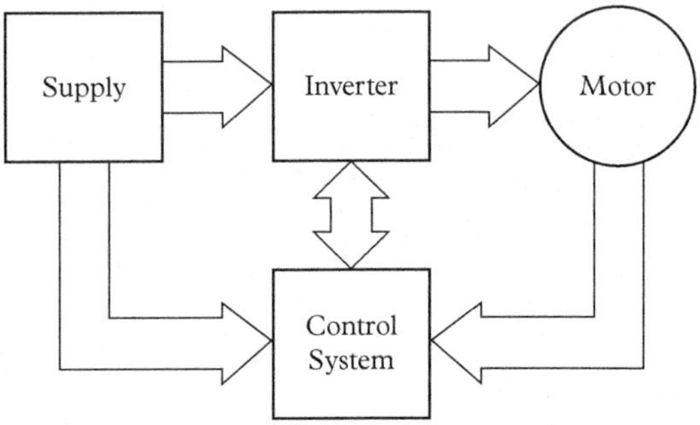

Figure 1.1 *A controlled PMS motor system including power supply, power converter, motor, and control system.*

must provide a fixed DC power. It can be generated by a DC source, such as a battery, or converted from a fixed voltage and fixed frequency source, such as the electricity network, or a stand-alone source of energy, such as a generating unit. In the case of rectification, a filtering stage is also necessary to smooth the rectifier output. This is commonly performed by a parallel capacitor bank. As a result, a low ripple voltage is provided. This is used in connection with the voltage source inverters (VSIs). It is also possible to use a series inductor for filtering. This is used in connection with the current source inverters (CSI).

The power converter in the controlled motor system is usually an inverter. The inverter receives the rectified power and inverts it to AC power. The instantaneous value of the inverter output consisting of the magnitude, frequency, and phase of the signal depends on the switching strategy of the inverter. VSIs, rather than CSIs, are far more common in AC motor drives due to their cost benefit and ease of control. The application of CSIs is mainly limited to high power motor drives. The domination of VSIs is more severe in PMS machines due to the limited use of these motors in high power ratings. The VSIs will be reviewed in the next section.

The motor, as the heart of the energy conversion system, converts electrical energy to mechanical energy. However, it may occasionally convert mechanical energy to electrical energy, to be delivered back to the power supply in a regenerative mode of operation, e.g., in the braking mode of an electric vehicle. PMS motors enjoy many desirable features, such as high efficiency, high power density, excellent controllability, etc. The application of PMS motors is growing rapidly and is expected to override the use of induction motors in major applications in future. PMS motors are usually supplied by power converters under the influence of appropriate control systems. The physical structure and operating principles of the motors will be elaborated on in this chapter.

The control system is the information processing part of a controlled motor system. It receives command signals from the applicant or the overhead control system, plus system signals from the motor, power converter, and power supply to generate driving signals for the power electronic switches of the inverter. The system signals are mainly obtained by the electrical and mechanical sensors. At least one current sensor on the link between the DC power supply and the inverter is an indispensable part of the system. However, two or three current sensors are usually used for measuring motor currents. Voltage, position, speed, and flux sensors may also be used. Modern control systems include a software part, consisting of signal processing routines, and a hardware part, which executes the routines. The signal processing is arranged according to the motor control methods as the focus of this book. The methods include vector, direct torque, predictive, deadbeat, and combined vector and direct torque controls. The software part may also include position and speed estimation and parameter estimation. The control methods will be reviewed briefly in this chapter and elaborated in detail later in the book.

1.2 Power electronic converter

Power electronic converters are electrical power processing systems that transfer the electrical power of the power supply to a variable amplitude and variable frequency supply. They may be regarded as power amplifiers when their output power is considered with respect to their reference signals. Traditionally, two types of power electronic converters are used for driving AC machines, i.e., single- and two-stage converters. The first one, known as a cyclo-converter, converts directly fixed amplitude and frequency electrical power to variable amplitude and frequency power, while the second type performs the same function by first converting the fixed AC power to DC power by a rectifier, then converting it to a variable magnitude and variable frequency AC power by an inverter.

1.2.1 Voltage source inverter

The inverter is either a VSI or a CSI. In PMS motor drives, the application of VSIs is much more common than CSIs or cyclo-converters, because the latter two converter types are mostly suitable for high power motors in the range of several megawatts (MWs) and above, where the application of PMS machines is not common in this power range yet. Also, multilevel inverters are being used in high power applications. Figure 1.2 schematically depicts a three-phase full-bridge

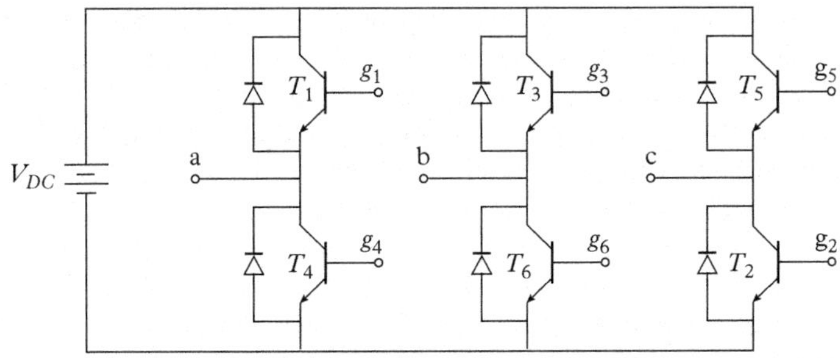

Figure 1.2 *Schematic view of a three-phase two-level voltage source inverter.*

two-level VSI as the most commonly used inverter in motor drives. The inverter switching states are shown in Table 1.1.

Table 1.1 *Inverter switching states.*

$S_a = 1$	T_1 ON	T_4 OFF
$S_a = 0$	T_1 OFF	T_4 ON
$S_b = 1$	T_3 ON	T_6 OFF
$S_b = 0$	T_3 OFF	T_6 ON
$S_c = 1$	T_5 ON	T_2 OFF
$S_c = 0$	T_5 OFF	T_2 ON

A VSI consists of DC voltage terminals and three legs with two power electronic switches installed on each leg. Freewheeling diodes are connected across the switches to facilitate the turn-off function of the switches. The switches are turned on and off with high frequency by driving circuits. The driving circuits receive the input signals from the control system and provide isolated gating outputs toward the switch gates.

As mentioned previously, a rectifier mainly provides the DC link voltage. The voltage is smoothed by a rather large voltage-smoothing capacitor, which is installed in parallel to the DC link. During the regenerating, the mean link current, I_{DC}, is reversed and charges the capacitor. This may cause an over-voltage DC link and harm the inverter switches. To prevent the over-voltage of the DC link, either a pulse width modulation (PWM) power-dissipating circuit is turned on by a switching transistor or the generated energy is fed back to the main supply. In the latter case, a power electronic converter with the capability of bidirectional power transfer substitutes for the rectifier.

Power electronic switches are mainly insulated-gate bipolar transistors (IGBTs). Power metal-oxide semiconductor field effect transistors (power MOSFETs) and MOS-controlled thyristors (MCTs) are also used.

A CSI, with a high impedance series inductor replacing the DC-link capacitor, may also be used, especially for driving high power PMS motors. However, the VSI is far more popular in practice due to physical and performance merits.

Any VSI inverter works in a switching mode in which the inverter switches can have two operating states: either ON, i.e., conducting state, or OFF, i.e., non-conducting state. The on and off states of the switches connect the output terminals of the inverter to V_{DC}, 0, $-V_{DC}$. The two switches of a leg must have opposite states in order to prevent a short circuit through the switches, which can be hazardous. The switching states of the inverter can be seen in Table 1.1.

Switching states are followed regularly so that a variable voltage variable frequency (VVVF) power can be produced at the inverter terminals according to the commanded voltage signals. Pulse wave modulation schemes control the switching so that the desired output voltages are delivered to the machine terminals. A PWM scheme is how commanded voltage signals, produced by the current controllers, are projected to the gating signals of the power electronic switches of the inverter. Different PWM schemes can do the job. Two of the most popular ones are the sinusoidal PWM and the space vector PWM.

1.2.2 Sinusoidal PWM

The sinusoidal PWM is realized by comparing the three commanded phase voltages of the machine, i.e., modulating signals, by a constant high-frequency triangular signal, i.e., carrier wave (Bose 2005). The crossing points of the modulating signals and the carrier wave determine the instances of applying gating signals to the inverter switches to modulate the DC link voltage and produce the output voltages of the inverter as seen in Fig. 1.3. As the frequency of carrier wave is much higher than that of the AC modulating signals, the latter ones are seen as flat lines in the figure. The crossing of each modulating signal with the carrier wave determines the on-period of

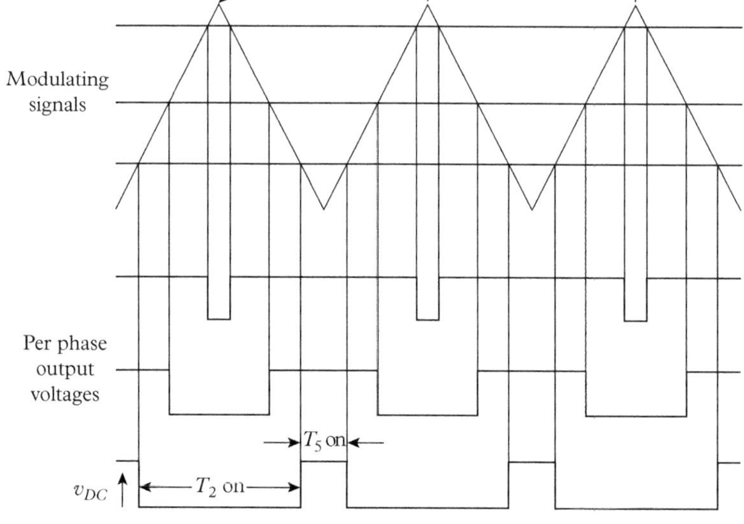

Figure 1.3 *Principle of sinusoidal pulse wave modulation (only a small fraction of a cycle for sine modulating signals is shown).*

two switches in a leg. A pair of such periods for switches T_2 and T_5 belonging to the inverter's right leg from Fig. 1.2 is shown in Fig. 1.3. It is clarified that the name of the modulating scheme stems from the fact that the modulating signals are sinusoidal at steady-state.

1.2.3 Space vector PWM

The space vector modulation (SVM) scheme is based on the concept of rotating space vectors (Rashid 2004). The voltage space vector of an AC machine stator is defined as the sum of three voltage vectors as

$$\overline{v}_s = v_a + v_b\, e^{j\frac{2\pi}{3}} + v_c\, e^{j\frac{4\pi}{3}}, \tag{1.2.1}$$

where v_a, v_a, and v_a are phase voltages. Also, $e^{j\frac{2\pi}{3}}$ and $e^{j\frac{4\pi}{3}}$ represent unit vectors aligned with the winding axes of phase "b" and "c," respectively, if the axis of phase "a" is regarded as the base. Therefore, the three terms of eqn (1.2.1) can be regarded as three vectors with magnitudes equal to v_a, v_b, and v_c, and space angles of 0°, 60°, and 120°, respectively. Using this concept, voltage space vector of an inverter is presented in terms of the DC bus voltage and the states of the inverter switches are presented as

$$\overline{v}_s = \frac{2}{3}\, V_{DC}\left(S_a + S_b\, e^{j\frac{2\pi}{3}} + S_c\, e^{j\frac{4\pi}{3}}\right), \tag{1.2.2}$$

where S_a, S_b, and S_c are used to determine the connection of output terminals of the inverter to DC bus voltage levels V_{DC} or 0. For instance, S_a is 1 when the upper transistor in the inverter's left leg of Fig. 1.2 is on and the lower transistor on the same leg is off. Thus, the inverter "a" terminal is connected to V_{DC}, and S_a is 0 when the states of the switches on the left leg are reversed and the "a" terminal is connected to zero level voltage. As S_a, S_b, and S_c are either 1 or 0, eqn (1.2.2) represents eight voltage vectors as depicted in Fig. 1.4.

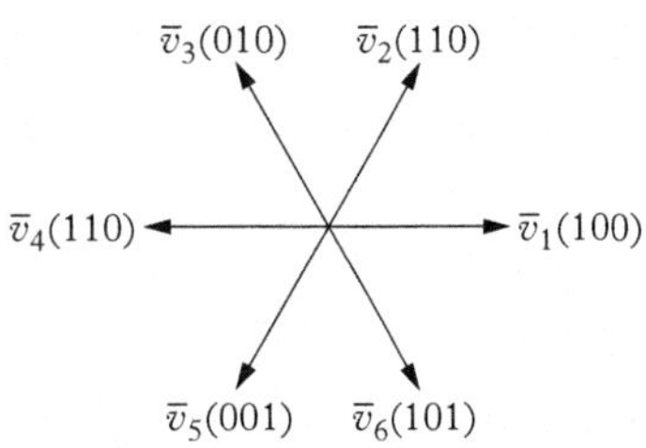

Figure 1.4 *Inverter voltage vectors.*

It is seen that there are six non-zero or active voltage vectors, 60° apart from each other, and two zero voltage vectors located at the center. The zero voltages are produced when the three upper switches are all on or all off; thus, the motor terminals are short circuited at the upper or lower DC bus terminal, i.e., V_{DC} or 0, respectively. Each voltage vector is determined by a three-digit number showing the connection of the inverter terminals to V_{DC} or 0, e.g., $\overline{v}_2(110)$. The non-zero vectors have the same magnitude and are 60° apart.

The inverter can effectively generate any voltage vector, other than the eight voltage vectors, by applying the two adjacent vectors for a portion of a sampling period. By using superposition, it is possible to write

$$\overline{v}_s = \overline{v}_k t_k + \overline{v}_{k+1} t_{k+1}, \tag{1.2.3}$$

in which $\overline{v}_k$ and $\overline{v}_{k+1}$ are two voltage vectors adjacent to $\overline{v}_s$. Also, T_s is the sampling period and t_k and t_{k+1} are the duration of applying $\overline{v}_k$ and $\overline{v}_{k+1}$, respectively, where

$$T_s = t_k + t_{k+1}. \tag{1.2.4}$$

The tip of such a vector moves on the hexagon locus, connecting the tips of the six non-zero voltages of the inverter shown in Fig. 1.5, when t_k and thus t_{k+1} change. The inverter can also generate any voltage vector inside the hexagon if t_k and t_{k+1} are scaled down, and a zero voltage, $\overline{v}_z$, is applied in the remaining portion of T_s. Then

$$\overline{v}_s = \overline{v}_k t_k + \overline{v}_{k+1} t_{k+1} + \overline{v}_z t_0, \tag{1.2.5}$$

$$T_s = t_k + t_{k+1} + t_0, \tag{1.2.6}$$

where t_0 is the duration of the zero voltage. This provides a linear or under-modulation mode of the inverter. The inverter output voltages are sinusoidal if the tip of voltage vector rotates on a circle inside the hexagon at a constant speed. The voltage vectors outside the hexagon correspond to the over-modulation mode of the inverter (Fig. 1.5).

SVM has found increasing application due to its superior performance characteristics, including lower harmonics (Bose 2005).

PWM can be implemented by separate microchips to avoid a software PWM and release the computation capability of the central processing unit. Application-specific digital signal processors (ASDSPs), for electrical motor control, embed more than one hardware PWM circuit.

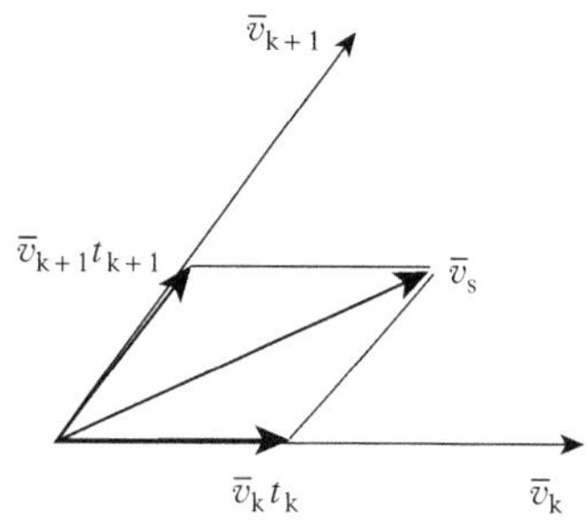

Figure 1.5 *Space vector modulation in terms of two inverter voltage vectors.*

1.3 PM materials

The salient features of PMS machines in comparison with the basic features of conventional electric machines all stem from the use of permanent magnet (PM) materials in their rotors. PMs are among the fastest growing materials of the past three decades. This is because of their superior characteristics, and intensive research and development that have been carried out to further enhance their features. The features have been extensively reported in numerous specialized references. An overview of major characteristics and a carefully selected set of features of PM materials, playing an important role in

PMS motors, are briefly reviewed in this section as an introduction to PMS machines. Also, a snapshot of the PM market is given by a few statistics to highlight the fast-growing field of PM materials and their applications. More emphasis is placed on the neodymium iron boron (NdFeB) magnet as the material of highest performance and application among other magnet materials.

1.3.1 Characteristics

PM materials have a special hysteresis loop that is central to most of their properties. In particular, the second quarter of the loop, referred to as the demagnetization curve, mainly characterizes the performance of materials. Also, the energy product of a material is an important measure of magnetic performance. Properties of magnetic materials depend, to a wide extent, on their temperature. These are briefly discussed later.

1.3.1.1 Demagnetization curve

B–H hysteresis loop demonstrates graphically the magnetic flux density, *B*, in a magnetic medium in terms of the magnetic field intensity, *H*, in the medium in a Cartesian coordinate system as in Fig. 1.6. The field intensity, in hard magnetic materials, is produced by an external magnetic field. As the field increases, *B* increases with an increasing *H* on an initial magnetization path until it saturates at a certain *H*, at

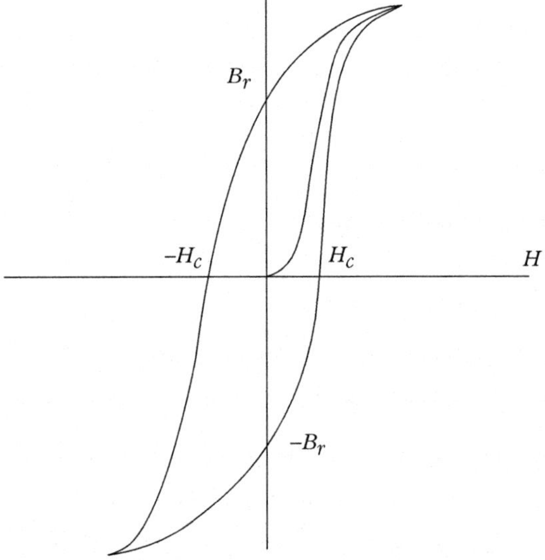

Figure 1.6 *B–H hysteresis loop of a typical hard magnetic material.*

the top of the loop. If the external field starts decreasing at the saturation point so do H and B, but on a demagnetization path with higher values of B with respect to the values of B on the initial magnetization path for the same values of H. As a result, when H reaches 0, B reaches a value called remanence, B_r. Now, if the external field starts increasing in the negative direction, H in the medium becomes negative in the second quadrant of the coordinate system until B becomes 0. This corresponds to a negative value of H called coercivity, H_c. The hysteresis loop reaches another saturation in the third quadrant and continues on a magnetizing path in the third and fourth quadrants, and closes itself at the first quadrant as seen in Fig. 1.6.

B–H hysteresis loop for permanent magnet materials is important mainly in the second quadrant, since PMs enjoy inherent field intensity and the corresponding flux density in the material. As a result, their performances are assessed by the demagnetization of the inherent flux density against opposing (negative) magnetic fields. Therefore, the demagnetization curve, rather than the entire *B–H* loop, is usually considered in PM materials. A demagnetization curve for a typical PM material is depicted with more detail in Fig. 1.7.

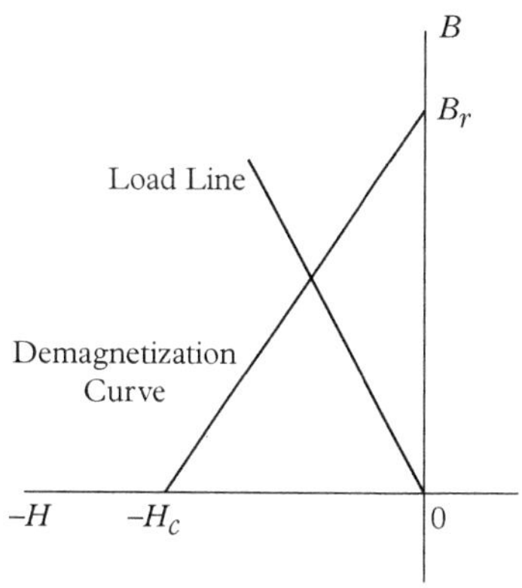

Figure 1.7 *Demagnetization curve of a typical PM material in a machine.*

Several specifications of PM materials are determined by the demagnetization curves. B_r and H_c are particularly crucial in defining PM materials. Bearing in mind the procedure described previously for achieving the hysteresis loop, B_r is the maximum flux density in the material at a particular temperature when the field intensity in the magnet vanishes after the material is magnetized to saturation. Also, H_c is the negative field intensity in the magnet caused by the external magnetic field to bring the flux density in the material to 0 from B_r.

A useful feature of most modern PM materials is that the demagnetization curve is a straight line, which is determined by B_r and H_c, with the slope of the line being the permeability of the material. However, the curves of low-grade PM materials experience a knee in the second quadrant. The operating point of the magnet material lies on the demagnetization curve somewhere between B_r and H_c. The operating point depends on the location of the magnet in the PMS machine. It is also influenced by the operating point of the machine. This is because the magnetic field in the PM depends on the permeance of the whole flux path and the ampere-turns that produce the flux.

Flux density of the magnet in operating point on the demagnetization curve is always less than the remanence, due to the excitation requirement of the flux path in the machine, particularly that needed by the air gap. It can be easily calculated by applying the Ampere law to the flux path, neglecting the reluctance of the iron and leakage flux. The line connecting the origin of the *B–H* coordinate system to the magnet operating point is the load line or the air gap line. The absolute value of load line slope, when normalized by μ_0, is called the permeance coefficient (PC), which depends on the geometric dimensions of the machine. The operating point slips down on the demagnetization curve under the influence of the demagnetization caused by flux weakening due to the motor current. As a result, the flux density and the field intensity reduce further from those corresponding to the load line. Consequently, the magnet load line shifts to the left in parallel with the original load line.

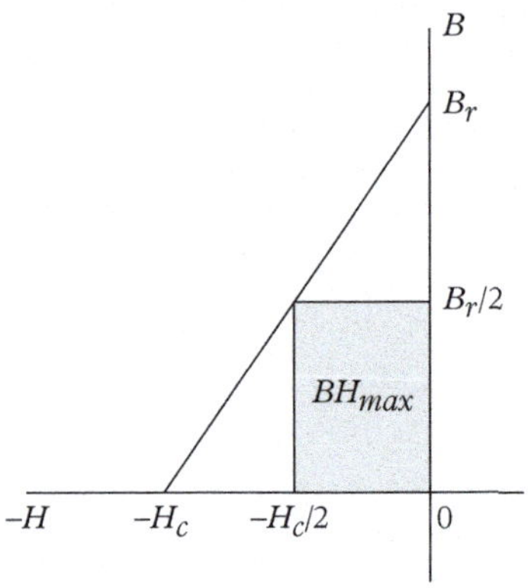

Figure 1.8 *Graphical representation of maximum energy product of a typical PM material in connection with its demagnetization curve.*

1.3.1.2 Energy product

An important parameter of a PM material is the energy product, which is the product of B and H. The maximum possible energy product, BH_{max}, is the biggest value of the product, corresponding to the area of the largest square that can be drawn under the demagnetization curve as seen in Fig. 1.8 (Coey 2012). This is a measure of the material remanence and coercivity combined. It shows how strong the flux density of magnet is, and to what extent it can resist against demagnetization by external magnetic fields. High value of maximum energy product leads to PMS machines with high torque and power characteristics as seen in Fig. 1.9 (Gutfleisch *et al.* 2011).

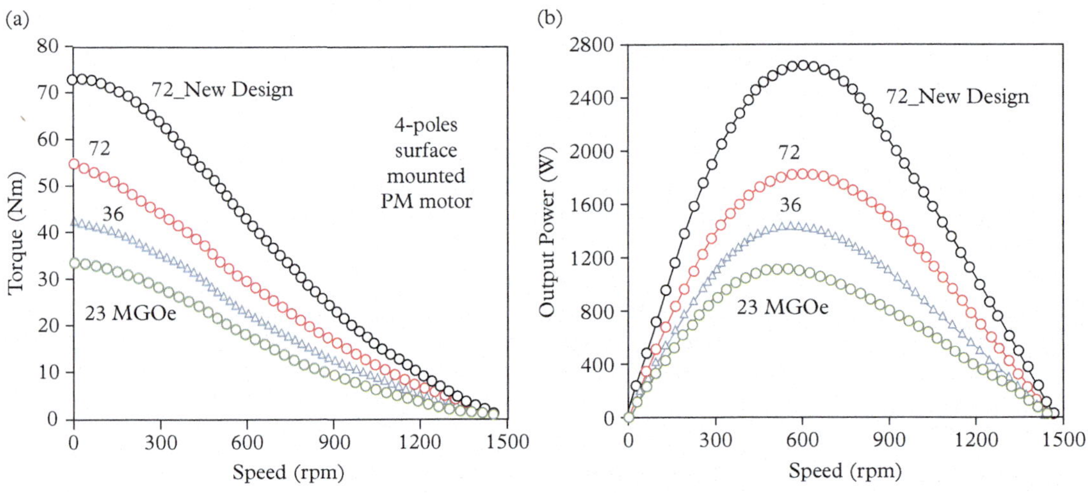

Figure 1.9 *The effect of the energy products of magnets with linear B–H demagnetization curves on a PM motor performance with three different energy product values and two different motor designs: torque vs speed (a) and output power vs speed (b) (Gutfleisch* et al. *2011).*

It is possible to calculate the maximum energy product by substituting for H in $B.H$ from the linear equation of the demagnetization curve to yield

$$B.H = -1/\mu B\,(B_r - B)\,, \tag{1.3.1}$$

where μ is the permeability of PM material. Then, solving the derivative of (1.3.1) with respect to B gives the value of flux density corresponding to BH_{max} as $B_r/2$. Substituting this value back into (1.3.1) results in BH_{max} in terms of B_r, and also in terms of H_c as

$$BH_{max} = -\frac{B_r^2}{4\mu} = -\frac{\mu H_c^2}{4}. \tag{1.3.2}$$

The energy product is maximum when the magnet operating point is on the demagnetizing curve at $B = B_r/2$ and $H = -H_c/2$, as seen in Fig. 1.8. Such an operating point is optimal from the point of view of PM material utilization and requires the PMS machine to work under considerable flux weakening by appropriate stator current control. Therefore, maintaining this PM operating point may not coincide with the required flux weakening from the motor point of view, as the current control usually has some objectives corresponding to optimal motor performance.

1.3.1.3 Temperature dependency

Exposure of hard magnetic materials to high temperatures for certain periods of time results in reversible or irreversible loss of magnetism in materials. The reversible demagnetization occurs at a limited temperature and is usually linear with the temperature. Therefore, it can be modeled by constant coefficients. Irreversible demagnetization occurs at a threshold temperature referred to as the Curie temperature, at which the magnetization vanishes permanently; i.e., it is not resumed by only lowering the temperature. However, it may be possible to remagnetize the material at normal temperature if metallurgical changes due to high temperatures have not taken place in the material. These changes begin to occur at temperatures below the Curie temperature in rare earth magnets and above that temperature in ferrite magnets. The temperature demagnetization causes the remanence and the coercivity to reduce. Therefore, the demagnetization curve moves downward as depicted in Fig. 1.10, making the maximum energy product smaller. As a result, the risk of magnet demagnetization, by the motor current, increases.

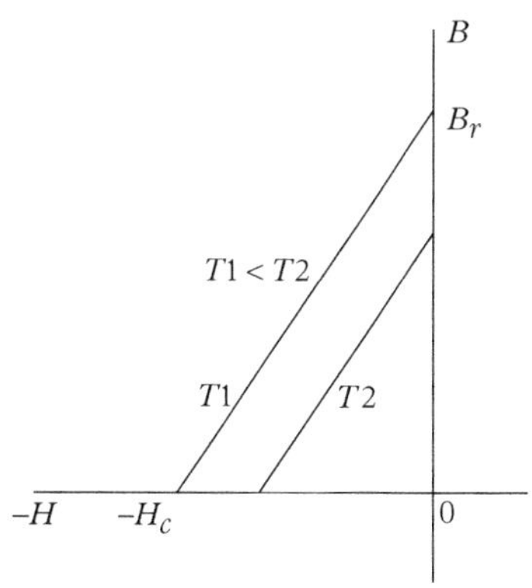

Figure 1.10 *Temperature effect on demagnetization curve of PM materials.*

A gradual demagnetization also occurs in PM materials as time goes on. This process depends on the magnet operating point and temperature. The demagnetization is negligible in high energy PM materials at normal temperature as it is well below 0.5% after 10 years.

1.3.2 Properties of PM materials

There are four types of permanent magnet materials, i.e., neodymium iron boron (NdFeB), SmCo, ferrite, and alnico. NdFeB magnets were discovered in 1983 and have linear demagnetization characteristics, as seen in Fig. 1.11, with very high remanence and coercivity, being 1.28 T and 900 kA m^{-1}, respectively, for a specific grade. As a result, the energy product of the material is highest among all hard magnetic materials, 300 kJ m^{-3} for a specific grade, resulting in corresponding PMS machines with high power and torque density. NdFeB is electrically conducting and magnetically anisotropic. It is usually magnetized in the direction of the thickness of the pieces. The material is mechanically strong and features good machinability. The main weakness of the material is low Curie temperature, which is around 310°C with metallurgical changes at about 200°C. The temperature increases because of the addition of dysprosium (DY), which is an expensive element. The NdFeB magnets are coated to prevent corrosion of the magnets and are not cheap. Nevertheless, they strongly dominate the electric machine market.

SmCo magnets were discovered in the 1960s, and are alloys of cobalt and samarium produced by the powder metallurgy process. They are magnetized before pressing. The material has a linear demagnetization characteristic as depicted in Fig. 1.11 with high remanence and coercivity, and therefore high maximum energy product. The material is electrically conducting and magnetically anisotropic, and is magnetized along the thickness of pieces. SmCo magnets are mechanically hard and brittle; therefore, they must be handled with care.

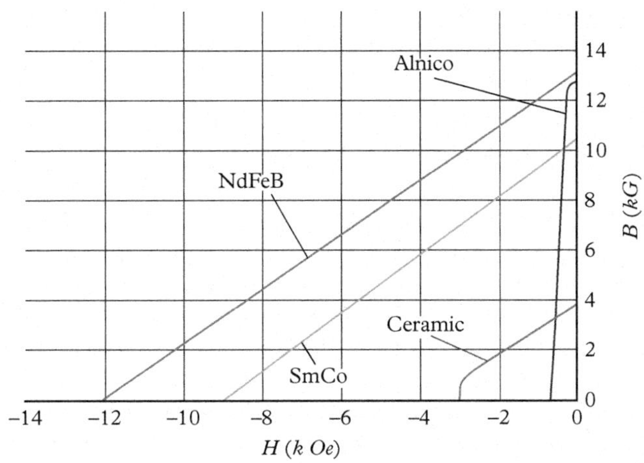

Figure 1.11 *Demagnetization characteristics of common PM materials.*

The material has a high Curie temperature of 700–800°C, depending on the grades, with metallurgical change at 300–350°C. The main deficiency of the material is the very high price, which restricts its use to a marginal market.

Ferrite magnets were produced in the form of sintered materials by the powder metallurgy process in the 1950s. They consist of compounds of ferric oxide and carbonate of barium or strontium. They feature a rather linear demagnetization characteristic, as seen in Fig. 1.11, with low remanence and coercivity, resulting in low maximum energy product. Therefore, they do not provide a compact motor design. Electrically, ferrite magnets are very good insulators and produce insignificant eddy current loss caused by external AC fields. The mechanical properties of the material are unique, being very hard and brittle. For this reason, they are known as ceramics. They are resistant against corrosion and are, by weight, the most used PM material in PMS motors, due to their low prices.

Alnico, the last magnetic material mentioned here, is the oldest commercially available PM. The alloy consists of aluminum, nickel, cobalt, and iron plus other elements, and is available in both sintered and cast forms. The material has non-linear demagnetizing characteristics, as seen in Fig. 1.11, with a very high remanence and a very low coercivity. As a result, alnico's energy product is low and can easily be demagnetized by external fields. It has very good thermal capability. Electrically, it is a good conductor and magnetically anisotropic with magnetic orientation along the length. Isotropic grades of alnico are also available for some applications. A comparison of the PM materials' main properties is presented in Table 1.2.

Research into improving the properties of existing PM materials and finding new materials is underway, with two particular areas being active today. The first is to improve the thermal performance of NdFeB magnets or to come up with similar materials having

Table 1.2 *Comparative properties of PM materials.*

PM material	*Remanence*	*Coercivity*	*Energy product*	*Curie temperature*	*Price*	*Applications*
NdFeB	*High*	*High*	*High*	*Low*	*High*	*Very high*
SmCo	*High*	*High*	*High*	*High*	*Very high*	*Low*
Ferrite	*Low*	*Low*	*Low*	*High*	*Low*	*High*
Alnico	*High*	*Very low*	*Low*	*High*	*Low*	*Low*

good thermal resistance. The second is to find cost-effective materials like ferrites, but with high energy products.

1.3.3 PM market

An overview of the PM market, regarding electric motors, can be delivered by considering the related supply and demand sides. It can roughly be presented by four aspects of matter, i.e., worldwide production in terms of material type and producing regions, worldwide consumption, and price. The information is reviewed briefly in the course of time to give a dynamic account of the market.

1.3.3.1 Production by type

The PM market started with alnico and then ferrite magnets. The market did not change a lot with the invention of SmCo magnets in the 1960s, due to their very high cost. However, NdFeB magnets changed the market dramatically, following their invention in the early 1980s. The global PM market in 2010 is estimated in terms of PM types in Fig. 1.12 (Gutfleisch *et al.* 2011). It can be seen that NdFeB magnets dominate the market with a share of about two-thirds, followed by ferrite magnets with a share of one-third. The shares of SmCo and alnico magnets are negligible, due to their high price and low performance, respectively. Magnet production by volume, however, shows a huge share of the market for ferrites and a low share for NdFeB. A chronological presentation of material sales during 1985–2020 is shown in Fig. 1.13 (Dent 2012). Total sales are predicted to exceed \$15 billion by 2020. It is evident from this figure that the market domination of NdFeB will be further consolidated in the future.

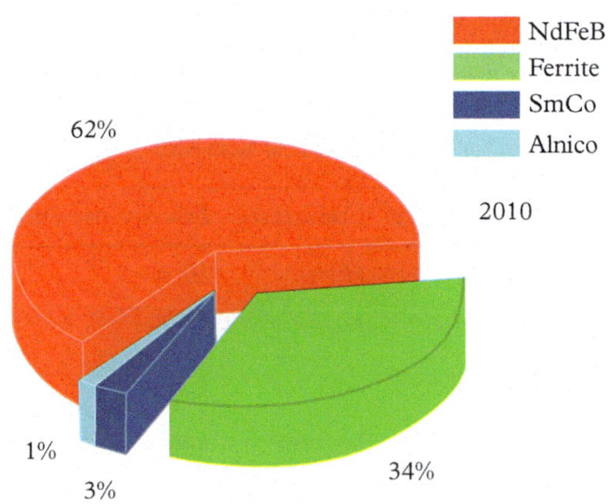

Figure 1.12 *Estimated breakdown of global permanent magnet value by type in 2010, total value: \$9 bn (Gutfleisch* et al. *2011).*

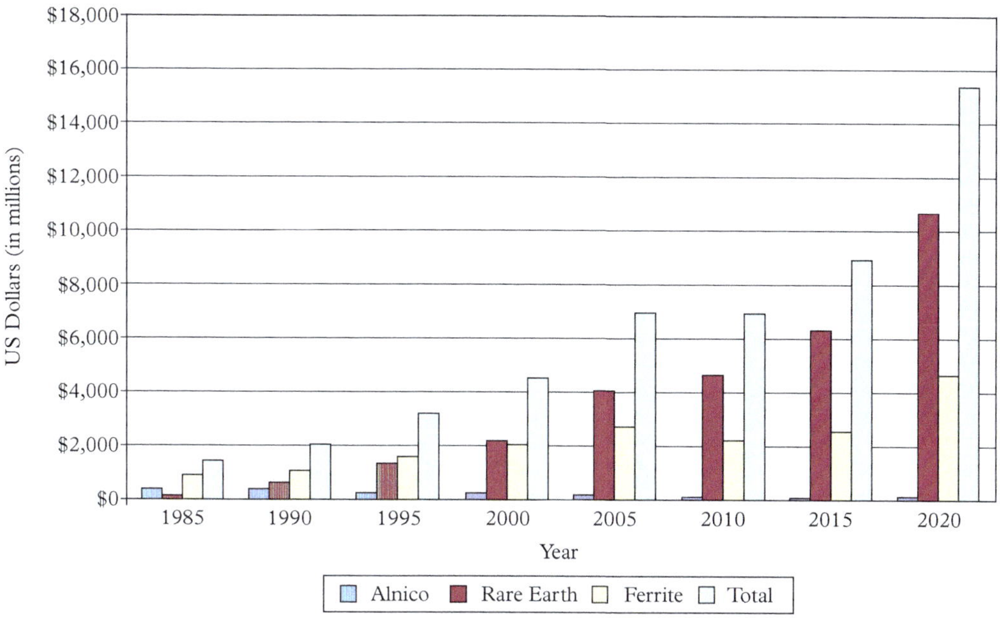

Figure 1.13 *Breakdown of global permanent magnet sales by type during 1985–2020 (Dent 2012).*

1.3.3.2 ***Production by region***

As mentioned previously, NdFeB is by far the dominant PM material in the market. Traditionally, the materials were produced mainly in Japan, Europe, and the United States, before China became the predominant producer in the mid-1980s. China now has about 85% of the worldwide share, which continues to increase. Figure 1.14 shows the trend of NdFeB production by regions during the period 2005–2020 (Benecki *et al.* 2010). With the exception of a slowdown in 2008–2009, due to the global recession, production has always been increasing. It is anticipated that production growth rate will increase in the near future as shown in Fig. 1.14.

1.3.3.3 ***PM consumption by application***

Permanent magnet materials have found widespread applications in many sectors. PM motors are a major sector as seen in Fig. 1.15 (Kara *et al.* 2010). The figure shows that PM motors' share in the consumption of NdFeB magnets has grown since 2003, reaching 26% by 2008, demonstrating that PM motors have a high rate of growth. Japan's share has grown even faster, from 20% in 1999 to 34% in 2003, increasing from 960 tons to 1785 tons within 4 years (Kozawa 2011).

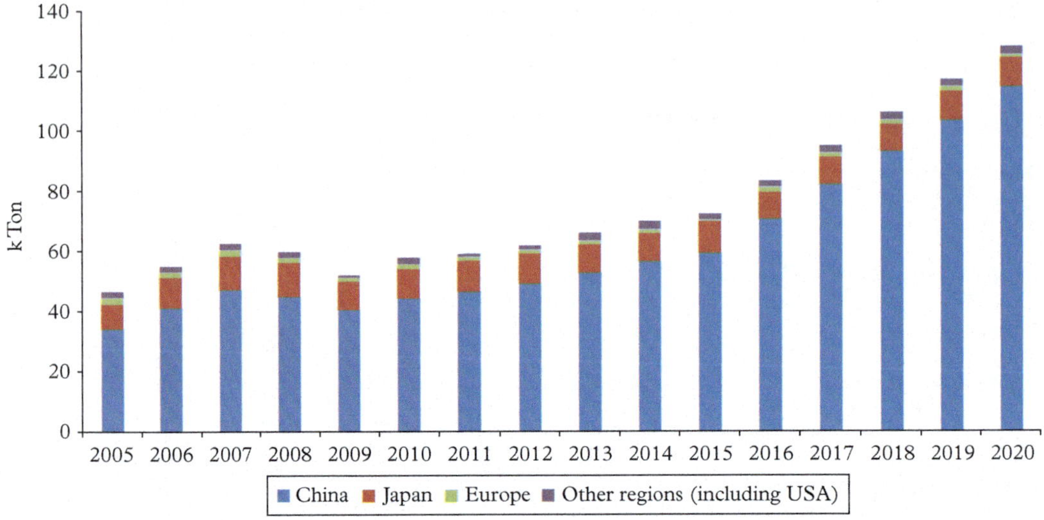

Figure 1.14 *Estimated and forecasted worldwide production of NdFeB material by region during 2005–2020 (Benecki* et al. *2010).*

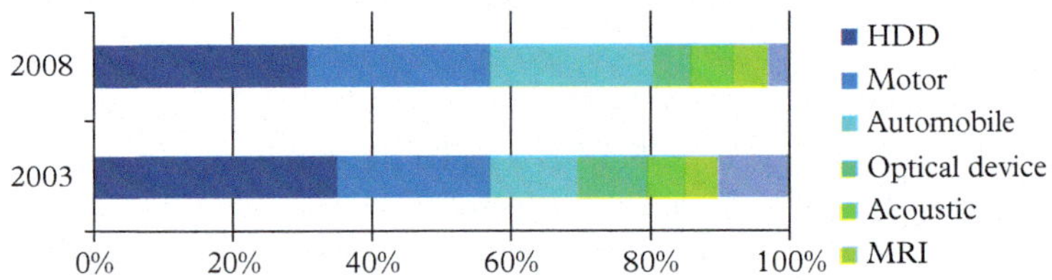

Figure 1.15 *The application of NdFeB magnets in terms of sectors in 2003 and 2008 (Kara* et al. *2010).*

Salient features of PMS motors, including high efficiency, high power density, and fast dynamics, drive market growth. As a result, the PMS motor market is anticipated to capture the traditional market of variable speed induction motor drive applications in sectors such as oil and gas, paper and pulp, metals and mining, fans and pumps, and compressors, chemicals, plastic, cement, power, and automotive, in order to save energy, and improve quality and productivity. At the same time, new applications such as electric and hybrid vehicles, train propulsion systems, and fly by wire that are suited to PMS motors are increasingly expanding the market. It is estimated that the market for PM motors will grow at an annual rate of 11.7% during 2014–2020, more than double that for induction motors (Allied Market Research 2015). The global PMS motor market was valued at USD 14.32 billion in 2014 and is expected to reach USD 25.18 billion by 2020 (Allied Market Research 2015).

1.3.3.4 Prices

The prices of permanent magnet materials differ considerably. Alnico and ferrite magnets are far cheaper than rare earth magnets. SmCo used to be the most expensive one, and NdFeB magnets were in between and closer in price to SmCo magnets. However, adding dysprosium, a rare earth element, to NdFeB magnets to protect against low temperature demagnetization makes them comparable in price with SmCo magnets. The high cost of rare earth magnets, compared with that of other PM materials, is partially compensated for by the lower volume of material needed to provide the same flux density in the machine air gap due to their high energy product, as seen in Fig. 1.16 for an example. The figure shows two interior permanent magnet (IPM) motors for automotive cooling fans with ferrite and NdFeB magnets. The motor with the NdFeB magnet consumes much less magnet volume (Ding 2013).

This, in turn, may reduce the rotor and stator core iron, increase the machine power and torque densities, and improve the motor dynamic behavior. The price of rare earth permanent magnet materials depends on the price of rare earth elements. NdFeB magnets experienced price fluctuations at the start of 2010 due to changes in the price of neodymium. The neodymium price rose 10-fold during a 4-year period, reaching a maximum in mid-2011, due to policies adopted by China as the dominant producer. Neodymium's high price influenced the cost of NdFeB magnets, and affected the PM motor market and its future predictions. Neodymium's price declined after this and dropped to about one-fifth of the maximum price in 2013, due to market responses and changes in policies (Ding 2013).

Figure 1.16 *IPM motors for automotive cooling fans, using ferrite (left) and NdFeB magnets (right) (Ding 2013).*

1.4 PMS machines

Permanent magnet synchronous machines emerged in the 1970s as the evolution of older machines at the crossroads of permanent magnet DC machines, line-start PM machines, and power converter-supplied induction machines. Thus, they feature many advantages over other types of machines, making them the main competitor to induction motors as the workhorse of today's industry. The cost of permanent magnet material in the rotor is the major material cost of a PMS motor. The magnet cost for the motor used in the hybrid electric vehicle Prius 09, for instance, formed 64% and 81% of the total material cost in 2005 and 2012 (Rahman 2014), respectively. The magnet cost, though adding to the primary motor cost, reduces the operating cost. It eliminates the need for any type of rotor winding or cage, resulting in no copper loss on the rotor. The passive rotor causes no heat and means the machine can operate at lower temperatures. This, in turn, contributes to fewer losses. In contrast to induction motors, there is no stator magnetizing current in PMS motors. The less stator current as a result reduces the copper loss in stator windings. Adding all these energy savings, the overall machine efficiency increases significantly. PMS motors usually work at an efficiency much higher than that of standard induction motors of the same rating. They also are more efficient than premium efficiency motors, as seen in Fig. 1.17 (US Department of Energy 2014). In fact, they can meet or exceed the IE4 efficiency level. The high efficiency of the motors compared to that of induction motors does not drop much at high speeds or under heavy loads, resulting in a higher overall energy saving. The high efficiency of a typical PMS

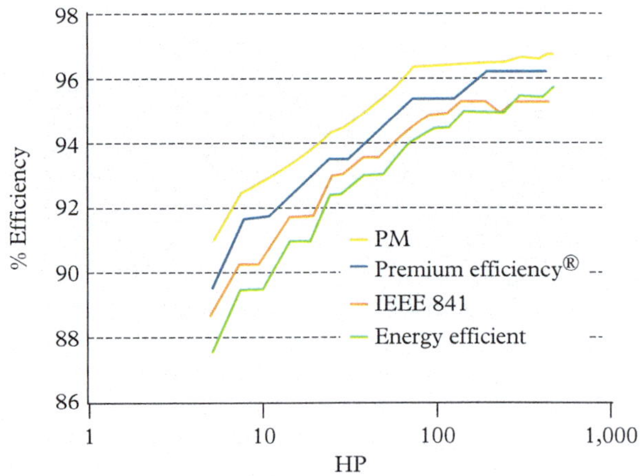

Figure 1.17 *Full load PMS motor efficiency in comparison with induction motor efficiency, illustrated from Baldor Electric (US Department of Energy 2014).*

motor results in a low operating cost for the motor that compensates for the high primary cost. Applications in which the machine has a high utilization factor of the machine results in an overall cost saving when PMS machines replace the induction machines. The saving value depends on the electricity price and is usually substantial when the electricity is delivered to the market with no subsidies.

As the global warming crisis necessitates environmental protection measures, a decisive factor—low emissions—must also be taken into account in assessing PMS motors with respect to their competitors. In this regard, the PMS motors are regarded as environmental friendly devices. This factor is accounted for by the lower carbon cost of PMS motors. Therefore, the overall cost of PMS motors is comparable to or less than that of induction motors.

Use of high energy PM materials results in high air gap flux density, with smaller motor size and weight leading to compact motor designs, with increased power and torque densities, as seen in Fig. 1.18 (Kang 2009). These features open the door to a wide range of applications, from electric vehicles to aerospace systems, which in addition to efficiency, weight, and volume are critical.

The higher torque density of modern PMS machines not only improves the steady-state operation of the machines, but also provides them with higher dynamic performances. This feature, in line with model simplicity and higher controllability of the motors, makes them the most appropriate option for high performance drive applications. The simplicity of the mathematical model stems from the fact that there is no electrical dynamics in the rotor. A simple motor model leads to a simple control system and/or higher control performance.

1.4.1 Structures

The stator of PMS machines is not different from that of wound rotor synchronous or induction machines. However, the main difference in

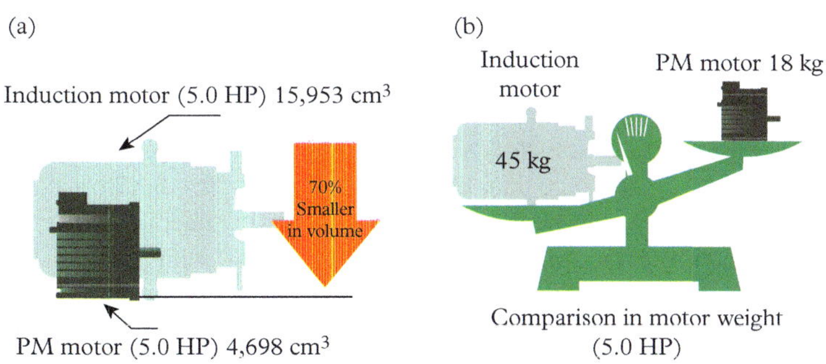

Figure 1.18 *Advantages of PMS motors over induction motors: (a) size comparison and (b) weight comparison (Kang 2009), with permission from Yaskawa America, Inc..*

the machines' basic structures lies in the rotor configuration, which influences motor performance in different aspects, including torque and power density, torque speed characteristic, flux weakening capability, and high-speed operation, among others.

1.4.1.1 Rotor configurations

Different rotor configurations stem from the PM pole shape and location in the rotor core, contributing to air gap flux density distribution. Using high energy rare earth magnets, low thickness poles are sufficient for establishing the needed air gap flux density.

Magnet location is the main criterion for categorizing PMS machines. There are two categories of PMS machines when looking at them from the control point of view: motors with and without magnetic saliency. The categories depend upon magnet location. The non-salient machines are surface-mounted magnet machines and the salient machines are interior magnet and inset magnet machines. Figure 1.19 shows schematically the cross-sections for some common rotor configurations, where direct- and quadrature-axes of the machines are also shown. The *d*-axis is usually depicted on the PM pole axis, while the *q*-axis is perpendicular to that in electrical angle. The cross-section of a surface-mounted magnet rotor is shown in Fig. 1.19(a), where the magnet poles are mounted on the surface of rotor core by strong adhesives as seen in Fig. 1.20 (Kikuchi 1997). The magnetic air gap is therefore large as the magnet permeability is almost the same as that of free air. This leads to the same reluctance in the magnet flux paths crossing the air gap along the *d*- and *q*-axes, providing a single machine inductance. However, the mechanical air gap is non-uniform, causing loss and audible noise as the motor works.

The salient machines are either inset magnet machines or interior PM machines. Inset magnet machines have magnet poles that are sandwiched inside the rotor iron, but their exterior surface is not covered by iron as seen in Fig. 1.19(b). These machines have uniform or non-uniform air gap, depending on the magnet poles being totally in the rotor iron or semi-projected. Interior magnet machines have PM poles totally buried inside the rotor iron, resulting in uniform air gap. However, the machine flux path in the rotor is not uniform as the *d*-axis flux path through poles experiences much higher reluctance than the *q*-axis flux paths through iron alone. This magnetic saliency is taken into account in machine modeling by introducing two winding inductances of different values. The saliency provides an extra torque known as reluctance torque. It also gives an excellent opportunity for effective flux weakening. It is recognized that the flux weakening, here and in many other

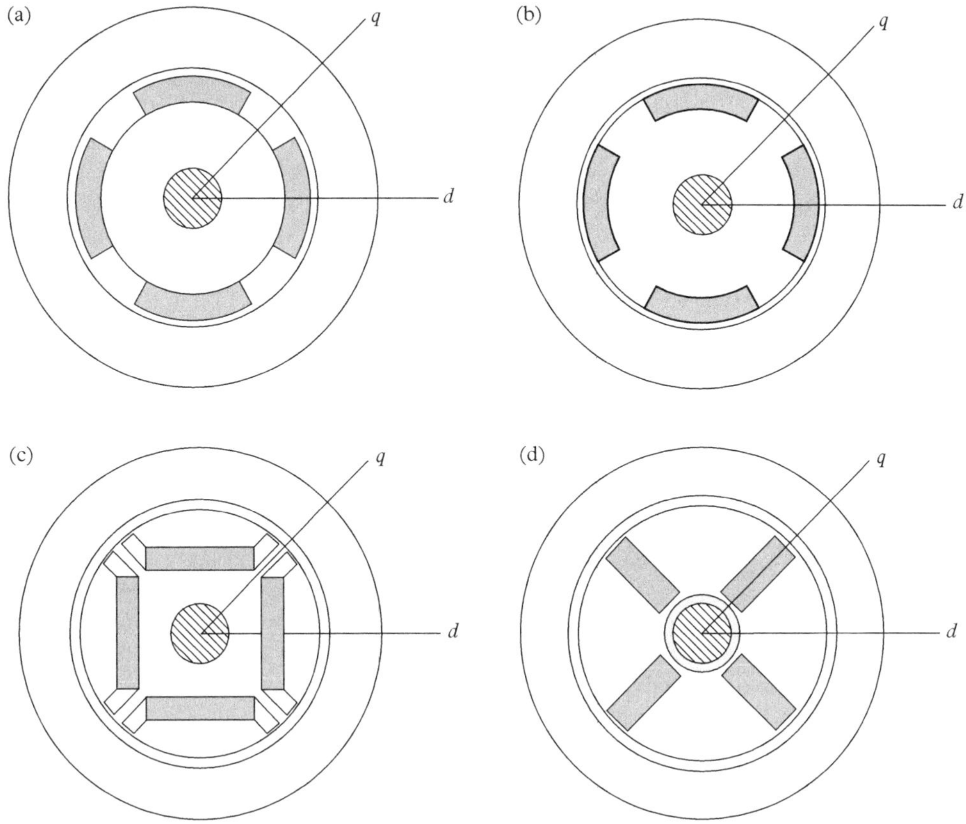

Figure 1.19 *PMS motor types based on the location of PM rotor poles: (a) surface-mounted, (b) inset, (c) interior, and (d) radial interior.*

machines, is necessary for high-speed operation and/or high efficiency, high power factor, etc. IPM machines are proposed in different configurations, depending on the magnet shape, location, and magnetization orientation. The rectangular shape is the most popular one, but the trapezoidal shape is also used. The magnets are usually placed close to the circumference of the rotor body as in Fig. 1.19(c). They may also be placed well inside the rotor. The IPM machines may also be built with radial poles as in Fig. 1.19(d). This pole configuration, known as flux concentrating, is usually appropriated for low energy magnet materials, e.g., ferrite and alnico magnets since it reinforces the air gap flux. The PM poles are usually magnetized along the rotor radius, especially in the shape depicted in Figs. 1.19(a)–(c). However, the magnetization may be oriented

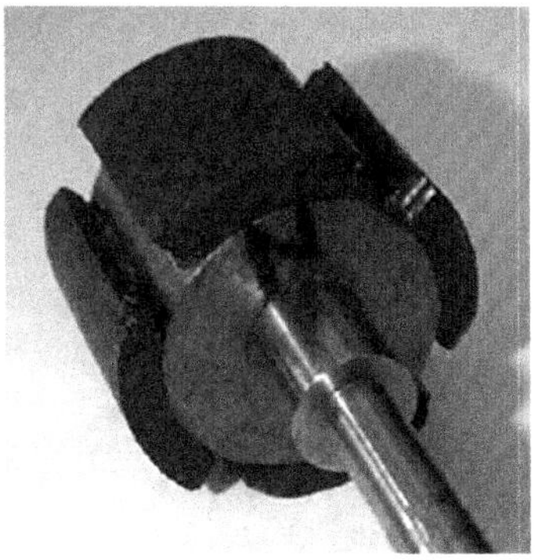

Figure 1.20 *Permanent magnet poles mounted on the surface of the rotor core (Kikuchi and Kenjo 1997).*

perpendicular to the radius as with the flux-concentrating configuration of Fig. 1.19(d). The thickness of the magnet poles along the magnetization orientation depends on the desired air gap flux. Therefore, high energy magnet poles are usually thin in that direction.

Many other rotor configurations for achieving different performance features have been suggested. A rotor consisting of eight PM poles of V shape as in Fig. 1.21, for instance, has been proposed for hybrid electric vehicles to yield higher torque and power densities.

Figure 1.21 *Rotor laminations of IPM motor with V-shaped poles used in 2010 Prius hybrid vehicle (Burress* et al. *2011).*

1.4.1.2 Pole magnetization orientation

A sinusoidal distribution of magnet flux density in the air gap contributes to a smooth developed torque. The poles shape may tend toward trapezoidal form by cutting their outer corners to produce a magnet flux density distribution in the air gap close to the sinusoidal form. However, the sinusoidal flux density distribution can be provided by magnetization orientation.

The magnetization of PM poles are oriented along the rotor radius or in parallel to the pole axis as shown in Figs. 1.22(a) and (b), respectively, for a surface-mounted PMS motor. In addition, radial sinusoidal amplitude magnetization and constant amplitude sinusoidal angle magnetization can be done as seen in Figs. 1.22(c) and (d), respectively. Magnetic flux density distribution along the circumferential of the air gap for the magnetization of Figs. 1.22(a) and (b) is shown in Fig. 1.23 (Shin-Etsu rare earth magnets). It is seen that the parallel magnetization provides a flux distribution closer to the ideal sinusoidal distribution. This is the most popular magnetization orientation.

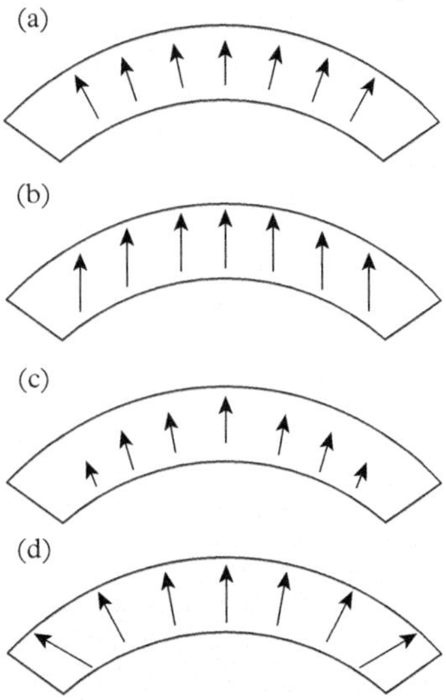

Figure 1.22 *Magnetization orientations of motor poles: (a) radial, (b) parallel, (c) radial sinusoidal, and (d) sinusoidal angle.*

Each type of magnetization needs special equipment and procedures to come up with the desired orientation that may be costly and complex.

1.4.1.3 Modular poles

Modulated poles with different PM materials may be used to provide sinusoidal flux density distribution as an alternative to complex magnetization systems. Figure 1.24 shows modular poles for surface-mounted and IPM motors (Isfahani *et al.* 2008). A PM module with stronger magnet material for the middle part and weaker magnet for the side parts provides a more sinusoidal flux distribution in the air gap, while saving the magnet cost.

Figure 1.25 compares the flux density distribution of a modular PM pole with those of conventional PM poles with weak and strong field intensity PM materials (Isfahani *et al.* 2008). In each case, the fundamental component of the flux distribution is presented for the

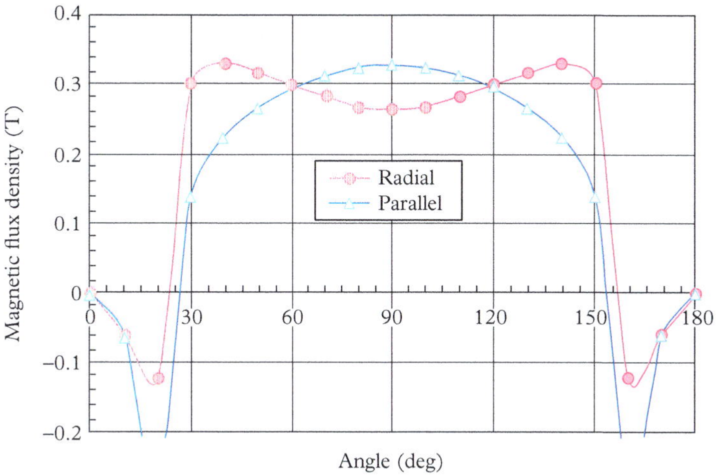

Figure 1.23 *Magnetic flux density distribution along the circumferential of the air gap for radial and parallel PM pole magnetizations (Shin-Etsu rare earth magnets).*

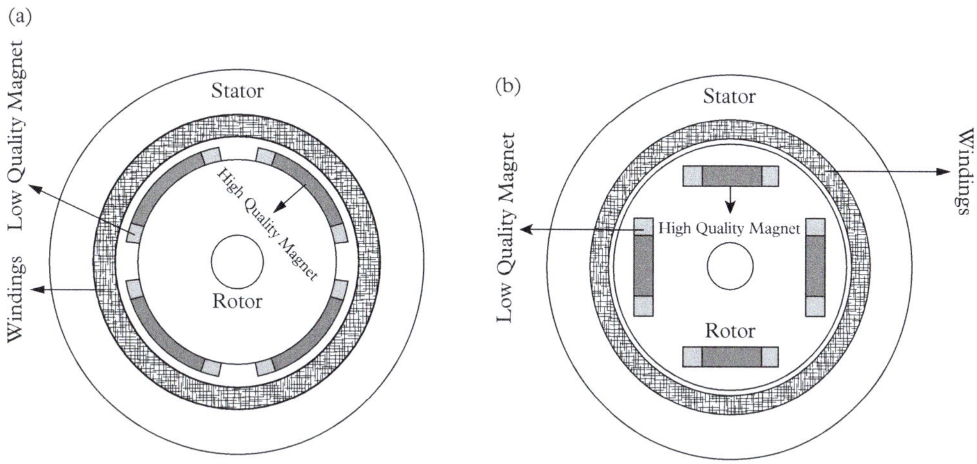

Figure 1.24 *Modulated PM poles: (a) surface-mounted poles and (b) interior type poles (Isfahani* et al. *2008).*

sake of comparison. It is seen that the flux distribution of the modular pole is closer to a sinusoidal shape than those of conventional PM poles.

1.4.2 Operating principles and characteristics

Torque production in PMS machines is the same as in wound rotor synchronous machines, except for the development of the excitation field by PM poles instead of a DC winding. At steady-state, a

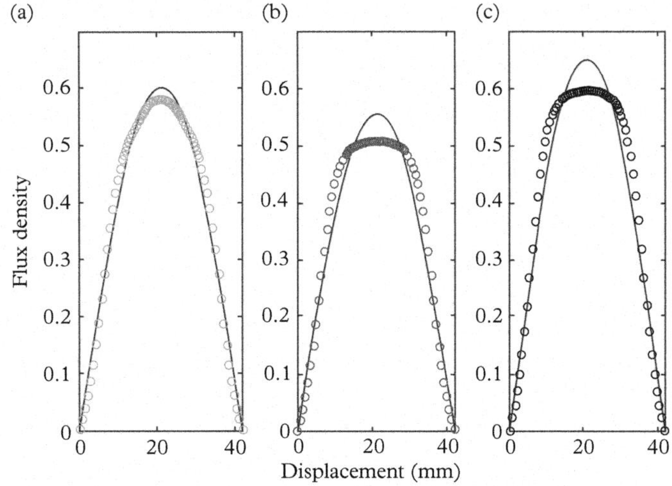

Figure 1.25 *Flux density distribution of the PM poles (circled line) together with their fundamental components (solid line): (a) modular pole with both weak and strong field intensity PM materials, (b) conventional pole with weak magnetic field intensity PM material, and (c) conventional pole with strong magnetic field intensity PM material (Isfahani* et al. *2008).*

synchronously rotating magnetic field with constant magnitude and speed is produced in the air gap as the stator phase windings are supplied by a balanced three-phase voltage source. The magnet poles also produce a sinusoidal distributed magnetic field with constant magnitude in the air gap. When the two fields rotate in the same speed, a constant torque known as the magnet torque or alignment torque is developed by their interactions. The machines with magnetic saliency, i.e., IPM, and inset machines, develop another torque in addition to the magnet torque known as reluctance torque. This torque is produced by the difference in the reluctances of flux paths along the *d*- and *q*-axes. As the rotor rotates, the stator windings examine a varying flux linkage due to the varying reluctances. This flux linkage in interaction with the stator current produces the reluctance torque. That is why salient machines have higher torque density.

The steady-state machine speed and torque depend on the stator current magnitude and frequency. At the transient state, e.g., at starting, the rotor and, thus, its magnetic field may not be synchronized with the stator-produced magnetic field. A main function of the motor control system is to provide synchronization during the transient state and maintain it during the steady-state. This is achieved by adjusting the stator current phase, in addition to its magnitude simply by positioning the rotor such that the stator and rotor fields are synchronized. Therefore, the machine rotor position is needed not only for position or speed control applications, but also for the normal operation of the machine. This is in contrast with the operation of induction machines where the information of rotor position is not required for basic operation. However, in drive applications the rotor position is usually needed for both induction and PMS motors.

Other than the super-premium efficiency, as mentioned previously, PMS motors enjoy high power factor due to the replacement of magnet flux linkage for the magnetization current as compared with induction motors. PMS motors, especially those with magnetic saliency (IPM motors), provide very desirable torque speed characteristics over wide speed ranges. Having higher torque density and lower rotor inertia, PMS motors demonstrate excellent dynamic performance. Non-active rotors result in a less complex mathematical model for PMS machines compared with that for most AC machines. This leads to better controllability and variable speed capability with less complicated control systems. A PMS motor works at a lower temperature, due to the lack of rotor copper loss and magnetizing current. This prolongs the motor lifetime and reduces the motor maintenance cost.

There are always non-ideal factors in the machine structure and power supply that result in torque ripples. These include:

- space harmonics in the rotating magnetic field, due to non-perfect sinusoidal distribution of windings around the air gap;
- the non-sinusoidal distribution of magnet flux in the air gap due to the magnet shape and magnetization orientation;
- the stator current noise due to the switching nature of the inverter operation; and
- the magnetic saturation of iron, and the cogging torque due to the attractive force between the magnet poles and stator slots.

Conventional solutions to torque ripples in other machines, such as skewing and chorded windings, in addition to optimized designs to reduce cogging torque are common in PMS machines (Jahns 1994). The inverter switching noise is filtered by machine impedance if the switching frequency is sufficiently high, as is the case in today's motor drives. Therefore, the ripple torque is very low and is not usually needed to use the available control techniques to reduce the effects of the previously mentioned factors. This extends the machine use in applications that need fast dynamics and smooth performance, such as machine tools.

1.4.3 Similar machines

PMS machines must be distinguished from other similar machines. Permanent magnet DC machines do not have much in common with

PMS machines in operating principle and construction. They work according to the principle of conventional DC machines, except that the magnetic field is developed by the magnet poles on the stator instead of electromagnets. Brushless DC (BLDC) machines are similar to PMS machines in their constructions, except for their concentrated stator windings and rectangular flux density in the air gap. As far as the motors' operation is concerned, they are supplied by DC sources, where the DC power is applied to two phases at a time, in contrast to the PMS machines, where they ideally have sinusoidal flux density distribution in the air gap and all phases are supplied by the AC sources. Line-start PM motors are PMS machines with a conducting rotor cage as the cage of induction motors. They are usually constant speed synchronous machines supplied by the main. They start as induction motors via the function of the cage and run as synchronous machines with the help of magnet poles.

Synchronous reluctance machines, while not using any magnet, have much in common in their construction and operating principles with IPM machines. They can be sought as IPM machines with no magnet poles but magnet hollow locations. Therefore, they have different d- and q-axis inductances and produce reluctance torque, as do the IPM machines. However, they do not develop the magnet torque specific to all PMS machines. Many of the materials on the modeling, analysis, and control of IPM machines can be adapted to synchronous reluctance machines if PM flux density is assumed zero.

1.5 Control system

PMS machines lend themselves to many types of high performance control methods more conveniently than induction and wound rotor synchronous machines. This is due to a simpler dynamic model of PMS machines, which stems from the replacement of PM poles for a conducting cage or a DC-supplied winding on the rotor. As a result, the machine model in different reference frames can readily be obtained. There is, in fact, no electrical dynamics in the rotor. Thus, the control system design focuses solely on the stator winding equations.

Vector control as the standard high performance control method of AC machines is applied to PMS machines in all possible reference frames. Direct torque control (DTC) as the second popular control method for AC machines has been widely investigated for PMS machines since the mid-1990s in many excellent varieties. In addition, less popular but more promising control methods, at least in the research stage, are also adapted to PMS machines. These include deadbeat control (DBC), predictive control, and combined control (CC). The control system shown in Fig. 1.1 may use any of the previously mentioned control methods.

The system may also include position and/or speed estimation schemes to provide sensorless control. Many types of estimation schemes common to other machines can be adapted to PMS machines. In addition, there are very effective position/speed estimation methods devoted solely to PMS machines. These methods mainly use the consequences of magnet poles or the magnetic saliency of the machines.

Finally, a control system may include parameter estimation schemes to provide accurate values of machine parameters under actual operating conditions. This is crucial in many control methods as their desirable performances rely on the accurate information of motor parameters. The control methods, position/speed estimation, and parameter estimation schemes are briefly reviewed later. The detailed elaboration of the methods and schemes is the focus of the following chapters.

1.5.1 Vector control

In electrical machines, a torque is developed due to the interaction of two magnetic fields. In a mathematical sense, the torque is proportional to the outer product of the two magnetic flux linkage vectors. In separately excited DC machines, the two magnetic fields are provided by two separate flux sources, i.e., stator or field winding and rotor or armature winding. Furthermore, in these machines a commutation system always locates the two magnetic fields spatially perpendicular to each other, providing the best space position of the two fields for the purpose of torque production. Alternatively, it is possible with regard to the machine torque proportional to an outer product of a magnetic flux linkage vector and a current vector according to Lorentz's law.

In AC machines, the two torque-contributing fields are neither independent nor orthogonal in general. Also, there is no commutation system to locate the fields in orthogonal positions with respect to each other. The main goal of a vector control (VC) method is to make the two magnetic fields independent and perpendicular as much as possible. Therefore, a separately excited DC machine may resemble the performance of an AC machine under VC, despite very different constructions of the two types of machines. VC is a means by which the stator phase currents of AC machines are transformed into a current vector of two perpendicular components, analogous to the field current and the armature current of a separately excited

DC motor, to control the torque and flux independently. The function of the commutation system in DC machines is carried out by the current transformation in vector-controlled AC machines to keep the current components in normal position with each other. Information on rotor position or flux linkage vector position is needed for the transformation.

A VC system is a fast control system since the perpendicular current components are instantaneously controlled. In scalar controls, e.g., a V/f control, only magnitude and frequency of the applied voltage to the machine are determined. In VC, however, the instantaneous value of the motor voltage is determined. It may be regarded that the phase of the applied voltage is also controlled through the control of a stator current vector, resulting in fast motor dynamics.

VC appears in different schemes and provides many control features, such as motor flux weakening, current and voltage limitation, unity power factor, and loss minimization operations, etc. It is accepted as the most common control method by the industry.

1.5.2 Direct torque control

It has been mentioned in connection with VC that the torque of an AC machine can be regarded as the outer product of rotor and stator flux linkages. In a PMS machine, the rotor flux linkage depends on the magnet poles and is fixed for a specific machine. Therefore, if the magnitude of the stator flux linkage is kept constant during the motor transient state, the torque response depends largely on the angle between the two flux linkages. Based on this reasoning, the DTC is a stator flux linkage VC, in contrast with the VC, which is a stator current VC. A fast torque response is provided by the DTC for an AC machine when the stator flux linkage vector is rotated with respect to the rotor flux linkage vector as fast as possible. At steady-state, the stator and rotor flux linkage vectors are synchronous. However, there is a slip speed during the torque transient. A high torque response can be provided if the speed difference becomes as high as possible. Bearing in mind that the rotor flux linkage rotates with the rotor speed and the rotor as a mechanical body has a large time constant, the rotor speed cannot change rapidly. As a result, if the speed of stator flux linkage is changed rapidly, the fast torque response is achieved.

It can be shown that the stator flux linkage vector rotates fast and its magnitude remains constant when a voltage vector perpendicular to the stator flux linkage vector is applied to the machine stator winding. This is the golden rule of DTC. However, the inverter is capable of producing a limited number of voltage vectors. Therefore, it is impossible to keep the applied voltage vector in perpendicular orientation to the flux linkage vector as the machine works. Nevertheless, it is possible to compromise to the best possible voltage vector among the inverter voltages in each inverter switching instance. Such a voltage vector must have an angle to the flux linkage vector as close as possible to 90° as a condition resulting in a fast torque response. An inverter switching logic, implemented as a predetermined look-up table, is an easy means for selecting the desired voltage vector command at each switching instance. The switching logic needs the errors between the commanded and the estimated values of flux linkage magnitude and torque in addition to the angle of the stator flux linkage vector. The estimated torque and flux linkage magnitude and angle are given by processing measured phase voltages and currents.

1.5.3 Predictive control

The model-based predictive control, as the most common version of predictive control, performs two main tasks when applied to a PMS motor. First, it uses a discrete-time model of the motor to predict the motor output variables for all possible inverter voltages, i.e., seven different voltage vectors for a two-level inverter. Second, an objective function is calculated for all the possible voltages to find the voltage, which minimizes the function. A representative of the errors between the predicted and the reference values of machine output variables is a part of the objective function in addition to other possible terms to optimize the machine performance as desired. The minimizing voltage is then applied to the machine by the inverter. The predictive control can be implemented in connection with VC or DTC. In the former case, the motor current components, and in the latter case, the motor torque and flux linkage are predicted. Comparing the predictive control with DTC, the former determines the desired voltage vector based on mathematical calculations, while DTC selects the desired voltage according to a heuristic approach. In addition, the predictive control uses a constant switching frequency, while DTC have a variable switching frequency.

1.5.4 Deadbeat control

Deadbeat control (DBC) can be regarded as a version of predicted control in which the desired voltage vector is determined, such that the controlled variable of the motor, e.g., the developed torque, reaches to its reference in one switching instance. Here, a discrete-time inverse model of the machine is used, instead of a conventional controller, to calculate the desired voltage to the machine. Therefore, an accurate model of the system is required.

Deadbeat control can be applied to motor drives as a current VC, or as a direct torque and flux control. In both schemes, the deadbeat controller calculates the desired voltage components to be applied to the machine in order to reach the reference signals in the next sampling interval. However, in the former scheme, the reference signals are the machine stator current components, while in the latter scheme the reference signals are the reference torque and flux linkage. In both schemes, a modulator like a sinusoidal PWM or SVM is needed to generate the switching signals for the inverter switches. Direct torque and flux control by DBC is much more interesting than the current control by DBC in motor drives.

1.5.5 Combined control

VC deals primarily with a machine current control, and controls the machine torque and flux linkages through the current control, while the DTC controls the machine flux linkage amplitude and torque directly without any current control. Also, the mathematical foundations, and the principles of VC and DTC, are far apart. However, the machine basic performances under these two methods are so close. This is because the two methods share a fundamental basis theoretically and practically despite their apparent differences. The basis shows how it is possible to replace current control with flux linkage control and vice versa. Therefore, consistent parts of VC and DTC are selected to build a CC system to combine performance advantages of the two methods. The CC combines a current control with a switching table. The torque development under the CC has less pulsation with respect to DTC and faster response with respect to VC.

1.5.6 Rotor position and speed estimation

Sensorless control is a common option in many motor drive products in the market. This is due to the mature technology, although a great deal of research is still focused on the topic. Motor position and speed controls are the most needed control loops in motor control applications. The control loops of these types need actual rotor position and/or speed signals. There are several means for sensing these variables for motor control systems. Using different laws and rules, they are classified under the mechanical sensors all together. Tachometers, resolvers, and encoders are among the most used devices of the kind. A wide variety of these devices is readily available in the market for sensing the rotor position or speed with varying degrees of accuracy, depending on the application requirements.

The sensorless motor control is very desirable in practice. It is also very challenging due to its demanding criteria, such as accuracy, robustness, swiftness, and capability of working over the entire range of motor operation. Consequently, the position and speed estimation has appeared as a popular research area in the past two decades, providing a very dense literature. This is also due to various opportunities available for tackling the problem. As a result, many position and speed estimation methods have been presented in the literature, many of them applied to PMS motors, or even specifically developed for these motors. The information of initial rotor position is needed, even for open-loop operation of PMS motors, which is not demanded by induction motors. This has brought a new dimension to the rotor position estimation of PMS machines and broadened the corresponding research agenda, compared with that of induction machines as the most practically available drive in the market.

It is common to categorize position estimation of PMS machines into two main groups of back electromotive force (EMF)-based methods and saliency-based methods. This is justified partly because of the distinct cause of the two groups. Back EMF depends on the motion of rotor (motor operation) and saliency is a motor structure property. However, as the research is expanding the methods, it is extremely difficult to stick to these traditional categories. Observer-based estimation is just an example that is capable of implementing both back EMF and saliency methods. Therefore, observer-based estimation has emerged as a major category of position and speed estimation. Hypothetical position estimation is another example. Online signal injection methods using saliency of the motor is a recently discovered commercial application. The position and speed estimation methods are presented in Chapter 6.

1.5.7 Motor parameter estimation

Motor parameters play an important role in motor modeling. Therefore, full information of the parameters is necessary for most motor control systems. They are also needed in rotor position and speed estimation. As a result, accurate determination of parameters is often regarded as a preparation step for motor control or, more commonly, as a part of control systems. The parameters vary with motor operating point and ambient conditions. The variations affect the motor performance and may deviate it from being optimal. Therefore, it is desirable to update the parameter values by parameter estimation schemes.

The PMS motor parameter estimation schemes are divided into two main methods, i.e., offline and online. Several offline schemes are proposed for estimation of all major motor parameters. The offline schemes can be divided into DC standstill and AC standstill tests, no-load test, load test, and VC scheme. These schemes use equivalent circuit equations or vector diagram equations in connection with measurement of voltage and current to calculate motor parameters for particular operating points or over a range of operating points. They may need special test arrangements for measuring machine voltage and current.

On the other hand, the online parameter estimation schemes use the online values of motor variables usually used in the control system to estimate motor parameters with the help of closed-loop observers. Many online estimation schemes are proposed for the parameters of PMS motor. They include closed-loop observers for estimation of motor inductances; model reference adaptive system (MRAS)-based estimation of λ_m and R_s, and also motor inductances; recursive least-squares (RLS) schemes for estimation of motor inductances and, for estimation of inductances and magnet flux linkage, extended Kalman filter (EKF) schemes, etc. The online schemes take into account the motor parameter variations caused by any sources. They are being increasingly implemented as a part of modern control systems in the market. The PMS motor parameter estimation is the focus of Chapter 7 in this book.

1.6 Summary

An overview of PMS motors and the related control system is presented in this chapter, as an introduction to the rest of the book. The interconnections of the control system to the power electronic inverter and the motor are emphasized, and the major parts of the system overviewed. Pulse width modulated VSIs, as the most commonly used power converter in PMS motor drives, is briefly discussed. PMS motors configurations and operating principles are also presented after considering characteristics of permanent magnet materials. Major PMS motor control methods including vector control, direct torque control, predictive control, deadbeat control, and combined control are briefly reviewed. Finally, several rotor position and speed estimation schemes, and offline and online parameter estimation methods are overviewed.

BIBLIOGRAPHY

Allied Market Research (2015). *Global permanent magnet motor market-size, industry analysis, trends, opportunities, growth and forecast, 2014–2020.* https://www.alliedmarketresearch.com/permanent-magnet-motor-market.

Atallah, K., Zhu, Z.Q., and Howe, D. (1998). Armature reaction field and winding inductances of slotless permanent-magnet brushless machines. *IEEE Trans. Magnet.* 34(5), 3737–3744.

Benecki, W.T., Claggett, T.K., and Trout, S.R. (2010). *Permanent magnets 2010–2020: a Comprehensive Overview of the Global Permanent Magnet Industry.* Walter T. Benecki LLC, Highland beach, FL.

Bose, B. (2005). *Modern Power Electronics and AC Drives.* Prentice-Hall of India, Delhi.

Boules, N. (1985). Prediction of no-load flux density distribution in permanent magnet machines. *IEEE Trans. Indust. Appl.* 21(3), 633–643.

Bradshaw, M.D. and Byatt, W.J. (1967). *Introductory Engineering Field Theory.* Prentice-Hall, Upper saddle River, NJ.

Breton, C., Bartolome, J., Benito, J., Tassinario, G., Flotats, I., Lu, C., and Chalmers, B. (2000). Influence of machine symmetry on reduction of cogging torque in permanent-magnet brushless motors. *IEEE Trans. Magnet.* 36(5), 3819–3823.

Burress, T.A., Campbell, S.L., Coomer, C., Ayers, C.W., Wereszczak, A.A., Cunningham, J.P., Marlino, L.D., Seiber, L.E., and Lin, H.T. (2011). *Evaluation of the 2010 Toyota Prius Hybrid Synergy Drive System* (No. ORNL/TM-2010/253). Oak Ridge National Laboratory (ORNL); Power Electronics and Electric Machinery Research Facility, Oak Ridge, TN.

Coey, J.M.D. (1996). *Rare-earth Iron Permanent Magnets* (No. 54). Oxford University Press, Oxford.

Coey, J.M.D. (2012). Permanent magnets: plugging the gap. *Scripta Materialia* 67(6), 524–529.

Dent, P.C. (2012). Rare earth elements and permanent magnets. *J. Appl. Phys.* 111(7), 07A721.

Ding, K. (2013). The rare earth magnet industry and rare earth price in China. *Joint European Magnetic Symposia*, http://www.epj-conferences.org.

Furlani, E. (1994). Computing the field in permanent-magnet axial-field motors. *IEEE Trans. Magnet.* 30(5), 3660–3663.

Gieras, J.F. and Wing, M. (2002). *Permanent Magnet Motor Technology.* Marcel Dekker, New York, NY.

Gutfleisch, O., Willard, M.A., Brück, E., Chen, C.H., Sankar, S.G., and Liu, J.P. (2011). Magnetic materials and devices for the 21st century: stronger, lighter, and more energy efficient. *Adv. Mater.* 23(7), 821–842.

Hanselman, D.C. (2003). *Brushless Permanent Magnet Motor Design.* The Writers Collective, Winnipeg, MB.

Hendershot, J.R. and Miller, T.J.E. (1994). *Design of Brushless Permanent-magnet Motors.* Magna Physics Publications and Oxford Science Publications, Oxford.

Howe, D. and Zhu, Z. (1992). The influence of finite element discretisation on the prediction of cogging torque in permanent magnet excited motors. *IEEE Trans. Magnet.* 28(2), 1080–1083.

Hwang, C-C. and Cho, Y. (2001). Effects of leakage flux on magnetic fields of interior permanent magnet synchronous motors. *IEEE Trans. Magnet.* 37(4), 3021–3024.

Iguchi, M. (1992). Market expansion programme of electric vehicles planned by the ministry of international trade and industry, Japan. In: *Proceedings of The Urban Electric Vehicle OECD-conference,* pp. 25–27. OECD, Stockholm.

Isfahani, A.H., Vaez-Zadeh, S., and Rahman, M.A. (2008). Using modular poles for shape optimization of flux density distribution in permanent-magnet machines. *IEEE Trans. Magnet.* 44(8), 2009–2015.

Islam, R., Husain, I., Fardoun, A., and McLaughlin, K. (2009). Permanent-magnet synchronous motor magnet designs with skewing for torque ripple and cogging torque reduction. *IEEE Trans. Indust. Appl.* 45(1), 152–160.

Jack, A., Mecrow, B., and Mitcham, A. (1992). Design and initial test results from a permanent magnet synchronous motor for a vehicle drive. In: *International Conference on Electrical Machine,* pp. 751–755. IEEE, Piscataway, NJ.

Jahns, T.M., Kliman, G.B., and Neumann, T.W. (1986). Interior permanent-magnet synchronous motors for adjustable-speed drives. *IEEE Trans. Indust. Appl.* 22(4), 738–747.

Jahns, T.M. (1994). Motion control with permanent-magnet AC machines. *Proc. IEEE* 82(8), 1241–1252.

Kang, J. (2009). *General Purpose Permanent Magnet Motor Drive Without Speed and Position Sensor.* Yaskawa Electric America, Inc, Oak Creek, WI. WP.AFD.05.

Kara, H., Chapman, A., Crichton, T., Willis, P., Morley, N., and Deegan, K. (2010). Lanthanide resources and alternatives. In: *A report for the Department for Transport and the Department for Business, Innovation and Skills.* Oakdene Hollins, Aylesbury, UK.

Kenjō, T. and Nagamori, S. (1985). *Permanent-magnet and Brushless DC Motors*. Oxford University Press, Oxford.

Kikuchi, T. and Kenjo, T. (1997). A unique desk-top electrical machinery laboratory for the mechatronics age. *IEEE Trans. Educ.* 40(4), 18 pp.

Kim, K.C., Lim, S.B., Koo, D.H., and Lee, J. (2006). The shape design of permanent magnet for permanent magnet synchronous motor considering partial demagnetization. *IEEE Trans. Magnet.* 42(10), 3485–3487.

Kim, K-C., Kim, K., Kim, H.J., and Lee, J. (2009). Demagnetization analysis of permanent magnets according to rotor types of interior permanent magnet synchronous motor. *IEEE Trans. Magnet.* 45(6), 2799–2802.

Kozawa, S. (2011). Trends and problems in research of permanent magnets for motors — addressing scarcity problem of rare earth elements. *Sci. Technol. Trends* 38(1), 40–54.

Krishnan, R. (2010). *Permanent Magnet Synchronous and Brushless DC Motor Drives*. CRC Press, Boca Raton, FL.

Miller, T.J.E. (1989). *Permanent Magnet and Reluctance Motor Drives*. Oxford Science Publications, Oxford.

Miller, T. and Rabinovici, R. (1994). Back-EMF waveforms and core losses in brushless DC motors. *IEE Proc. Elect. Power Appl.* 141(3), 144–154.

Nasar, S.A. (1987). *Handbook of Electric Machines*. McGraw-Hill, New York, NY.

Nasar, S.A., Unnewehr, L.E., and Boldea, I. (1993). *Permanent Magnet, Reluctance, and Self-synchronous Motors*. CRC Press, Boca Raton, FL.

Qishan, G. and Hongzhan, G. (1985). Effect of slotting in PM electric machines. *Elect. Mach. Power Syst.* 10(4), 273–284.

Parker, R.J. (1990). *Advances in Permanent Magnetism*. Wiley-Interscience, Hoboken, NJ.

Petrov, I. and Pyrhonen, J. (2013). Performance of low-cost permanent magnet material in PM synchronous machines. *IEEE Trans. Indust. Electron.* 60(6), 2131–2138.

Rahman, M.A. (2013). History of interior permanent magnet motors. *IEEE Indust. Appl. Mag.* **19** (1), 10–15.

Rahman, M.A. (2014). Status review of advances in hybrid electric vehicles. *General Meeting, IEEE PES*, http://docplayer.net/26036618-Status-review-of-advances-in-hybrid-electric-vehicles.html.

Rashid, M.H. (2004). *Power Electronics*, 3rd edn. Prentice-Hall, Upper Saddle River, NJ.

Rasmussen, K.F., Davies, J.H., Miller, T., McGelp, M., and Olaru, M. (2000). Analytical and numerical computation of air-gap magnetic fields in brushless motors with surface permanent magnets. *IEEE Trans. Indust. Appl.* 36(6), 1547–1554.

Roshen, W. (1991). Ferrite core loss for power magnetic components design. *IEEE Trans. Magnet.* 27(6), 4407–4415.

Sebastian, T. (1995). Temperature effects on torque production and efficiency of PM motors using NdFeB magnets. *IEEE Trans. Indust. Appl.* 31(2), 353–357.

Shin-Etsu Rare Earth Magnets. *Magnetic circuit analysis, SPM motor.* http://www.shinetsu-rare-earth-magnet.jp/e/circuit/index.html

Sitapati, K. and Krishnan, R. (2001). Performance comparisons of radial and axial field, permanent-magnet, brushless machines. *IEEE Trans. Indust. Appl.* 37(5), 1219–1226.

Slemon, G.R. and Liu, X. (1990). Core losses in permanent magnet motors. *IEEE Trans. Magnet.* 26(5), 1653–1655.

Toliyat, H.A. and Kliman, G.B. (Eds). (2004). *Handbook of Electric Motors.* CRC Press, Boca Raton, FL.

US Department of Energy (2014). *Premium efficiency motor selection and application guide*, DOE/GO-102014-4107. Washington DC.

Yang, Z., Shang, F., Brown, I.P., and Krishnamurthy, M. (2015). Comparative study of interior permanent magnet, induction, and switched reluctance motor drives for EV and HEV applications. *IEEE Trans. Transport. Electrif.* 1(3), 245–254.

Yeadon, W. H., and Yeadon, A. W. *Handbook of Small Electric Motors.* McGraw-Hill.

Zhu, J. G. and Ramsden, V. S. (1998). Improved formulations for rotational core losses in rotating electrical machines. *IEEE Trans. Magnet.* **34** (4), 2234–2242.

Zhu, Z. and Howe, D. (1992). Analytical prediction of the cogging torque in radial-field permanent magnet brushless motors. *IEEE Trans. Magnet.* 28(2), 1371–1374.

Machine Modeling

2

In this chapter, after a brief introduction to modeling in general, different approaches of modeling of PMS machines are recalled and the necessity of dynamic modeling of the machines for control purposes is described. The machine physical model that acts as a mediator between an actual machine and a mathematical model is presented. A range of PMS machine dynamic models in different reference frames (RFs), including stationary phase variable, stationary two-axis, rotor, and stator flux RFs, are presented in a rather concise, but self-sufficient manner. The space vector model of the machines in different RFs is also considered in connection with the previously presented models. Finally, iron losses and saturation are then taken into account.

2.1 Modeling

Modeling, in general, is a familiar concept for human beings in different contexts and notations. It can be defined in a broad sense as presenting anything by an analogous thing. It is usually intended by modeling to represent actual things by simpler and more understandable things, while preserving as much as original underlying specifications. A model may thus be regarded as a hypothesized abstraction of a past or present reality, or a future phenomenon or event to better convey the basic principles and main functionality of the thing it represents.

The main feature of modeling in the real world is in getting enough information about the construction and performance of something without having examined it or even having access to it. A model can never be the same as the thing it is modeling, since every model has its own assumptions and limitations. Therefore, the accuracy of a model rests in its closeness to the thing it is intended to present. A good model must also be capable of providing a deep insight to the real things, and help to investigate potentials and possibilities. In this sense, a model is an instrument for exploring the reality beyond the present knowledge. It may be regarded in this sense as a means of knowledge creation.

Control of Permanent Magnet Synchronous Motors. Sadegh Vaez-Zadeh. © Sadegh Vaez-Zadeh 2018.
Published in 2018 by Oxford University Press. DOI 10.1093/oso/9780198742968.001.0001

Modeling is carried out for many purposes in different disciplines by a wide variety of methods. Emulation, prototyping, and simulations may include some kinds of modeling. Modeling in engineering is necessary today to meet the specifications needed, increase quality, and reduce the cost of products. It increases the chance of getting a product right or fully fixes a problem with minimum practical effort on a real system. Engineer modeling, in connection with computer simulation, helps to study new ideas without a possible hardware malfunction; thus, it reduces the risk of failure in real systems.

Mathematical modeling of systems is an essential part of engineering today. It is carried out by providing a physical model first, which resembles the real system in terms of its basic construction and performance. A physical model is usually represented by a shape, its dimensions, and properties of used materials. The system performance is the state of one or more aspects of the system, e.g., the system electrical aspect. Depending on the aspect, the basic variables and parameters of the system are determined. Then, the variables and parameters are related by algebraic and deferential equations governing the behavior of the system based on scientific laws. A mathematical model is either a static or a dynamic model. A static model represents the system in steady-state, while a dynamic model represents the time variations of the system state. The static models are presented by algebraic equations, but dynamic models include differential equations, too.

An electrical machine can be modeled mathematically by accounting for different aspects of the machine. Electrical, mechanical, magnetic, and thermal modeling of electrical machines have found applications in analysis, design, and control of machines. Dynamic modeling of the electromechanical aspect of machines is essential for machine control and dynamic performance analysis. A dynamic model of the machines consists of a set of algebraic and differential equations, which relate voltages, currents, flux linkages, torque, speed, etc. Dynamic performance of electric machines is very rapid with respect to most dynamic systems due to small time constants of the machine variables. Therefore, the model should be solved delicately to simulate the machine dynamic performance. PMS machine dynamic modeling is elaborated here, after taking a quick look at the physical model of the machines.

2.2 Physical model of PMS machines

A PMS machine is a complex system having many characteristics and specifications. Ideally, one may think of considering all these characteristics and specifications in designing a control system for the machine. However, this is neither necessary for achieving desirable performances nor practical in realizing such a system. The reason is obvious—one should not overdo the control system design. In fact, essential characteristics and specifications of the system should be taken into account, while others may be overlooked. Therefore, wise simplification of the machine structure and performance must be assumed to keep the fundamental behavior of the machine accurate, and leave unimportant aspects out of the modeling. By doing this, a physical model as a schematic description of the machine is obtained. This is closely related to the actual machine to meet the accuracy of the modeling on one hand, while it is related to the mathematical model to keep it manageable and solvable, on the other.

2.2.1 Machine schematic view

A schematic view of two popular PMS machines with four surface-mounted and interior permanent magnet poles is depicted in Figs. 2.1(a) and (b), respectively, to show the machine stator, rotor, and air gap. The stator windings of PMS machines are distributed sinusoidally in the stator slots around the periphery of the stator core with 120° displacements, producing a smooth rotating magnetic field in steady-state operation. Winding axes are, therefore, shown in the

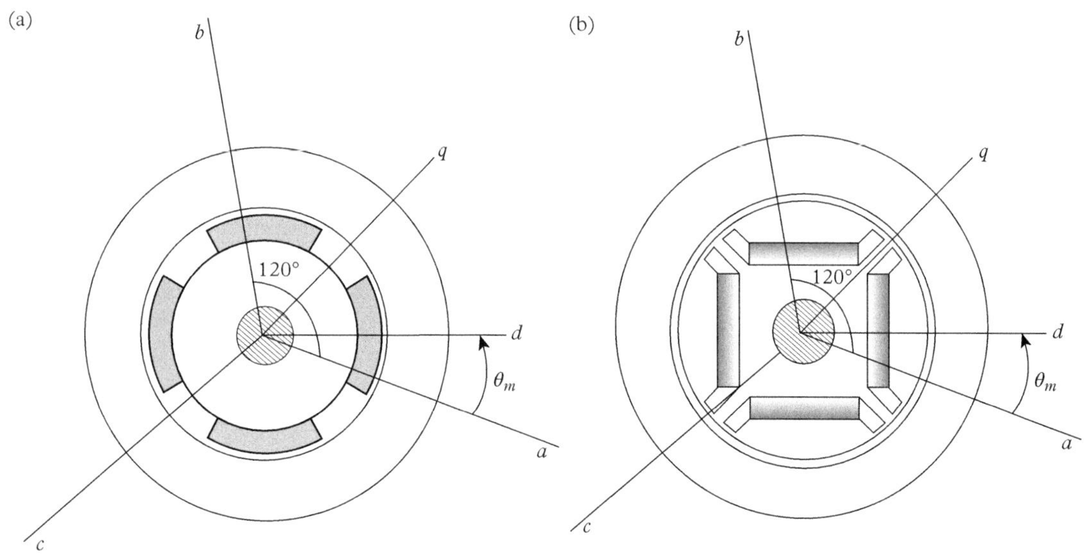

Figure 2.1 *Schematic view of PMS machines with (a) surface mounted poles and (b) interior poles.*

physical model as three stationary axes fixed to the winding centers and are 120° apart as noted by *a*, *b*, and *c* in Fig. 2.1. Also, the axis of a permanent magnetic pole, as the direct axis denoted by "*d*," is shown in the figure for surface-mounted PM and interior PM machines. This axis is placed on the point of maximum flux density of the pole through its length, which is at the center of the pole. A quadrature axis, which is 90° (electrical) apart from the *d*-axis and denoted by "*q*," is also shown in the figure. The axes are fixed to the rotor body and rotate with the rotor. The two sets of *a*–*b*–*c* and *d*–*q* axes are referred to as phase variable RF and rotor RF, respectively, and play an important role in the dynamic modeling of PMS machines.

2.2.2 Modeling assumptions

The physical model is derived by considering the following assumptions:

1. The stator windings of PMS machines are distributed sinusoidally in the stator slots around the stator core's periphery to produce a smooth rotating magnetic field in steady-state operation. However, stator slot harmonics distort the magnetic flux produced by the machine windings. They are ignored in the physical model. This means that the stator winding of each phase in the physical model is assumed to be distributed continuously and sinusoidal, such that a rotating magnetic field with fixed magnitude and constant speed is produced by the winding currents under the balanced steady-state operation. This assumption is justified by the fact that the slot harmonics have no influence on the fundamental dynamics of the machine due to their high frequency.
2. Switching power supplies used in variable speed drives provide switching harmonics in the machine currents and distort magnetic flux produced by the windings. It is assumed, however, in the physical model that the machine is supplied by a drive system with no switching harmonics. In fact, the frequency of switching harmonics is also much higher than the angular frequency of the machine, such that they may be considered as noises that are filtered by winding inductance.
3. A sinusoidal flux density distribution around the periphery of the machine air gap produced by permanent magnet poles mounted on the rotor surface or buried inside the rotor core is assumed in the physical model despite an almost trapezoidal pattern of flux distribution in most actual machines

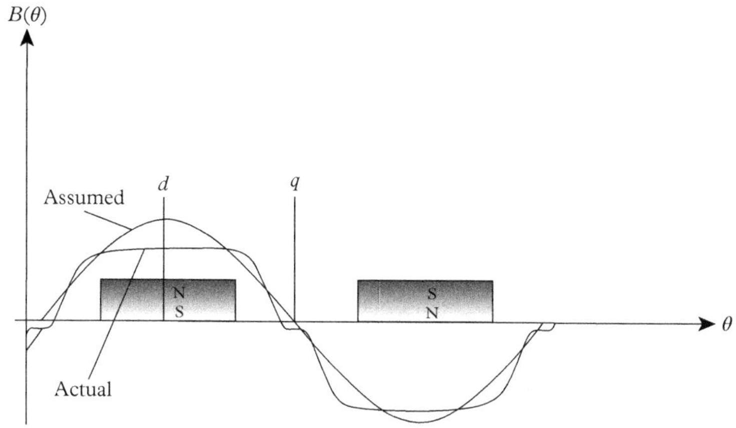

Figure 2.2 *Actual and assumed patterns of air gap flux density distribution produced by a pair of PM poles.*

(Hendershot and Miller 2010). The sinusoidal flux density distribution may in fact be the fundamental frequency component of the actual distribution. Figure 2.2 shows the two flux density distributions for a pair of poles. The *d*- and *q*-axes are also shown in the figure.

4. Iron losses occur in a PMS machine due to hysteresis and eddy currents phenomena. Both factors depend on the magnitude of magnetic flux in the iron and the supply frequency. Therefore, in high speed and high flux machines the losses behave against the essential high efficiency of the machines provided by the lack of slip loss and rotor winding loss with respect to induction machines and wound rotor synchronous machines. Conventional control methods of PMS machines ignore the iron losses to focus on the basic control issues. Therefore, the physical model primarily ignores the iron losses. However, loss minimization control of PMS motor drive is needed for high efficiency drives. This type of control needs machine models with iron losses. Such a model is thus presented separately in this chapter.

5. A PMS machine that is optimally designed is subject to magnetic saturation in some part of machine stator and rotor iron cores in particular operating conditions. The extent and occasions of saturation depend on many factors, including the rotor configuration, type of permanent magnet material used for the rotor poles, etc. The chance of saturation is high in PMS machines with high energy product permanent magnets. Saturation may have undesirable consequences in dynamic performance of PMS machines under a conventional type of

control system. However, the physical model generally ignores the magnetic saturation. Nevertheless, a machine model that considers saturation is presented separately in this chapter.

6. Stray losses are ignored in the modeling through this chapter. This is justified by negligible size of the losses and the complications involved in their modeling.
7. Bearing in mind the earlier assumptions, sinusoidal induced EMFs in machine-phase windings are assumed in the physical model under the balanced steady-state operating conditions. This develops a constant and smooth torque under the conditions mentioned previously (Pillay and Krishnan 1989).

The earlier-mentioned assumptions, although more or less different from the reality, should be considered for the sake of simplifying mathematical modeling, and maintaining and reflecting the basic construction and fundamental performances of PMS machines.

2.3 Phase variable model

The mathematical model of PMS machines can be derived from the model of conventional wound rotor synchronous machines with some modifications. In PMS machines, the excitation is provided by PMs instead of a DC field winding. Therefore, the field equations are removed from the machine model. There are also no damper windings in inverter-fed PMS machines. As a result, the voltages and currents of PMS machines are restricted to those of the stator windings and there are no rotor equations in the model.

2.3.1 Phase variable reference frame

The stator windings of machine are distributed in the stator slots around the periphery of the stator core with 120° displacements. Taking this fact into account, a convenient way of machine modeling is to consider any set of machine phase variables, i.e., phase voltages, currents, or flux linkages on a three-axis RF, with its axes being the magnetic axes of stator windings *a*, *b*, and *c*. This stationary RF can be regarded as *a–b–c* RF, stator RF, or phase variable RF, and is shown in Fig. 2.3 (Krause 1986).

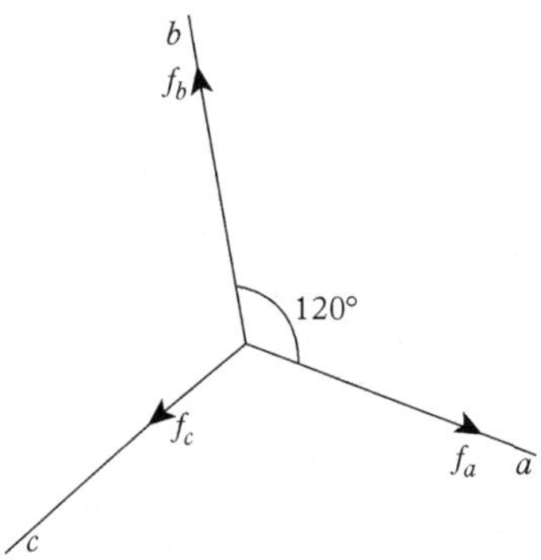

Figure 2.3 *Presentation of machine phase variables in stator RF.*

Three machine variables in time domain f_a, f_b, and f_c are shown in the RF. Here, f may represent a phase voltage, a phase current, or a phase flux linkage. It must be emphasized that f_a, f_b, and f_c do not represent vectors of any type, e.g., space vectors. They should not be confused with phasors. According to this type of representation, voltage, current, and flux linkage of each phase, all are depicted on the corresponding phase axis. Also, in general, they are not sinusoidal signals, but signals of the same quantity of machine three phases, with arbitrary variations in time. In a special case, when a machine operates under steady-state conditions, however, f_a, f_b, and f_c represent a symmetrical three-phase sinusoidal time-varying system of variables.

2.3.2 Machine equations in phase variable reference frame

The voltage equations of a three-phase PMS motor, supplied by an arbitrary three-phase power source, in contrast to a wound rotor synchronous machine model, are restricted to the stator equations only and are represented in machine variables *a–b–c* stationary RF by (Krause *et al.* 2013; Pillay and Krishnan 1989)

$$\begin{bmatrix} v_a \\ v_b \\ v_c \end{bmatrix} = \begin{bmatrix} R_s & 0 & 0 \\ 0 & R_s & 0 \\ 0 & 0 & R_s \end{bmatrix} \begin{bmatrix} i_a \\ i_b \\ i_c \end{bmatrix} + p \begin{bmatrix} \lambda_a \\ \lambda_b \\ \lambda_c \end{bmatrix} \tag{2.3.1}$$

where p is a derivative operator, R_s is a stator phase winding resistance, and v_a, v_b, v_c, i_a, i_b, i_c, and λ_a, λ_b, λ_c are the phase winding voltages, currents, and flux linkages, respectively. These equations are generally time-varying. A three-phase equivalent circuit of the machine, based on eqn (2.3.1), can be drawn as shown in Fig. 2.4 where derivatives of the phase flux linkages are substituted by the equivalent induced voltages e_a, e_b, and e_c in the phases.

Each of the three stator phase flux linkages is provided by two independent flux sources, i.e., the stator currents and the magnet poles. Thus, the flux linkages can be presented in a matrix form as

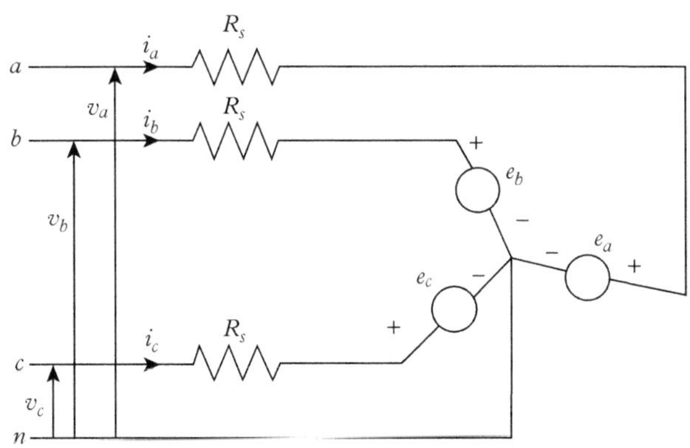

Figure 2.4 *A three-phase equivalent circuit model of PMS machines.*

$$\begin{bmatrix} \lambda_a \\ \lambda_b \\ \lambda_c \end{bmatrix} = \begin{bmatrix} L_{aa} & M_{ab} & M_{ac} \\ M_{ba} & L_{bb} & M_{bc} \\ M_{ca} & M_{cb} & L_{cc} \end{bmatrix} \begin{bmatrix} i_a \\ i_b \\ i_c \end{bmatrix} + \lambda_m \begin{bmatrix} \cos\theta_r \\ \cos\left(\theta_r - \frac{2\pi}{3}\right) \\ \cos\left(\theta_r + \frac{2\pi}{3}\right) \end{bmatrix}, \tag{2.3.2}$$

where λ_m represents the maximum flux linkage of a phase, produced by the magnet poles only—it depends on the magnet properties and motor structure. Therefore, it is constant for a specific machine. The self-inductances, L_{aa}, L_{bb}, L_{cc}, and the mutual inductances, M_{ab}, M_{bc}, ... M_{ca}, depend on the rotor position, θ_r , defined as the electrical angle of rotor pole axis or rotor d-axis, from the axis of stator phase winding a where

$$\theta_r = P \cdot \theta_m, \tag{2.3.3}$$

and P is the number of pole pairs, and θ_m is the mechanical angle of rotor pole axis or rotor d-axis, from the axis of stator phase winding a, as shown in Fig. 2.1. The self and the mutual inductances are given by (Kulkarni and Ehsani 1992)

$$L_{aa} = L_1 + L_2 \cos(2\theta_r), \tag{2.3.4}$$

$$L_{bb} = L_1 + L_2 \cos\left(2\theta_r + \frac{2\pi}{3}\right), \tag{2.3.5}$$

$$L_{cc} = L_1 + L_2 \cos\left(2\theta_r - \frac{2\pi}{3}\right), \tag{2.3.6}$$

$$M_{ab} = M_{ba} = -\frac{1}{2}L_1 + L_2 \cos\left(2\theta_r - \frac{2\pi}{3}\right), \tag{2.3.7}$$

$$M_{bc} = M_{cb} = -\frac{1}{2}L_1 + L_2 \cos(2\theta_r), \tag{2.3.8}$$

$$M_{ac} = M_{ca} = -\frac{1}{2}L_1 + L_2 \cos\left(2\theta_r + \frac{2\pi}{3}\right), \tag{2.3.9}$$

where L_1 is the inductance component due to the space fundamental air gap flux linkage and L_2 is the inductance component due to rotor position-dependent flux linkage. Thus, the inductances are time varying in general when the rotor is not stalled. However, the inductance variations do not have any connection with the time variation of applied arbitrary voltages.

Substituting eqn (2.3.2) into eqn (2.3.1) yields the machine voltage equations in terms of the stator currents and the magnet flux linkages as

$$\begin{bmatrix} v_a \\ v_b \\ v_c \end{bmatrix} = \begin{bmatrix} R_s & 0 & 0 \\ 0 & R_s & 0 \\ 0 & 0 & R_s \end{bmatrix} \begin{bmatrix} i_a \\ i_b \\ i_c \end{bmatrix} + p \begin{bmatrix} L_{aa} & M_{ab} & M_{ac} \\ M_{ba} & L_{bb} & M_{bc} \\ M_{ca} & M_{ca} & L_{cc} \end{bmatrix} \begin{bmatrix} i_a \\ i_b \\ i_c \end{bmatrix}$$

$$+ \lambda_m p \begin{bmatrix} \cos\theta_r \\ \cos(\theta_r - \frac{2\pi}{3}) \\ \cos(\theta_r + \frac{2\pi}{3}) \end{bmatrix}. \qquad (2.3.10)$$

The instantaneous electrical input power to the motor is calculated as

$$P_{in} = v_a i_a + v_b i_b + v_c i_c. \qquad (2.3.11)$$

A small part of the input power is dissipated as electrical losses consisting of copper, iron, and stray losses. The stray loss is ignored here, as it is mentioned in physical modeling with quite acceptable approximation. The stator iron losses will also be ignored for the moment, but will be considered later in this chapter. Therefore, taking the motor windings loss, P_{Cu}, into account, the electromagnetic power is expressed as

$$P_{el} = P_{in} - P_{Cu}. \qquad (2.3.12)$$

Ignoring the rotor iron losses as in the physical model, the power mentioned is totally converted to mechanical power, since there is no copper loss in the rotor. Consequently, the motor electromagnetic torque is given by

$$T_e = \frac{P_{el}}{\omega_m}, \qquad (2.3.13)$$

where ω_m is the mechanical angular speed of a machine, which corresponds to the electrical angular speed, ω_r, and the number of motor pole pairs, P, by

$$\omega_m = \omega_r / P. \qquad (2.3.14)$$

2.4 Stationary two-axis model

In three-phase AC machines, the machine windings perform identical roles under balanced operating conditions. Also, in contrast to DC machines, both stator and rotor often contribute to the air gap flux. Therefore, the phase variable model of the machine may not be the best model to separate torque control from flux control. Fortunately, the two-axis theory of AC machines in stationary and rotating RFs has been developed to do the job. The machine model in the stationary RF is presented in this section, while the rotating RF will be presented in other sections.

2.4.1 Two-axis stationary reference frame

An equivalent two-winding machine of three-winding machines can be derived mathematically (Krause 1986). Such a fictitious two-winding machine as is presented schematically in Fig. 2.5 is particularly useful in the machine analysis and control system design.

The equivalent machine has the same performance as the original three-phase PMS machine. Also, the machine rotor is the same as the rotor of the original machine. However, there are two windings in the stator of the fictitious machine distributed sinusoidally in the stator slots with their axes 90° apart. The two windings may be named as stationary direct and quadrature windings. The fictitious machine can be modeled in a two-axis RF fixed to the machine stator together with the magnetic axes of the two windings. Therefore, the axes are perpendicular to each other and referred to as the *D*- and *Q*-axes as shown in Fig. 2.5.

2.4.2 Model transformation to stationary two-axis reference frame

A *D*–*Q* RF is depicted together with an *a*–*b*–*c* reference in Fig. 2.6. Referring to Fig. 2.6, the phase variables f_a, f_b, and f_c can be mapped to the *D*- and *Q*-axes to obtain the fictitious variables f_D and f_Q, respectively. These variables in general may represent arbitrary

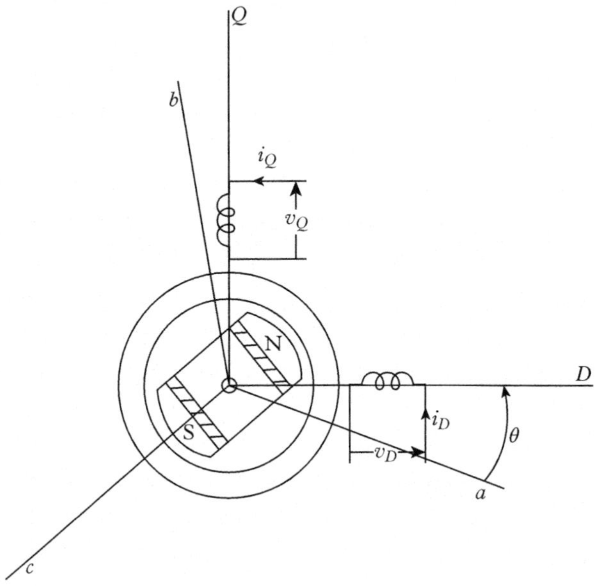

Figure 2.5 *A schematic view of a two-winding fictitious PMS machine.*

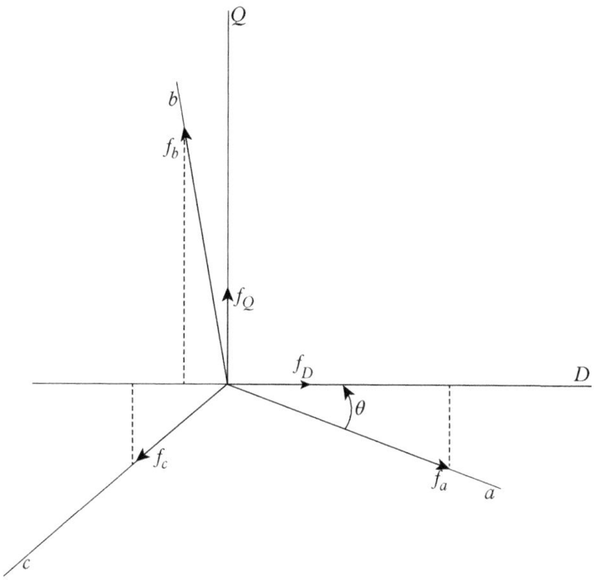

Figure 2.6 *System transformation from three-axis stationary (a–b–c) to two-axis stationary (D–Q) RF.*

time-varying voltages, currents, or flux linkages of a two-winding fictitious machine. The time variations of D and Q variables depend on the time variations of f_a, f_b, and f_c. Figure 2.6 shows in particular the procedure of mapping a–b–c variables on the D-axis to get f_D. It is obtained by projecting f_a, f_b, and f_c on the D-axis, and then adding together the projected components. The same can be done to get f_Q.

This RF transformation requires an accurate knowledge of the angle between the three-phase stationary RF and the two-axis stationary RF, i.e., the angle from the a-axis to D-axis, θ. Knowing this angle, the mathematical transformation may be done by the matrix form equation

$$\begin{bmatrix} f_D \\ f_Q \\ f_O \end{bmatrix} = \boldsymbol{K} f_{abc} = \frac{2}{3} \begin{bmatrix} \cos\theta & \cos\left(\theta - \frac{2\pi}{3}\right) & \cos\left(\theta + \frac{2\pi}{3}\right) \\ -\sin\theta & -\sin\left(\theta - \frac{2\pi}{3}\right) & -\sin\left(\theta + \frac{2\pi}{3}\right) \\ \frac{1}{2} & \frac{1}{2} & \frac{1}{2} \end{bmatrix} \begin{bmatrix} f_a \\ f_b \\ f_c \end{bmatrix}. \tag{2.4.1}$$

A fictitious variable, f_O, is also used in the transformation for the sake of matrix symmetry in order to make possible a reverse transformation from D–Q RF back to a–b–c RF by inverting the 3×3 transformation matrix in eqn (2.4.1). This is sometimes called the zero component and may be regarded as perpendicular to the plane of D–Q RF. The zero component is independent from θ. This component will be 0 if

$$f_a + f_b + f_c = 0, \tag{2.4.2}$$

as is the case for the variables of a star-connected three-phase winding with no neutral connection or a delta-connected three-phase winding. The true stationary two-axis RF is then is established.

A reverse transformation from D–Q RF to a–b–c RF can be done by

$$\begin{bmatrix} f_a \\ f_b \\ f_c \end{bmatrix} = \boldsymbol{K}^{-1} f_{DQO} = \begin{bmatrix} \cos\theta & -\sin\theta & 1 \\ \cos(\theta - \frac{2\pi}{3}) & -\sin(\theta - \frac{2\pi}{3}) & 1 \\ \cos(\theta + \frac{2\pi}{3}) & -\sin(\theta + \frac{2\pi}{3}) & 1 \end{bmatrix} \begin{bmatrix} f_D \\ f_Q \\ f_O \end{bmatrix}. \tag{2.4.3}$$

It is common to choose a D-axis in line with the a-axis, i.e., $\theta = 0$ in eqn (2.4.1). In fact, this special case is often used in a–b–c to D–Q transformation to simplify the calculations.

If the general variable f throughout eqn (2.4.1) is replaced by a voltage symbol v, then a voltage transformation from a–b–c RF to D–Q reference will be done by (2.4.1). This transformation can be applied to the voltage equations of eqn (2.3.1) by multiplying both sides of the equation to the matrix form factor of phase variables of eqn (2.4.1), i.e., $\boldsymbol{K}$. This provides a pair of voltage equations in two-axis stationary RF as

$$v_D = R_s\, i_D + p\lambda_D, \tag{2.4.4}$$
$$v_Q = R_s\, i_Q + p\lambda_Q, \tag{2.4.5}$$

where all voltage, current, and flux linkage components are in D–Q RF.

A two-part equivalent circuit can be drawn, based on eqns (2.4.4) and (2.4.5), to visualize the D–Q model as in Fig. 2.7.

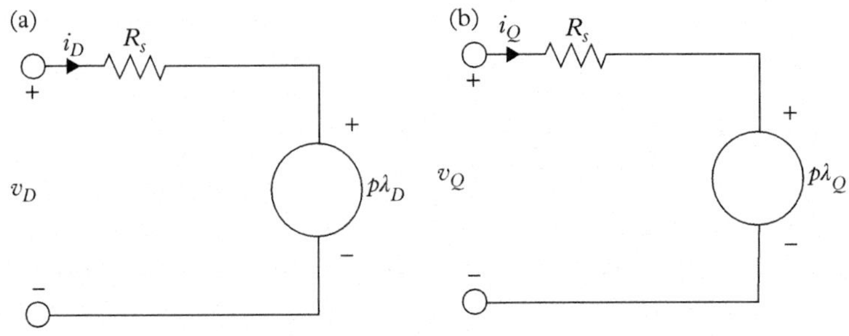

Figure 2.7 *Equivalent circuit model of PMS machines in two-axis stationary RF. (a) D-axis circuit, (b) Q-axis circuit.*

This is an interesting feature of the equivalent circuit in this RF that the D- and Q-axis circuits are separately drawn as for a separately exited DC machine in which the field and armature circuits are separated circuits. Nevertheless, in this derivation, each flux linkage component depends on both current components. Therefore, a coupling exists between the two circuits.

The machine instantaneous input power in D–Q RF must be equal to the instantaneous input power of eqn (2.3.11). This is obtained by

$$P_{in} = \frac{3}{2}(v_D i_D + v_Q i_Q + 2v_O i_O), \tag{2.4.6}$$

where the D- and Q-axis voltage and current components are obtained by eqn (2.4.3). The third term vanishes if eqn (2.4.2) is held.

With assumptions the same as those considered in finding the electromagnetic power in the a–b–c RF, the electromagnetic power in the D–Q RF can be obtained by referring to Fig. 2.7. Setting aside the copper loss occurring in R_s in the circuits of Fig. 2.7, the remaining power is converted to electromagnetic power through an energy conversion process. Therefore, the electromagnetic power is represented by

$$P_{el} = \frac{3}{2}(i_Q p\lambda_Q + i_D p\lambda_D). \tag{2.4.7}$$

Referring to eqns (2.4.4) and (2.4.5) in steady-state, it can be shown that

$$P_{el} = \frac{3}{2}\,\omega_r\left[\lambda_D\, i_Q - \lambda_Q\, i_D\right]. \tag{2.4.8}$$

Using eqns (2.3.13) and (2.3.14), the electromagnetic torque is given as

$$T_e = \frac{3}{2}\,P(\lambda_D\, i_Q - \lambda_Q\, i_D). \tag{2.4.9}$$

Although eqn (2.4.9) is obtained here by using eqns (2.4.4) and (2.4.5) in a steady-state, it can be shown that the torque equation is true under transient conditions, too.

Under the steady-state operation of the machine, the phase variables form a balanced three-phase sinusoidal system with a constant magnitude of F and a constant angular frequency of ω_e as

$$f_a = F\cos\omega_e t, \tag{2.4.10}$$

$$f_b = F\cos(\omega_e t - 2\pi/3), \tag{2.4.11}$$

$$f_c = F\cos(\omega_e t + 2\pi/3). \tag{2.4.12}$$

Then, the D and Q components are sinusoidal signals of the same magnitude and angular frequency, but with a phase difference of $\pi/2$ as

$$f_D = F\cos(\omega_e t + \theta), \tag{2.4.13}$$

$$f_Q = -F\sin(\omega_e t + \theta). \tag{2.4.14}$$

In a special case, where there is no angle between the two RFs, i.e., $\theta = 0$, then (2.4.13) and (2.4.14) reduce to

$$f_D = F\cos(\omega_e t), \tag{2.4.15}$$

$$f_Q = -F\sin(\omega_e t). \tag{2.4.16}$$

2.5 Rotor reference frame model

The machine model in the stationary two-axis stationary RF, like the phase variable model, suffers from time variation of parameters. The variations complicate the control system design. Time variations can be compensated by transforming the model to a two-axis model in a rotating RF fixed to the machine rotor (Krause 1986). The model parameters in this RF will appear to be constant as a major feature of the model, along with other model features.

2.5.1 Transformation to rotor reference frame

The rotor RF has two orthogonal axes fixed to the rotor, including a longitudinal axis (d-axis) aligned with the magnetic axis of a rotor permanent magnet pole, which is normally at the center of the magnet pole with a positive direction from the south to north pole of the magnet inside the magnet and a quadrature axis (q-axis) perpendicular to d-axis, i.e., 90° ahead of the d-axis in the direction of rotor rotation as mentioned in relation to the physical model of Section 2.2.

Figure 2.8 shows the transformation from a two-axis D–Q stationary RF to the rotor RF by visualizing the procedure through which f_d is determined. This is done by first projecting f_D and f_Q to the d-axis. Then, an algebraic summation of the two components projected on this axis gives the value of f_d. Following a similar procedure regarding the q-axis gives f_q. The space angle θ_d in Fig. 2.8 is the angle between the stationary two-axis stationary RF and the rotor RF in the direction of rotor rotation in electrical degree. It depends on the rotor speed and the angle initial value.

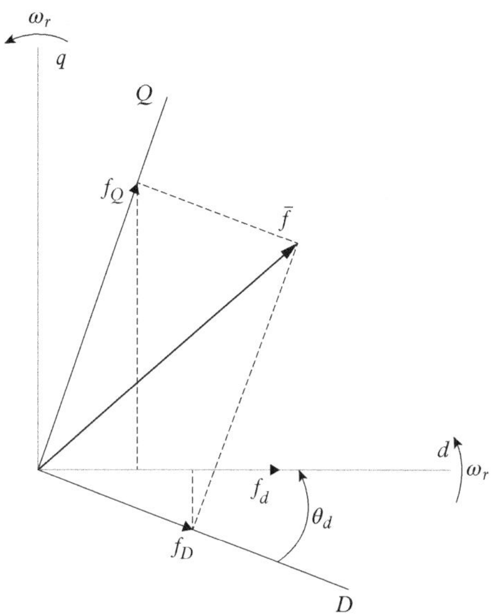

Figure 2.8 *System transformation from two-axis stationary (D–Q) RF to rotor (d–q) RF.*

The transformation is carried out mathematically by the matrix equation

$$\begin{bmatrix} f_d \\ f_q \end{bmatrix} = \begin{bmatrix} \cos\theta_d & \sin\theta_d \\ -\sin\theta_d & \cos\theta_d \end{bmatrix} \begin{bmatrix} f_D \\ f_Q \end{bmatrix}, \quad (2.5.1)$$

where

$$\theta_d = \int_0^t \omega_r dt + \theta_{d0}, \quad (2.5.2)$$

and θ_{d0} is the initial angle between the two RFs. The inverse transformation is as follows:

$$\begin{bmatrix} f_D \\ f_Q \end{bmatrix} = \begin{bmatrix} \cos\theta_d & -\sin\theta_d \\ \sin\theta_d & \cos\theta_d \end{bmatrix} \begin{bmatrix} f_d \\ f_q \end{bmatrix}. \quad (2.5.3)$$

An alternative transformation is to transform the machine model to the two-axis rotating RF directly from the three-axis stationary *a–b–c* RF. Figure 2.9 shows the transformation by visualizing the procedure by which f_d is calculated. This is done by first projecting f_a, f_b, and f_c on to the *d*-axis. Then, an algebraic summation of all three components projected onto this axis gives the value of f_d. A similar approach on to the *q*-axis gives f_q. The space angle θ_r is the electrical angle

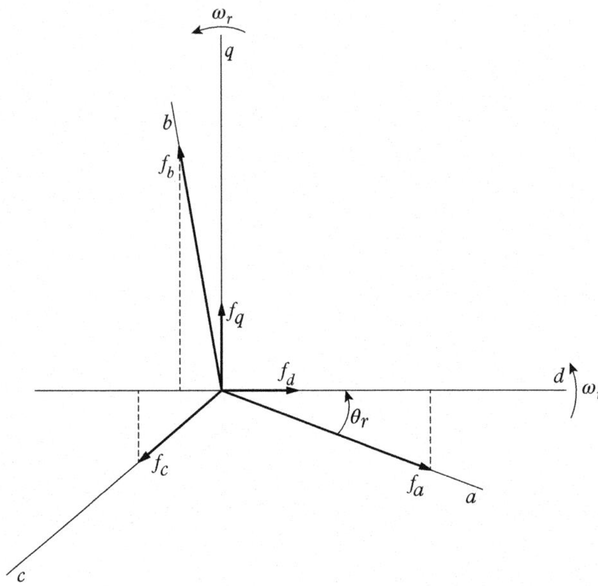

Figure 2.9 *System transformation from three-axis stationary (a–b–c) RF to rotor (d–q) RF.*

between phase "*a*" of the stationary *a–b–c* RF and the *d*-axis of the rotor RF in electrical degree. It depends on the rotor speed and the angle's initial value.

The well-known Park's transformation with little modifications is used to carry out mathematically the transformation procedure described in connection with Fig. 2.9 (Krause 1986; Rahman and Zhou 1996). It is done in matrix form as follows:

$$\begin{bmatrix} f_d \\ f_q \\ f_o \end{bmatrix} = \mathbf{P} f_{abc} = \frac{2}{3}\begin{bmatrix} \cos\theta_r & \cos(\theta_r - \frac{2\pi}{3}) & \cos(\theta_r + \frac{2\pi}{3}) \\ -\sin\theta_r & -\sin(\theta_r - \frac{2\pi}{3}) & -\sin(\theta_r + \frac{2\pi}{3}) \\ \frac{1}{2} & \frac{1}{2} & \frac{1}{2} \end{bmatrix}\begin{bmatrix} f_a \\ f_b \\ f_c \end{bmatrix}, \tag{2.5.4}$$

where f can be a phase voltage, a phase current, or a phase flux linkage and the 3×3 matrix is the transformation matrix. The angle θ_r is in electrical degrees and can be obtained as

$$\theta_r = \int_0^t \omega_r dt + \theta_0, \tag{2.5.5}$$

where ω_r is the instantaneous electrical rotor speed and θ_0 is the initial value of θ_r at $t = 0$ in electrical degrees. Since a reverse transformation from the *d–q* RF to the *a–b–c* RF needs to be carried out, a third axis as f_o is considered in eqn (2.5.4), normal to the plane of *d–q* axes and relates to f_{abc} by a third row of the transformation matrix of the same elements. By this way, the transformation matrix would be a square matrix and, therefore, an invertible one.

For arbitrary *a*, *b*, and *c* signals, a transformation from the *d–q* RF to the *a–b–c* RF can be performed via an inverse transformation by

$$\begin{bmatrix} f_a \\ f_b \\ f_c \end{bmatrix} = \boldsymbol{P}^{-1} f_{dqo} = \begin{bmatrix} \cos\theta_r & -\sin\theta_r & 1 \\ \cos(\theta_r - \frac{2\pi}{3}) & -\sin(\theta_r - \frac{2\pi}{3}) & 1 \\ \cos(\theta_r + \frac{2\pi}{3}) & -\sin(\theta_r + \frac{2\pi}{3}) & 1 \end{bmatrix} \begin{bmatrix} f_d \\ f_q \\ f_o \end{bmatrix}. \tag{2.5.6}$$

In a three-phase star connected circuit with no neutral ground or a delta connected circuit which are common in practice, there is

$$f_a + f_b + f_c = 0. \tag{2.5.7}$$

Therefore, the normal or the zero component, f_o, vanishes.

It is emphasized that the *d* and *q* variables, in general, are time varying signals, when *a*, *b*, and *c* variables are not balanced three-phase sinusoidal signals. However, interesting results are provided under the steady-state operation of the machine where the *d* and *q* variables reduce to DC signals as will be discussed at the end of this section. This is an important feature of the rotor RF. But now, the model under general mode of operation is presented first.

2.5.2 Machine equations in rotor reference frame

By multiplying the transformation matrix to both sides of the stator voltage equation (2.3.1) and using eqn (2.5.4), the following voltage equations in the rotor RF are obtained (Pillay and Krishnan 1989):

$$v_d = R_s\, i_d + p\lambda_d - \omega_r\, \lambda_q, \tag{2.5.8}$$

$$v_q = R_s\, i_q + p\lambda_q + \omega_r\, \lambda_d, \tag{2.5.9}$$

where v_d and v_q are the *d*- and *q*-axis stator voltages, i_d and i_q are the *d*- and *q*-axis stator currents, and λ_d and λ_q are the *d*- and *q*-axis stator flux linkages, respectively. Comparing the voltage equations in the rotor RF with those of the two-axis stationary RF, eqns (2.4.4) and (2.4.5), it is seen that the speed voltages, $\omega_r\lambda_q$ and $\omega_r\lambda_q$, are specific to the former set of equations. They are developed due to the relative rotation between the stator windings and the RF in which the windings voltage equations are presented. They are induced along each axis by the flux linkage component of the other axis. In fact, the speed voltages are 90° apart with their developing flux linkages as expected.

The *d*- and *q*-axis flux linkage components are obtained during the detailed derivation of eqns (2.5.8) and (2.5.9) as

$$\lambda_d = L_d i_d + \lambda_m, \tag{2.5.10}$$

$$\lambda_d = L_q i_q, \tag{2.5.11}$$

where L_d and L_q stand for the direct- and the quadrature-axis inductances, respectively. The inductances appear to be constant. This can be proved mathematically if eqns (2.5.8)–(2.5.9) are fully expanded. In a qualitative sense, it can be described by the fact that the variations of L_{aa}, L_{bb}, L_{cc}, and M_{ab}, M_{bc}, . . . , and M_{ca} in *a–b–c* RF are due to the rotor rotation and the *d–q* RF itself is rotating with the rotor as well. Therefore, the variations of the inductances in *a–b–c* RF are canceled out during the RF transformation by rotation of the rotor RF. In other words, the transformation cancels out the inductance variations, by compensating the cause of their variations which is a relative rotation between the rotor and *a–b–c* RF in which the inductances are being presented.

Looking back at eqns (2.5.10) and (2.5.11), the *d*- and *q*-axis inductances are given by the inductance components L_1 and L_2 of eqns (2.3.4)–(2.3.9) as

$$L_d = \frac{3}{2}\left(L_1 + L_2\right), \tag{2.5.12}$$

$$L_q = \frac{3}{2}\left(L_1 - L_2\right). \tag{2.5.13}$$

It is seen that each component of a flux linkage is produced by the current component of its own axis. More specifically, λ_d depends on i_d and λ_m, while λ_q depends on i_q only. A vector diagram of the machine in *d–q* RF is depicted in Fig. 2.10 showing the machine stator voltage, current, and flux linkage vectors, and their components.

The machine voltage equations in rotor RF in terms of *d*- and *q*-axis current components can be obtained by substituting the flux linkage components from eqns (2.5.10) and (2.5.11) into eqns (2.5.8) and (2.5.9), which yields

$$v_d = R_s\, i_d + L_d p i_d - \omega_r\, L_q i_q, \tag{2.5.14}$$

$$v_q = R_s\, i_q + L_q\, p i_q + \omega_r\, L_d i_d + \omega_r \lambda_m. \tag{2.5.15}$$

Based on these equations, an equivalent circuit model of the machine in rotor RF is shown in Fig. 2.11 (Sebastian *et al.* 1986).

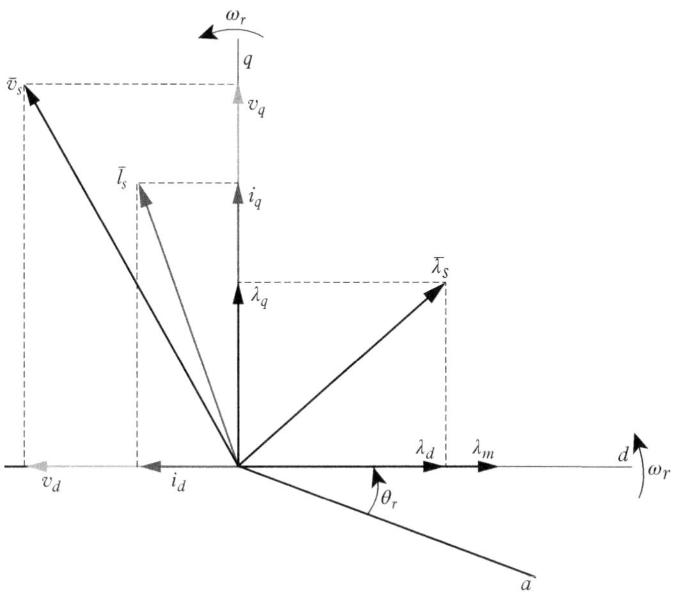

Figure 2.10 *A vector diagram of PMS machines.*

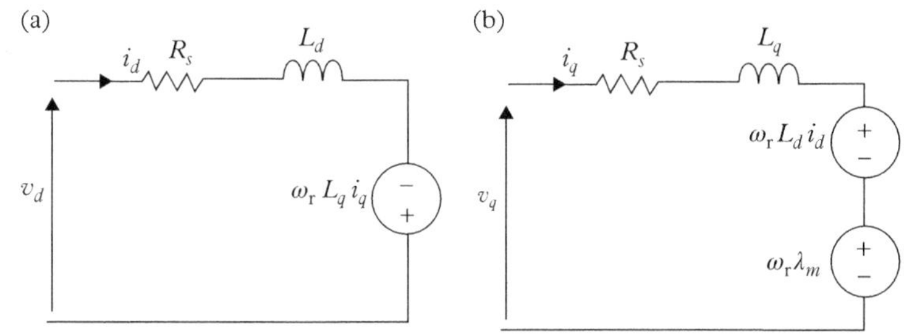

Figure 2.11 *Equivalent circuit model of PMS machines in rotor RF (a) d-axis circuit and (b) q-axis circuit.*

It is evident from Fig. 2.11 that the machine model in rotor RF enjoys the feature of having two separate circuits corresponding to the d- and q-axes of the RF. This is similar to the equivalent circuit of separately exited DC machines in which the field and armature circuits are apart. However, in the PMS model, the d- and q-axis circuits experience a mutual coupling through speed voltages as seen in voltage eqns (2.35)–(2.36) in the sense that a voltage component along an axis depends not only on the current component along its corresponding axis, but also on the current component along the other axis.

The instantaneous input power of PMS machines in motoring operation in the rotor RF can be obtained by substituting the phase variables in eqn (2.3.11) with their corresponding d- and q-axis components. This yields

$$P_{in} = \frac{3}{2}(v_d i_d + v_q i_q + 2v_o i_o), \tag{2.5.16}$$

where the third term vanishes if eqn (2.5.7) is held. Setting aside the copper loss occurring in R_s, the remaining power is converted into electromagnetic power through an energy conversion process. Therefore, the electromagnetic power is represented by the product of speed voltages in d- and q-axis equivalent circuits and the corresponding circuit currents as

$$P_{el} = \frac{3}{2}\omega_r(\lambda_d i_q - \lambda_q i_d). \tag{2.5.17}$$

Note that the 3/2 factor appears in this equation, too. Also, the mechanical power, P_m, is equivalent to the electromagnetic power, less the rotor electromagnetic losses. If there are no losses in the rotor iron and in the permanent magnet material of the poles, or if these losses are considered in the total iron losses as will be discussed later in this chapter, then $P_m = P_{el}$. The electromagnetic power can be regarded as the product of an electromagnetic torque and motor mechanical speed. Thus, using eqns (2.3.13)–(2.3.14), the electromagnetic torque is given as

$$T_e = \frac{3}{2}P(\lambda_d\, i_q - \lambda_q\, i_d)\,. \tag{2.5.18}$$

Substituting for d- and q-axis flux linkages in eqn (2.5.18) from eqns (2.5.10) and (2.5.11) yields

$$T_e = \frac{3}{2}P\,[\lambda_m + (L_d - L_q)i_d]\, i_q\,. \tag{2.5.19}$$

This is, in a sense, the most important equation of PMS motors and deserves detailed consideration. First, it depends on d- and q-axis currents and four motor parameter, i.e., λ_m, L_d, L_q, and P, and not the speed. Also, it consists of two parts, i.e., a magnet torque and a reluctance torque. The magnet torque is presented by

$$T_m = \frac{3}{2}P\lambda_m i_q, \tag{2.5.20}$$

which is a linear function of i_q and independent of d- and q- axis inductances. The reluctance torque is presented by

$$T_r = \frac{3}{2}P\,(L_d - L_q)i_d\, i_q, \tag{2.5.21}$$

which is, in general, a non-linear function, depending on both i_d and i_q. Also, the inductance difference of the machine which is a consequence of magnetic saliency of the machine plays an important role in this torque component. In PMS machines with non-surface-mounted magnet poles there is $L_q > L_d$, resulting in a positive reluctance torque if i_d is negative. A positive i_d, however, contributes to the reduction of electromagnetic torque and usually is avoided in normal motor operation. For this reason the saliency ratio of the machine, defined as

$$\rho = \frac{L_q}{L_d}, \tag{2.5.22}$$

plays an important role in machine performance and is regarded as a prime motor design parameter: the bigger ρ is, the bigger the reluctance torque is. The mechanical torque is given by

$$T_m = \frac{P_m}{\omega_m}. \tag{2.5.23}$$

This is the same as the electromagnetic torque if there are no rotor electromagnetic losses. Now let's present the steady-state model of the PMS machines in rotor RF. Recall that under steady-state operation, PMS machine phase variable quantities are as those in eqns (2.4.10)–(2.4.12). Also, the rotor speed does not have any transients and is constant at a synchronous speed equal to the angular frequency of the supplied voltage, i.e.:

$$\omega_r = \omega_e. \tag{2.5.24}$$

As a result, the transformation provides constant d and q components as

$$f_d = F\cos\theta_0, \tag{2.5.25}$$

$$f_q = -F\sin\theta_0. \tag{2.5.26}$$

In fact, the sinusoidal time variation of a–b–c variables is compensated by the rotation of the d–q RF due to eqn (2.5.4). Furthermore, in a special case where there is no initial angle between the two RFs, i.e., $\theta_0 = 0$ at $t = 0$, then eqns (2.5.25) and (2.5.26) reduce to

$$f_d = F, \qquad f_q = 0. \tag{2.5.27}$$

This is an interesting result, denoting that any symmetrical three-phase system of signals in a–b–c RF reduces to a single DC signal in d–q RF rotating with the rotor by a synchronous speed. This simplifies the machine steady-state model enormously. In practice, the machine steady-state voltage equations are obtained when the time variations

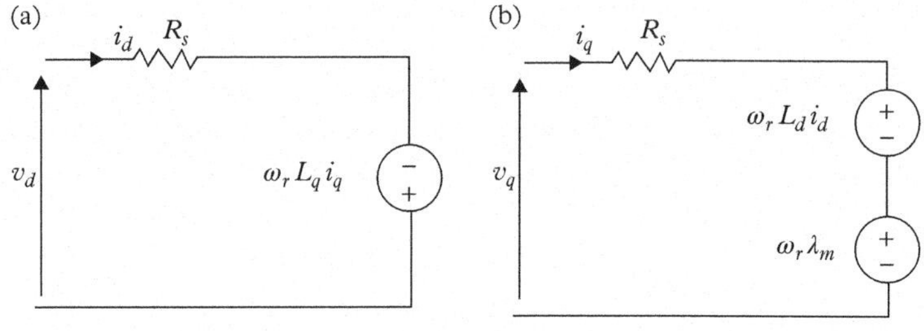

Figure 2.12 *Steady-state equivalent circuit model of PMS machines in rotor RF. (a) d-axis circuit and (b) q-axis circuit.*

of flux linkage and current components in eqns (2.5.8)–(2.5.9) are removed. As a result, the following equations are obtained:

$$v_d = R_s\, i_d - \omega_r\, \lambda_q i_q, \tag{2.5.28}$$

$$v_q = R_s\, i_q\, \omega_r \lambda_m. \tag{2.5.29}$$

The flux linkage eqns (2.5.10) and (2.5.11) are still valid. Therefore, using the previously mentioned equations, the steady-state voltage components are given in terms of current components as

$$v_d = R_s\, i_d - \omega_r\, L_q i_q, \tag{2.5.30}$$

$$v_q = R_s\, i_q + \omega_r\, L_d i_d + \omega_r \lambda_m. \tag{2.5.31}$$

The equivalent circuit model is given, based on eqns (2.5.30)–(2.5.31) as in Fig. 2.12 (Sebastian *et al.* 1986).

The power and torque equations (2.5.16)–(2.5.19) are valid in steady-state, too. Also, the machine power factor in obtained as

$$PF = \cos(\gamma - \alpha) = \frac{1}{\sqrt{1 + \left[\frac{L_q i_q^2 + \lambda_m i_d + L_d i_d^2}{\lambda_m i_q + (L_d - L_q) i_d i_q}\right]^2}}, \tag{2.5.32}$$

where γ and α are the angles of the voltage and current vectors, respectively.

2.6 Stator flux reference frame model

The machine model may be represented in rotating RFs other than the rotor RF. One such frame is a two-axis rotating RF fixed to the stator flux linkage space vector.

2.6.1 Transformation to stator flux linkage reference frame

The stator flux linkage vector is presented in the rotor RF in terms of its d- and q-axis components, λ_d and λ_q, by eqns (2.5.10) and (2.5.11), respectively. These are the components of stator flux linkage space vector, $\bar{\lambda}_s$, as depicted in a rotor RF in Fig. 2.13. The modulus, λ_s, and the angle of the vector with respect to the d-axis, δ, are given as

$$\lambda_s = \sqrt{\lambda_d^2 + \lambda_q^2}, \qquad (2.6.1)$$

$$\delta = tag^{-1} \frac{\lambda_q}{\lambda_d}. \qquad (2.6.2)$$

This angle, which is, in fact, the angle between the stator flux linkage vector and the rotor PM flux linkage or the so-called load angle, is constant only under steady-state conditions, when both the rotor and the stator flux linkages vector rotate with the same synchronous speed. It must be noted that the stator flux linkage is provided by both the stator winding currents and the rotor PM poles.

The stator flux linkage RF is now introduced as x–y RF with its direct axis, x, in line with the stator flux linkage space vector and its quadrature axis, y, perpendicular to the x-axis. By these assumptions, it is evident that in this RF

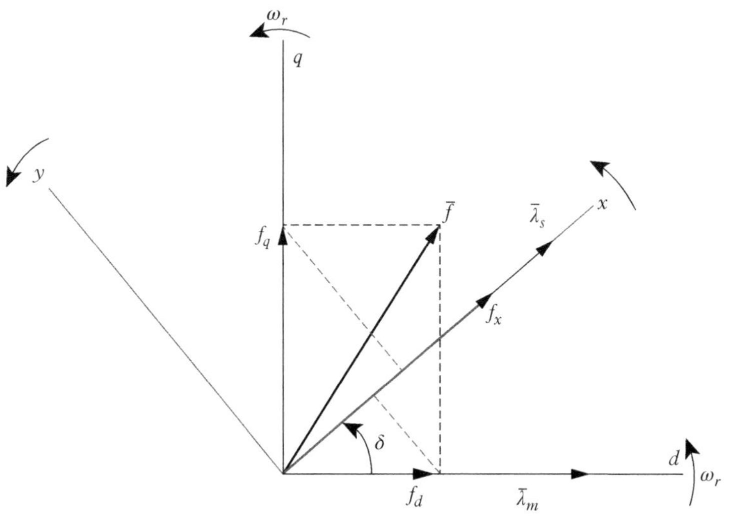

Figure 2.13 *System transformation from rotor (d–q) RF to stator flux linkage (x–y) RF.*

$$\lambda_x = \lambda_s \tag{2.6.3}$$

$$\lambda_y = 0. \tag{2.6.4}$$

Having known δ, it is possible to map any d- and q-component of a signal like voltage, current, or flux linkage from d–q RF to x–y RF. It is simply done by projecting both the d- and q-components onto each x- and y-axis to get the x- and y-components of the signal. The procedure for carrying out a d–q to x–y RF transformation is shown in Fig. 2.13 for getting the x-axis component only. The y-axis is obtained similarly. The transformation can be mathematically carried out by a simple matrix calculation as

$$\begin{bmatrix} f_x \\ f_y \end{bmatrix} = \begin{bmatrix} \cos\delta & \sin\delta \\ -\sin\delta & \cos\delta \end{bmatrix} \begin{bmatrix} f_d \\ f_q \end{bmatrix}, \tag{2.6.5}$$

where f is any machine variable, e.g., voltage, current, or flux linkage. The 2×2 transformation matrix in eqn (2.6.5) is obtained by following the projection of d- and q- components onto the x- and y-axes as described previously.

2.6.2 Machine equations in stator flux linkage reference frame

By applying the RF transformation mentioned earlier to the machine voltage equations in the d–q RF, the following voltage equations in the x–y RF are obtained:

$$v_x = R_s\, i_x + p\lambda_x\,, \tag{2.6.6}$$

$$v_y = R_s\, i_y + \omega_s\, \lambda_x, \tag{2.6.7}$$

where v_x and v_y are the x- and y-axis stator voltages, i_x and i_y are the x- and y-axis stator currents, and λ_x is the x-axis stator flux linkage and ω_s is defined as

$$\omega_s = p\delta_s, \tag{2.6.8}$$

where

$$\delta_s = \delta + \theta_r. \tag{2.6.9}$$

The presence of δ_s in the model is because the x–y RF is always apart from the stator windings by δ_s and not by θ_r, as is the case with the rotor RF. Comparing the voltage equations in the x–y RF with those of the rotor RF, eqns (2.5.8) and (2.5.9), it is seen that there is not a speed voltage along the x-axis and there is not a dynamic term along the y-axis. These are both due to the fact that $\lambda_y = 0$ as expressed previously. Therefore, voltage equations are even less involved in this RF with respect to those in the rotor RF. However, a speed voltage is induced along the y-axis by the flux linkage component of the other axis. In fact, the speed voltage is 90° apart from its developing flux linkages as expected.

A reverse RF transformation from the x–y RF back to the d–q RF can be performed by multiplying both sides of eqn (2.6.5) by the inverse of the transformation matrix, which yields

$$\begin{bmatrix} f_d \\ f_q \end{bmatrix} = \begin{bmatrix} \cos\delta & -\sin\delta \\ \sin\delta & \cos\delta \end{bmatrix} \begin{bmatrix} f_x \\ f_y \end{bmatrix}. \quad (2.6.10)$$

Torque equation (2.5.19) can be presented in the x–y RF by substituting for d- and q-axes current components in terms of the x- and y-axes components following the above-mentioned general transformation to yield

$$T_e = \frac{3}{2}P\left[\lambda_d(i_x \sin\delta + i_y \cos y) - \lambda_q(i_x \cos\delta - i_y \sin\delta)\right]. \quad (2.6.11)$$

Also, from Fig. 2.13,

$$\cos\delta = \frac{\lambda_d}{\lambda_s}, \sin\delta = \frac{\lambda_q}{\lambda_s}. \quad (2.6.12)$$

Substituting for $(\cos\delta)$ and $(\sin\delta)$ from eqn (2.6.12) into eqn (2.6.11) yields

$$T_e = \frac{3}{2}P\lambda_s i_y. \quad (2.6.13)$$

The voltage and torque equations in the x–y RF can also be developed by transforming the machine equations directly from the a–b–c RF into the x–y RF by a transformation similar to eqn (2.5.4) in which θ_r is replaced with δ_s. This way, further insight is given, in particular, into the development of the torque equation. In fact, the torque equation can be written similar to eqn (2.5.18) as

$$T_e = \frac{3}{2}P(\lambda_x i_y - \lambda_y i_x), \quad (2.6.14)$$

which reduces to eqn (2.6.13) due to eqns (2.6.3) and (2.6.4).

It can also be written as

$$T_e = \frac{3}{2}P\lambda_x i_y. \quad (2.6.15)$$

This is an interesting presentation for the torque equation, since it compactly expresses the machine torque as a product of a flux linkage component and a current component where the components are perpendicular to each other. This feature provides some conveniences with respect to eqn (2.5.19) when it comes to motor control.

Also, it is possible to calculate the y-axis stator current component in terms of λ_s and δ by first determining it from eqn (2.6.5) in terms of i_d and i_q, and then using eqns (2.5.10) and (2.5.11) to determine i_y in terms of λ_d and λ_q. Finally, replacing for λ_d and λ_q from eqn (2.6.12) and taking into account eqns (2.6.3) and (2.6.4) to yields

$$i_y = \frac{1}{2L_dL_q}[2\lambda_m L_q \sin\delta + \lambda_s(L_d - L_q)\sin 2\delta]. \tag{2.6.16}$$

Now, substituting eqn (2.6.16) into eqn (2.6.13) yields

$$T_e = \frac{3P\lambda_s}{4L_dL_q}[2\lambda_m L_q \sin\delta + \lambda_s(L_d - L_q)\sin 2\delta]. \tag{2.6.17}$$

This equation shows that the electromagnetic torque in a PMS machine can be regulated by controlling the modulus and the angle of stator flux linkage.

The d- and q-axis flux linkages can be transformed into x–y components by matrix calculations. The d- and q-axis flux can be written in matrix form as

$$\begin{bmatrix} \lambda_d \\ \lambda_q \end{bmatrix} = \begin{bmatrix} L_d & 0 \\ 0 & L_q \end{bmatrix} \begin{bmatrix} i_d \\ i_q \end{bmatrix} + \begin{bmatrix} \lambda_m \\ 0 \end{bmatrix}. \tag{2.6.18}$$

Substituting for d- and q-axis flux linkage components and d- and q-axis current components in terms of x- and y-axis components by using eqn (2.6.10) yields

$$\begin{bmatrix} \cos\delta & -\sin\delta \\ \sin\delta & \cos\delta \end{bmatrix} \begin{bmatrix} \lambda_x \\ \lambda_y \end{bmatrix} = \begin{bmatrix} L_d & 0 \\ 0 & L_q \end{bmatrix} \begin{bmatrix} \cos\delta & -\sin\delta \\ \sin\delta & \cos\delta \end{bmatrix} \begin{bmatrix} i_x \\ i_y \end{bmatrix} + \begin{bmatrix} \lambda_m \\ 0 \end{bmatrix}. \tag{2.6.19}$$

Multiplying both sides of eqn (2.6.19) by the inverse of the 2×2 matrix on the left-hand side of eqn (2.6.19) and doing some calculations gives

$$\begin{bmatrix} \lambda_x \\ \lambda_y \end{bmatrix} = \begin{bmatrix} L_d\cos^2\delta + L_q\sin^2\delta & (L_q - L_d)\sin\delta\,\cos\delta \\ (L_q - L_d)\sin\delta\,\cos\delta & L_d\sin^2\delta + L_q\cos^2\delta \end{bmatrix} \cdot \begin{bmatrix} i_x \\ i_y \end{bmatrix} + \lambda_m \begin{bmatrix} \cos\delta \\ -\sin\delta \end{bmatrix}. \tag{2.6.20}$$

This is the flux – current relationship in the *x–y* RF. It is seen that the relationship is not as straightforward as in the *d–q* RF. Nevertheless, the *x*-axis current can be found in terms of the *y*-axis current from the second row of eqn (2.6.20) by considering that $\lambda_y = 0$ as mentioned in eqn (2.6.4). This results in

$$i_x = \frac{\lambda_m \sin\delta - (L_d \sin^2\delta + L_q \cos^2\delta)\, i_y}{(L_q - L_d)\sin\delta\,\cos\delta}\,. \tag{2.6.21}$$

From eqns (2.6.21) and (2.6.16) the following equations are obtained:

$$i_x = \left(\frac{\sin^2\delta}{L_q} + \frac{\cos^2\delta}{L_d}\right)\lambda_s - \frac{\lambda_m}{L_d}\cos\delta\,, \tag{2.6.22}$$

$$i_y = \left(\frac{1}{L_q} - \frac{1}{L_d}\right)\frac{\sin 2\delta}{2}\lambda_s + \frac{\lambda_m}{L_d}\sin\delta\,. \tag{2.6.23}$$

Equation (2.6.23) is very useful in the design of a control system for PMS machines as will be presented in Section 3.7. When i_x from eqn (2.6.22) is inserted into eqn (2.6.11), the same result as in eqn (2.6.13) is obtained.

2.7 Space vector models

Space vector modeling is a compact form of representing AC electric machines in steady-state, as well as in transient state. It gives a physical sense in addition to a mathematical formulation of the system. The modeling is very much related to the two-axis and phase variable modeling methods of the machines. However, its compact form of representation and its ease of RF transformation provide it with extra futures. Nevertheless, its use is less common with respect to other types of modeling methods presented above. Therefore, the aim of this section is to discuss it rather comprehensively and self-sufficiently, and at the same time concisely.

2.7.1 Space vectors

There are different approaches to presenting space vector theory for AC machines (Boldea and Nasar 1992; Vas 1993). We adopt a straightforward approach, as a follow-up to the modeling methods presented earlier.

It is common to present any vector in a two-axis complex coordinate system as

$$\bar{f} = f_r + j f_i\,, \tag{2.7.1}$$

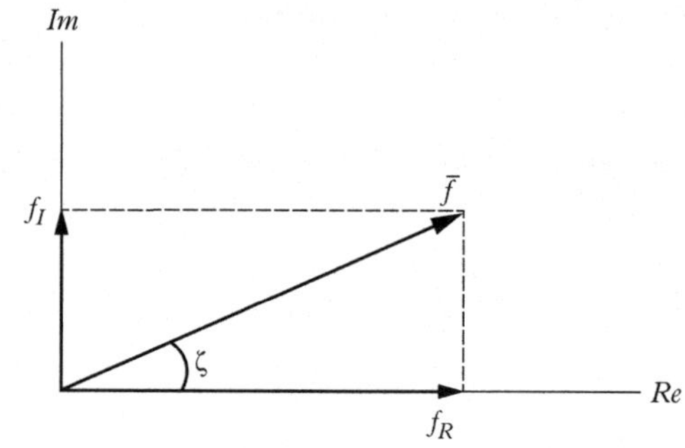

Figure 2.14 *Presentation of a vector variable in a complex coordinate system.*

where $\bar{f}$ is a vector, and f_r and f_i are its real and imaginary components, respectively. It is possible to show the vector in a complex two-axis coordinate as in Fig. 2.14, where the real and the imaginary axes are marked by *Re* and *Im*, respectively. It is known that $\bar{f}$ can also be represented in the form of

$$\bar{f} = fe^{j\varsigma}, \tag{2.7.2}$$

where f is the vector modulus and ς is the vector phase angle, which is the angle of the vector with respect to the *Re*-axis. The modulus and the phase angle are calculated in terms of the vector components as

$$f = \left(f_r^2 + f_i^2\right)^{1/2}, \tag{2.7.3}$$

$$\varsigma = \tan^{-1}\frac{f_i}{f_r}. \tag{2.7.4}$$

It is also possible to present the vector in a trigonometric form in terms of its modulus and phase angle by using the Euler formula as

$$\bar{f} = f\,(\cos\varsigma + j\sin\varsigma), \quad f_r = f\cos\varsigma, \quad f_i = f\sin\varsigma. \tag{2.7.5}$$

Recall that f_r and f_i represent $\bar{f}$ in a Cartesian coordinate system, while f and ς represent $\bar{f}$ in a polar system. Thus, the space vector modeling is closely related to the polar coordinate system. As such, the space vector modeling is not an alternative to other RF modelings, but is a method that can be applied to any RF. In other words, any stationary or rotating RF can be represented by space vector notation.

2.7.2 Machine equations in space vectors

Using the above theory of space vectors, the machine equations can be presented in stationary or rotating RFs. The vector magnitude is the same in all RFs. However, their space angles differ, depending on the RF. Any two-axis RF can be regarded as a complex coordinate as the one shown in Fig. 2.14. Therefore, the direct and quadrature axes of the RF resemble the real and imaginary axes of the coordinate. As such, a space vector can be defined for any machine variable, where its real and imaginary components are, in fact, the direct and quadrature axes components of the variable, respectively.

It is evident that a space vector is different from the conventional machine variables in the sense that it is a fictitious variable, having mathematical meaning only. However, it can be clearly defined, calculated, drawn, and even shown online via the computer monitor. Recall that a space vector in general is not a constant vector, but a time varying one, with changing modulus and phase angle. The stator current, voltage, and flux linkage space vectors are thus represented by $\bar{i}_s$, $\bar{v}_s$, and $\bar{\lambda}_s$, respectively.

2.7.2.1 Space vector models in stationary reference frames

The space vector of a machine variable in a stationary two-axis RF can be presented as

$$\bar{f} = fe^{j\varsigma_s}, = f_r + jf_i = f_D + jf_Q. \tag{2.7.6}$$

Using this presentation, $\bar{i}_s$, $\bar{v}_s$, and $\bar{\lambda}_s$ are given by their modulus, phase angles, and components in the stationary two-axis RF as

$$\bar{i}_s = i_s e^{j\alpha_s} = i_D + ji_Q, \tag{2.7.7}$$

$$\bar{v}_s = v_s e^{j\gamma_s} = v_D + jv_Q, \tag{2.7.8}$$

$$\bar{\lambda}_s = \lambda_s e^{j\delta_s} = \lambda_D + j\lambda_Q. \tag{2.7.9}$$

The modulus, i_s, v_s, and λ_s, are given by eqn (2.7.3), where the *Re* and the *Im* components are replaced by *D* and *Q* components, respectively. Also, their angles, α_s, δ_s, and γ_s, are given by eqn (2.7.4). Space vectors may also be presented in terms of *a–b–c* variables. It can be done by substituting *D*- and *Q*-axis components in eqns (2.7.7)–(2.7.9) by phase variables through a transformation of the stationary two-axis RF, *D–Q*, to the phase variable RF, *a–b–c*. This yields

$$\bar{f} = 2/3\,(f_a + af_b + a^2 f_c), \tag{2.7.10}$$

where a and a^2 are spatial operators, i.e., vectors of unity modulus having, respectively, 120° and 240° phase angles with respect to the axis of phase "*a*" of machine winding, i.e.,

$$a = 1e^{j2\pi/3}, \quad a^2 = 1e^{j4\pi/3}. \tag{2.7.11}$$

It is recalled that f_a, f_b, and f_c in Fig. 2.3 are not vectors, but arbitrary time-varying variables on three axes of 120° apart. Now, if the spatial feature is added to f_a, f_b, and f_c, then phase variable vectors of f_a, f_b, and f_c are introduced as

$$\bar{f}_a = f_a, \quad \bar{f}_b = a f_b, \quad \bar{f}_c = a^2 f_c. \tag{2.7.12}$$

They are varying vectors in fixed directions of phase winding axes. Thus, the space vector of a machine variable can be obtained by adding together the space vectors of phase variables and multiplying the result by 2/3 as seen in eqn (2.7.10). The notation of space vector for $\bar{f}$ can now be justified by bearing in mind that $\bar{f}$ is, in fact, a resultant of three space vectors f_a, f_b, and f_c as seen in Fig. 2.15. It changes in the space of phase axes, when three-phase vectors change their modulus. In other words, the time variations of f_a, f_b, and f_c are translated into the space variation of $\bar{f}$.

Using the vector notation, the machine voltage eqn (2.3.1) in stationary RF is represented by

$$\bar{v}_s = R_s \bar{i}_s + p\bar{\lambda}_s. \tag{2.7.13}$$

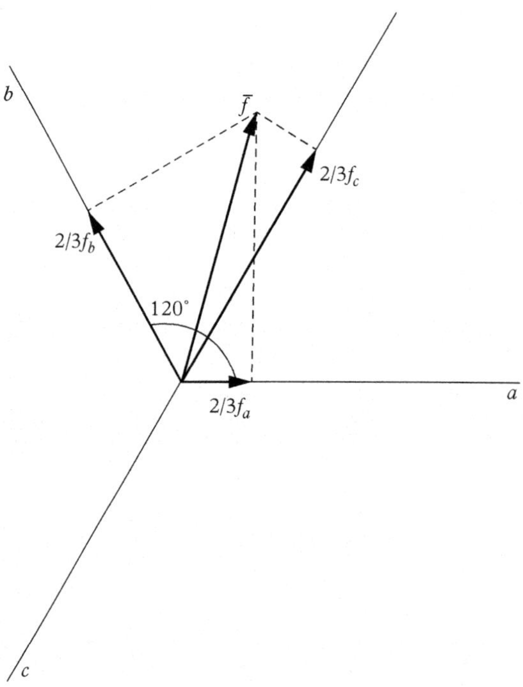

Figure 2.15 *The vector variable in terms of space vectors of phase variable components.*

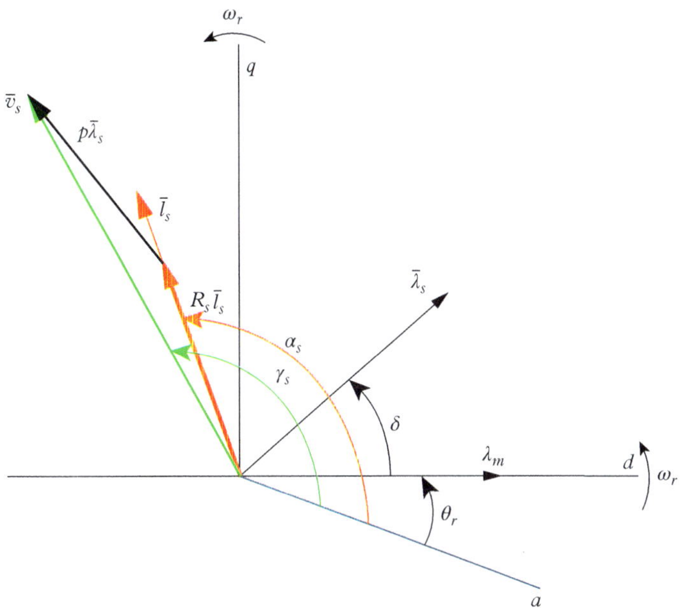

Figure 2.16 *A space vector diagram of PMS machines.*

A vector diagram of the machine based on this equation is shown in Fig. 2.16. An equivalent circuit model of the machine can also be drawn, based on eqn (2.7.13) as seen in Fig. 2.17 where the induced voltage vector is defined as

$$\bar{e}_i = p\bar{\lambda}_s. \tag{2.7.14}$$

Recall that the voltage, current, and flux linkage in this figure are all space vectors, and not the real machine variables.

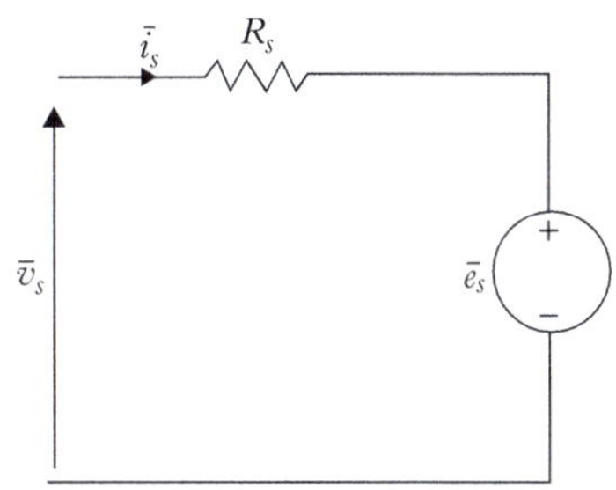

Figure 2.17 *A space vector equivalent circuit model of PMS machines in stationary RF.*

2.7.2.2 Space vector models in rotating reference frames

The vectors in eqns (2.7.7)–(2.7.9) are in the stationary RF. A space vector transformation can be performed to have the voltage, current, and flux linkage space vectors in a two-axis rotating RF. This can be done through a vector rotation as follows:

$$\bar{f} = fe^{j\varsigma_s}.\ e^{-j\theta_r} = fe^{j(\varsigma_s-\theta_r)} = f_d + jf_q, \tag{2.7.15}$$

where θ_r is the rotor angle with respect to the angle of phase "a" of stator winding or the angle between the stationary RF and the rotating RF defined by eqn (2.5.5), where the rotor rotation is in the direction shown in Fig. 2.18. It must be noted that $\bar{f}$ in the above equation is

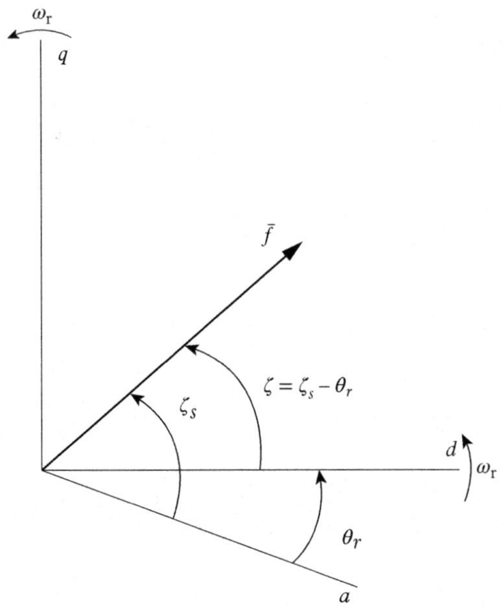

Figure 2.18 *A stationary RF to rotor RF transformation by space vector rotation.*

not the same as $\bar{f}$ in a stationary RF. In fact, the former vector is equal to the latter one in terms of modulus, but different in terms of angle. The vector rotation of (2.7.15) is the same as the RF transformation of eqn (2.5.4), but in space vector notation.

Space vector theory also simplifies the RF transformation among two-axis RFs through vector rotation. A d–q to x–y RF transformation of current vector, for instance, can be done by

$$\bar{i}_s = i_s e^{j\alpha} . e^{-j\delta} = i_s e^{j(\alpha-\delta)} = i_x + ji_y. \tag{2.7.16}$$

Again, it must be noted that $\bar{i}_s$ in this equation is the same as $\bar{i}_s$ in eqn (2.7.7) in terms of modulus, but not in terms of angle. A rotor RF to stator flux RF transformation is carried out in Fig. 2.19. A stator flux RF to rotor RF transformation in vector form can be done by multiplying a vector in the former RF to a unity vector with an angle of δ as

$$\bar{i}_s = i_s e^{j(\alpha-\delta)} . e^{j\delta} = i_s e^{j\alpha} = i_d + ji_q. \tag{2.7.17}$$

In order to transform eqn (2.7.13) to the rotor RF, first, the equation is rewritten by adding a double subscript "s" to the vectors to emphasize that they are presented in the stationary RF as

$$\bar{v}_{ss} = R_s \bar{i}_{ss} . + p\bar{\lambda}_{ss}. \tag{2.7.18}$$

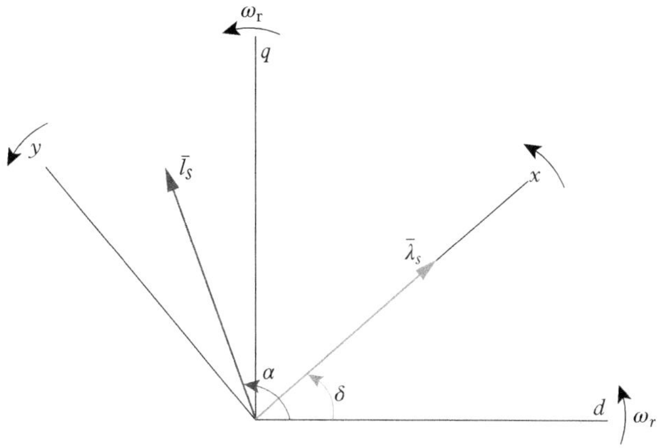

Figure 2.19 *A rotor RF to stator flux RF transformation by space vector rotation.*

Both sides of eqn (2.7.18) are then multiplied by a unit vector of $e^{-j\theta_r}$, where θ_r is the rotor angle with respect to the axis of phase "a" of stator winding to get

$$\bar{v}_{ss}.e^{-j\theta_r} = R_s\,\bar{i}_{ss}.\,e^{-j\theta_r} + (p\bar{\lambda}_{ss}).\,e^{-j\theta_r}. \tag{2.7.19}$$

Now, the voltage and the current vectors in the stationary RF are transformed to the rotor RF and the stator flux linkage in the stationary RF is substituted in terms of its vector in the rotor RF, where a double subscript "r" denotes the vectors being in the rotor RF:

$$\bar{v}_{sr} = R_s\,\bar{i}_{sr} + (p(\bar{\lambda}_{sr}\,e^{j\theta_r})).\,e^{-j\theta_r}. \tag{2.7.20}$$

It can be written by applying the derivative operator as

$$\bar{v}_{sr} = R_s\,\bar{i}_{sr} + (e^{j\theta_r}p\bar{\lambda}_{sr} + je^{j\theta_r}\lambda_{sr}p\theta_r).\,e^{-j\theta_r}. \tag{2.7.21}$$

Finally, a voltage equation in the rotor RF is obtained as

$$\bar{v}_{sr} = R_s\,\bar{i}_{sr} + p\bar{\lambda}_{sr} + j\omega_r\bar{\lambda}_{sr}. \tag{2.7.22}$$

It is seen that the speed voltage components in eqns (2.5.8)–(2.5.9) are presented in eqn (2.7.22) as the speed voltage vector.

The machine electromagnetic torque or the developed torque in terms of space vectors is given by (Vas 1992)

$$T_e = \frac{3}{2}P(\bar{\lambda}_s \times \bar{i}_s), \tag{2.7.23}$$

where $\bar{i}_s$ and $\bar{\lambda}_s$ must be in any same RF and × denotes an outer vector product. This can further be elaborated in the stationary RF as

$$T_e = \frac{3}{2} P \lambda_s \, i_s \, \sin(\alpha_s - \delta_s). \tag{2.7.24}$$

The torque can also be represented in the rotor RF as

$$T_e = \frac{3}{2} P \lambda_s \, i_s \, \sin(\alpha - \delta). \tag{2.7.25}$$

By substituting for the stator flux linkage vector and the current vector in terms of their d- and q-axis components in eqn (2.7.23), the torque eqn (2.5.18) is obtained. Expressing the vector product of eqn (2.7.23) in the stator flux RF and substituting for the stator flux linkage vector and the current vector in terms of their x- and y-axis components, the torque of eqn (2.6.13) is obtained.

2.8 Machine model considering iron losses

Under sinusoidal flux density of varying magnitude and frequency, the iron losses consisting of eddy current loss and hysteresis loss can be presented empirically by (Slemon and Liu 1990)

$$P_{Fe} = P_e + P_h = k_e \omega_e^2 B_{\max}^2 + k_h \omega_e B_{\max}^n, \tag{2.8.1}$$

where P_e and P_h are eddy current loss and hysteresis loss, respectively, ω_e and $B_{\max}$ are the flux frequency and maximum flux density, and k_e, k_h, and n are constants.

In practice $n < 2$. However, considering $n \approx 2$, an approximate iron loss equation can be obtained as

$$P_{Fe} = (k_e + k_h/\omega_e)(\omega_e B_{\max})^2 . \tag{2.8.2}$$

This equation suggests an iron loss resistance in parallel with the total induced voltages in d- and q-axis circuits of the steady-state model in Fig. 2.12 (Slemon and Liu 1990). Putting this resistance in the circuits as R_c, the equivalent circuit model of Fig. 2.20 is obtained. The circuit is justified since

$$P_{Fe} = \frac{3}{2}\frac{(\omega_e B_{\max})^2}{R_c} = \frac{3}{2}\left(\frac{e_d^2}{R_c} + \frac{e_q^2}{R_c}\right), \tag{2.8.3}$$

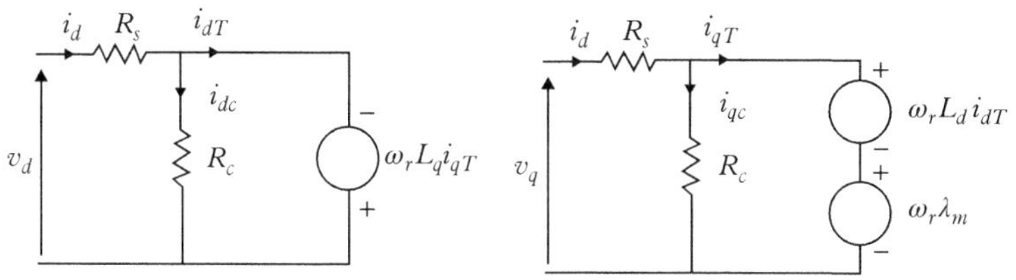

Figure 2.20 *Steady-state equivalent circuit model of PMS machines in rotor RF including iron losses: (a) d-axis circuit and (b) q-axis circuit.*

where e_d and e_q are the total induced voltage in the d- and q-axis directions, and R_c is obtained from eqns (2.8.2)–(2.8.3) as (Vaez *et al.* 1997)

$$R_c = \frac{2\pi^2}{k_e + k_h/f}. \tag{2.8.4}$$

The iron losses are then calculated in terms of R_c as

$$P_{Fe} = \frac{3}{2} R_c (i_{dc} + i_{qc})^2, \tag{2.8.5}$$

where i_{dc} and i_{qc} are iron loss current components. It is seen in eqn (2.8.4) that R_c depends on the machine supply frequency and thus depends on the synchronous speed. In fact, as the motor speed increases, the iron loss resistance decreases. This, in turn, increases the iron loss current components. As a result, the iron losses increase with increasing speed and become noticeable at high speeds, especially in machines with high energy permanent magnet materials in which flux density is high. It must be mentioned that despite eqn (2.8.4), the use of a constant R_c is also reported in many references (Slemon and Liu 1990; Morimoto *et al.* 1994).

Recall that eqn (2.8.4) gives R_c under steady-state conditions. However, a parallel R_c with constant value is also used in the dynamic simulation of PMS machines (Morimoto *et al.* 1994; Vaez *et al.* 1999). The voltage equations under these conditions are given in terms of the torque-producing components of i_q and i_d, i.e., i_{qT} and i_{dT}, respectively, as follows:

$$v_d = R_s i_{dT} + \left(\frac{R_s}{R_c} - 1\right) \omega_e L_q i_{qT}, \tag{2.8.6}$$

$$v_q = R_s i_{qT} + \left(\frac{R_s}{R_c} + 1\right) \omega_e L_d i_{dT} + \left(\frac{R_s}{R_c} + 1\right) \omega_e \lambda_m. \tag{2.8.7}$$

Also, the electromagnetic torque is given by

$$T_e = \frac{3}{2} P \left[\lambda_m + (L_d - L_q)\, i_{dT}\right] i_{qT}. \tag{2.8.8}$$

The current components i_{qT} and i_{dT} can be obtained in terms of i_q and i_d from a simple circuit analysis in steady-state as

$$i_{qT} = \frac{1}{a}(i_q - bi_d - c), \tag{2.8.9}$$

$$i_{dT} = i_d + d(i_q - bi_d - c), \tag{2.8.10}$$

where the coefficients a, b, c, and d are defined in terms of machine parameters. The torque equation in terms of i_q and i_d is obtained by substituting eqns (2.8.9)–(2.8.10) into eqn (2.8.8) as

$$T_e = \alpha i_d^2 + \beta i_q^2 + \gamma i_d i_q + \lambda i_d + \eta i_q + \sigma, \tag{2.8.11}$$

where the coefficients α, β, γ, λ, η, and σ are given in terms of machine parameters (Vaez-Zadeh 2001). If the iron loss currents are neglected in eqn (2.8.11), it is reduced to the conventional torque equation (2.5.19).

2.9 Magnetic saturation of iron core

Performance of PMS machines is influenced by magnetic saturation to varying degrees, depending on the machine design and control system employed. In machines with high flux densities, during normal operation, the chance of machine being saturated is high. Modern PMS machines with high energy PM materials like NdFeB have usually high flux density in air gap, and stator and rotor iron cores. Particular sections of cores where high flux follows through narrow passes are more vulnerable to saturation. Saturation, in general, tends to reduce machine inductances and the induced voltages in the stator windings. This is due to an increase in the reluctance of some flux paths. As a result, the machine performance may deteriorate under saturation. Saturation worsens the machine performance under the control. This happens particularly when the machine parameters drift from their nominal values due to the saturation. Therefore, saturation should be a concern when a control system is designed. This is usually done by taking the saturation into account in the modeling of machines and using such a model in the control system design. A machine model with extra parameters can be derived to include the saturation effects (Sneyers *et al.* 1985). However, it may complicate the control system design. Also, it is usually difficult to measure the parameters under varying operating conditions. Another option is to use conventional models with minor modifications. In fact, the saturation is caused by flux-producing currents. Therefore, it can be represented in terms of the currents. In PMS machines, the saturation mainly influences the machine d- and q-axis inductances, and the magnitude of magnet flux linkage. Therefore, it is a common practice to model L_d, L_q, and λ_m as functions of current components, i_d and i_q (Parasiliti and Poffet 1989; Mellor *et al.* 1991; Chalmers 1992). This makes the inductances vary according to the machine operating point. This method is particularly capable of taking saturation into account under high load conditions when machines draw high currents. First-order functions are usually sufficient for the modeling. The functions and their coefficients are found by measuring L_d, L_q, and λ_m over a range of currents, drawing the measured values against current components, i_d and i_q, and finally carrying out curve fitting. The outcome would be in the following form, for instance (Mellor *et al.* 1991):

$$L_d = L_d^0\,(1 - k_d^+ i_d), \qquad i_d \geq 0 \tag{2.9.1}$$

$$L_d = L_d^0\,(1 - k_d^- i_d), \qquad i_d < 0 \tag{2.9.2}$$

$$L_q = L_q^0\,(1 - k_q^+ i_q), \qquad i_q \geq 0 \tag{2.9.3}$$

$$L_q = L_q^0\,(1 - k_q^- i_q), \qquad i_q < 0 \tag{2.9.4}$$

$$\lambda_m = \lambda_m^0, \qquad i_q < 0 \tag{2.9.5}$$

$$\lambda_m = \lambda_m^0 - k_m(i_q - I_q), \qquad i_q \geq I_q \tag{2.9.6}$$

where L_d^0, L_q^0, and λ_m^0 are the nominal values of the machine parameters and k_d^+, k_d^-, k_q^+, and k_d^- are constant coefficients. The above functions are then used in the conventional machine model instead of constant machine parameters to have the saturation considered in the model. Using the model in control system design meets the requirement of accounting for the saturation in the control system.

2.10 Modeling PMS machines with surface-mounted poles

In PMS machines with inset or berried magnets in the rotor, there is always $L_q > L_d$. Therefore, saliency ratio, ρ, greater than unity results. However, in PMS machines with surface-mounted poles the inductances are equal, i.e.:

$$L_d = L_q = L_s, \tag{2.10.1}$$

where L_s denotes the synchronous inductance. As a result,

$$\rho = \frac{L_q}{L_d} = 1. \tag{2.10.2}$$

This unity ratio has important implications for machine operation and control since the machine model becomes simpler than that for salient machines. The machine voltage equations are obtained in a stationary two-axis RF for surface-mounted (non-salient) PMS machine by letting $L_2 = 0$ in eqns (2.3.4)–(2.3.9), since it is related to the rotor magnetic saliency. This results in constant self and mutual inductances, and simplifies the voltage equation (2.4.6) to

$$\begin{bmatrix} v_D \\ v_Q \end{bmatrix} = \begin{bmatrix} R_s & 0 \\ 0 & R_s \end{bmatrix} \begin{bmatrix} i_D \\ i_Q \end{bmatrix} + \begin{bmatrix} L_1 & 0 \\ 0 & L_1 \end{bmatrix} p \begin{bmatrix} i_D \\ i_Q \end{bmatrix} + \omega_r \lambda_m \begin{bmatrix} -\sin\theta_r \\ \cos\theta_r \end{bmatrix}. \tag{2.10.3}$$

The voltage equations in the rotor RF can be obtained by substituting (2.10.1) into (2.5.14) and (2.5.15). The voltage equations are then reduced to

$$v_d = R_s i_d + L_s(pi_d - \omega_r i_q), \tag{2.10.4}$$

$$v_q = R_s i_q + L_s(pi_q + \omega_r i_d) + \omega_r \lambda_m. \tag{2.10.5}$$

More importantly, the reluctance torque vanishes as $L_d - L_q = 0$. The machine torque as a result reduces to the magnet torque only, i.e.:

$$T_e = \frac{3}{2} P \lambda_m i_q. \tag{2.10.6}$$

The torque therefore depends on one variable, i_q, only. The torque resembles the torque of a DC machine in which the field and the armature are separated. In fact, the q-axis current in PMS machines plays the role of armature current in DC machines. Furthermore, the field is constant which makes the machine torque a linear function of i_q and provides an easier control system design. The flux linkage components in the stator flux linkage RF also reduce to

$$\begin{bmatrix} \lambda_x \\ 0 \end{bmatrix} = \begin{bmatrix} L_s & 0 \\ 0 & L_s \end{bmatrix} \begin{bmatrix} i_x \\ i_y \end{bmatrix} + \lambda_m \begin{bmatrix} \cos\delta \\ -\sin\delta \end{bmatrix} \tag{2.10.7}$$

according to (2.6.20) in connection with (2.10.1). As a result, the y-axis component of the stator current vector is easily obtained from (2.10.7) as

$$i_y = \frac{1}{L_s} \lambda_m \sin\delta. \tag{2.10.8}$$

Substituting (2.10.8) into (2.6.13) yields

$$T_e = \frac{3}{2} \frac{P}{L_s} \lambda_m \lambda_s \sin\delta = \frac{3}{2} \frac{P}{L_s} \lambda_m \lambda_q. \tag{2.10.9}$$

The above torque equation can also be easily obtained from eqn (2.6.17) in connection with eqn (2.10.1).

In a PMS motor with surface-mounted PM poles, the stator flux linkage in space vector form is

$$\bar{\lambda}_s = \bar{\lambda}_m + L_s \bar{i}_s . \tag{2.10.10}$$

Using this equation in connection with the torque equation (2.7.23), one can get

$$T_e = \frac{3}{2}P(\bar{\lambda}_m \times \bar{i}_s) = \frac{3}{2}P\lambda_m i_s \sin\alpha = \frac{3}{2}P\lambda_m i_q , \tag{2.10.11}$$

which is the same as eqn (2.10.6).

2.11 Dynamic equation of PMS machines

A PMS machine connected to a mechanical load forms a dynamic mechanical system as shown in Fig. 2.21. The dynamics of the machine's mechanical part is governed by the equation

$$T_e = T_L + B\omega_m + Jp\omega_m, \tag{2.11.1}$$

where T_L is the load torque, B is the viscous coefficient of the machine bearings, J is the motor and load inertia, and p is a derivation operator. It is also possible to obtain the equation in terms of the rotor position, θ_r, by replacing the machine speed by

$$\omega_m = p\theta_m. \tag{2.11.2}$$

The dynamic equation is then represented by

$$T_e = T_L + Bp\theta_m + Jp^2\theta_m, \tag{2.11.3}$$

where p^2 is the double derivation operator.

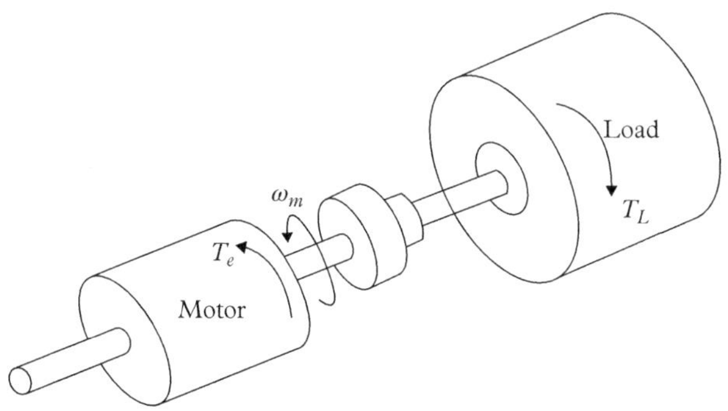

Figure 2.21 *A dynamic system consisting of a PMS machine connected to a mechanical load.*

2.12 Summary

This chapter presents mathematical modeling of PMS machines. A physical model of the machines is developed first by considering several simplifying assumptions. The machine modeling is carried out with the help of RFs. A machine model in terms of phase variables is then presented in a three-axis stationary RF as the most straightforward model. However, the model does not lend itself to control system design due to its complexity that arises from the time-varying machine parameters. An equivalent two-winding fictitious machine can be assumed to simplify machine analysis. The equivalent machine is modeled in a two-axis stationary RF with perpendicular direct and quadrature axes. It is obtained by an RF transformation from the three-axis stationary RF to the two-axis stationary RF with a fixed angle between the frames. However, the problem of time-varying machine parameters persists. A transformation to a two-axis rotating RF, fixed to the rotor, solves the problem and results in fixed parameters in this RF, where the direct axis is aligned with the magnetic axis of a permanent magnet rotor pole. A transformation to a two-axis rotating RF where the direct axis is aligned with the stator flux linkage vector is also presented. This RF particularly provides a compact torque equation for interior permanent magnet machines as the product of a stator flux linkage component and a stator current component. A concise treatment of modeling, based on space vector theory, is finally presented. The theory is based on the definition of a fictitious space vector for every machine variable and provides a deeper insight into the machine performance. Space vector modeling reveals a different aspect of RF transformation in which only the vector angle changes from one RF to another one. Accordingly, space vector models of PMS machines in different RFs are elaborated on. Equivalent circuits are drawn based on the mathematical models where appropriated. Iron losses and iron saturation are also taken into the models. The modeling in this chapter is primarily treated under the dynamic operation. However, steady-state models are also derived in each RF. The chapter ends with a brief presentation of the dynamic equation of PMS machines' mechanical part.

PROBLEMS

P.2.1. What are the differences between a space vector and a complex phasor, which is used in circuit analysis of sinusoidal signals in balanced systems?

P.2.2. Draw a system of balanced three-phase voltages with a maximum voltage of $V_s = 220\sqrt{2}$ and its transformed system of D and Q voltages versus time, where the a–b–c and D–Q RFs are spatially 30° apart.

P.2.3. Investigate a stationary RF of more than three axes for analysis of three-phase PMS machines. Carry out the transformation from an a–b–c RF to such an RF and find out the corresponding transformation matrix.

P.2.4. Develop an RF transformation of a PMS machine equation directly from a–b–c RF to x–y RF in matrix notation and show the transformation matrix.

P.2.5. Present the PMS machine stator voltage equation in rotor RF in terms of the space vector of the rotor permanent magnet flux linkage, $\bar{\lambda}_m$.

P.2.6. Find out a formula for the power factor of PMS machines taking into account the iron losses.

P.2.7. Try to simplify the iron losses by wise approximation to have more straightforward machine equations than eqns (2.8.10)–(2.8.15).

P.2.8. Draw a space vector diagram for PMS machines expressing voltage and flux linkage equations in the rotor RF.

P.2.9. Find out operating conditions of PMS machines under which the magnet torque is equal to the reluctance torque. Then calculate the total torque in terms of the saliency ratio.

P.2.10. Find out the optimum saliency ratio for having equal magnet torque and reluctance torque at the rated current. In addition, for a general saliency ratio, determine the ratio of i_d/i_q in terms of machine parameters to have equal magnet torque and reluctance torque.

P.2.11. Develop a relationship between i_d and i_q for achieving minimum machine copper loss under steady-state operation.

P.2.12. Consider the influences of a very big saliency ratio on the machine model and performance.

P.2.13. Suggest an equivalent circuit with an iron loss parameter in series with the d- and q-axis current components. What would be the relation of such a circuit with the conventional circuit of Fig. 2.20?

P.2.14. What would become the transformation matrix if the angle between the stationary and rotating RFs is assumed to be the angle between the a-axis and the q-axis (instead of the d-axis assumed throughout the chapter)?

P.2.15. Investigate the PMS machine torque in terms of current vector modulus when d- and q-axis current components are equal. Determine the current vector modulus in terms of machine parameters under this condition.

P.2.16. Determine an optimum δ to have a maximum torque with a constant i_s. Then calculate this maximum torque when $L_d = L_q$ and $L_q > L_d$. Draw T_e as a function of δ under both conditions in a single graph and comment on the differences.

P.2.17. Calculate the current coefficients $a, b, c,$ and d in eqns (2.8.13)–(2.8.14) in terms of machine parameters.

P.2.18. Calculate the torque coefficients α, β, γ, λ, η, and σ in eqn (2.8.15) in terms of machine parameters.

BIBLIOGRAPHY

Binns, K. and Wong, T. (1984). Analysis and performance of a high-field permanent-magnet synchronous machine. *IEE Proc. B-Electric Power Appl.* 131(6), 252–258.

Boldea, I. and Nasar, S.A. (1992). *Vector Control of AC Drives*. CRC Press, Boca Raton, FL.

Bose, B.K. (1988). A high-performance inverter-fed drive system of an interior permanent magnet synchronous machine. *IEEE Trans. Ind. Appl.* 24(6), 987–997.

Bracikowski, N., Hecquet, M., Brochet, P., and Shirinskii, S.V. (2012). Multiphysics modeling of a permanent magnet synchronous machine by using lumped models. *IEEE Trans. Ind. Electron.* 59(6), 2426–2437.

Chalmers, B. (1992). Influence of saturation in brushless permanent-magnet motor drives. *IEE Proc. B-Electric Power Appl.* 139(1), 51–52.

Chen, Z., Tomita, M., Doki, S., and Okuma, S. (2003). An extended electromotive force model for sensorless control of interior permanent-magnet synchronous motors. *IEEE Trans. Ind. Electron.* 50(2), 288–295.

Consoli, A. and Raciti, A. (1991). Analysis of permanent magnet synchronous motors. *IEEE Trans. Ind. Electron.* 27(2), 350–354.

Consoli, A. and Renna, G. (1989). Interior type permanent magnet synchronous motor analysis by equivalent circuits. *IEEE Trans. Energy Convers.* 4(4), 681–689.

Dehkordi, A.B., Gole, A.M., and Maguire, T.L. (2005). Permanent magnet synchronous machine model for real-time simulation. In: *International Conference on Power Systems. Transients (IPST'05)*, pp. 19–23. IPST, Montreal.

De La Ree, J. and Boules, N. (1989). Torque production in permanent-magnet synchronous motors. *IEEE Trans. Ind. Electron.* 25(1), 107–112.

Hendershot, J.R. and Miller, T.J.E. (2010). *Design of Brushless Permanent-Magnet Machines*, 2nd edn. Oxford University Press, Oxford.

Holm, S.R., Polinder, H., and Ferreira, J.A. (2007). Analytical modeling of a permanent-magnet synchronous machine in a flywheel. *IEEE Trans. Magnet.* 43(5), 1955–1967.

Honsinger, V. (1980). Performance of polyphase permanent magnet machines. *IEEE Trans. Power Apparat. Syst.* 99(4), 1510–1518.

Jannot, X., Vannier, J.C., Marchand, C., Gabsi, M., Saint-Michel, J., and Sadarnac, D. (2011). Multiphysic modeling of a high-speed interior permanent-magnet synchronous machine for a multiobjective optimal design. *IEEE Trans. Energy Convers.* 26(2), 457–467.

Li, J., Abdallah, T., and Sullivan, C.R. (2001). Improved calculation of core loss with nonsinusoidal waveforms. In: *Conference Record of the 2000 IEEE Industry Applications Conference 36th IAS Annual Meeting*, pp. 2203–2210. IEEE, Piscataway, NJ.

Krause, P.C. (1986). *Analysis of Electric Machinery*. McGraw-Hill, New York, NY.

Krause, P.C., Wasynczuk, O., Sudhoff, S.D., and Pekarek, S. (2013). *Analysis of Electric Machinery and Drive Systems*. John Wiley and Sons, Chichester.

Kulkarni, A.B., and Ehsani, M. (1992). A novel position sensor elimination technique for the interior permanent-magnet synchronous motor drive. *IEEE Trans. Ind. Appl.* 28(1), 144–150.

Mellor, P., Chaaban, F., and Binns, K. (1991). Estimation of parameters and performance of rare-earth permanent-magnet motors avoiding measurement of load angle. *IEEE Proc. B-Electric Power Appl.* 138(6), 322–330.

Mi, C., Slemon, G.R., and Bonert, R. (2003). Modeling of iron losses of permanent-magnet synchronous motors. *IEEE Trans. Ind. Appl.* 39(3), 734–742.

Miller, T.J.E. (1989). *Permanent Magnet and Reluctance Motor Drives*. Oxford Science Publications, Oxford.

Morimoto, S., Tong, Y., Takeda, Y., and Hirasa, T. (1994). Loss minimization control of permanent magnet synchronous motor drives. *IEEE Trans. Ind. Appl.* 41(5), 511–517.

Ojo, O. and Cox, J. (1996). Investigation into the performance characteristics of an interior permanent magnet generator including saturation effects. In: *Conference Record of the 1996 IEEE Industry Applications 31st IAS Annual Meeting*, San Diego, pp. 533–540. IEEE, Piscataway, NJ.

Parasiliti, F. and Poffet, P. (1989). A model for saturation effects in high-field permanent magnet synchronous motors. *IEEE Trans. Energy Convers.* 4(3), 487–494.

Pillay, P. and Krishnan, R. (1988). Modeling of permanent magnet motor drives. *IEEE Trans. Ind. Electron.* 35(4), 537–541.

Pillay, P. and Krishnan, R. (1989). Modeling, simulation, and analysis of permanent-magnet motor drives. I. The permanent-magnet synchronous motor drive. *IEEE Trans. Ind. Appl.* 25(2), 265–273.

Rabinovici, R. (1994). Eddy current losses of permanent magnet motors. *IEEE Proc. B-Electric Power Appl.* 141(1), 7–11.

Rahman, M., Little, T., and Slemon, G. (1985). Analytical models for interior-type permanent magnet synchronous motors. *IEEE Trans. Magnet.* 21(5), 1741–1743.

Rahman, M.A. and Zhou, P. (1994). Field-based analysis for permanent magnet motors. *IEEE Trans. Magnet.* 30(5), 3664–3467.

Rahman, M. A., and Zhou, P. (1996). Analysis of brushless permanent magnet synchronous motors. *IEEE Trans. Ind. Electron.* 43(2), 256–267.

Schiferl, R. and Lipo, T.A. (1988). Power capability of salient pole permanent magnet synchronous motors in variable speed drive applications. In: *Conference Record of the 1988 Industry Applications Society Annual Meeting*, pp. 23–31. IEEE, Piscataway, NJ.

Schifer, R. and Lipo, T. (1989). Core loss in buried magnet permanent magnet synchronous motors. *IEEE Trans. Energy Convers.* 4(2), 279–284.

Sebastian, T., Slemon, G., and Rahman, M. (1986). Modelling of permanent magnet synchronous motors. *IEEE Trans. Magnet.* 22(5), 1069–1071.

Sebastian, T. (1995). Temperature effects on torque production and efficiency of PM motors using NdFeB magnets. *IEEE Trans. Ind. Appl.* 31(2), 353–357.

Sebastiangordon, T. and Slemon, G.R. (1987). Operating limits of inverter-driven permanent magnet motor drives. *IEEE Trans. Ind. Appl.* 23(2), 327–333.

Slemon, G.R. and Liu, X. (1990). Core losses in permanent magnet motors. *IEEE Trans. Magnet.* 26(5), 1653–1655.

Sneyers, B., Novotny, D.W., and Lipo, T.A. (1985). Field weakening in buried permanent magnet ac motor drives. *IEEE Trans. Ind. Appl.* 21(2), 398–407.

Stumberger, B., Stumberger, G., Dolinar, D., Hamler, A., and Trlep, M. (2003). Evaluation of saturation and cross-magnetization effects in interior permanent-magnet synchronous motor. *IEEE Trans. Ind. Appl.* 39(5), 1264–1271.

Tseng, K-J. and Wee, S-B. (1999). Analysis of flux distribution and core losses in interior permanent magnet motor. *IEEE Trans. Energy Convers.* 14(4), 969–975.

Urasaki, N., Senjyu, T., and Uezato, K. (2000). An accurate modeling for permanent magnet synchronous motor drives. In: *15th Annual IEEE Applied Power Electronics Conference and Exposition, APEC 2000*, Vol. 1, pp. 387–392. IEEE, Piscataway, NJ.

Urasaki, N., Senjyu, T., and Uezato, K. (2004). Relationship of parallel model and series model for permanent magnet synchronous motors taking iron loss into account. *IEEE Trans. Energy Convers.* 19(2), 265–270.

Vaez, S., John, V., and Rahman, M. (1997). Energy saving vector control strategies for electric vehicle motor drives. In: *Proceedings of the Power Conversion Conference*, pp. 13–18. IEEE, Piscataway, NJ.

Vaez, S., John, V.I., and Rahman, M.A. (1999). An on-line loss minimization controller for interior permanent magnet motor drives. *IEEE Trans. Energy Convers.* 14(4), 1435–1440.

Vaez-Zadeh, S. (2001). Variable flux control of permanent magnet synchronous motor drives for constant torque operation. *IEEE Trans. Power Electron.* 16(4), 527–534.

Vas, P. (1992). *Electrical Machines and Drives, a Space-Vector Theory Approach.* Oxford University Press, Oxford.

Wijenayake, A.H. and Schmidt, P.B. (1997, May). Modeling and analysis of permanent magnet synchronous motor by taking saturation and core loss into account. In: *Proceedings of the 1997 International Conference on Power Electronics and Drive Systems*, Vol. 2, pp. 530–534. IEEE, Piscataway, NJ.

3 Vector Control

Vector control (VC) as the first high performance motor control method has been the control method most widely used by the industry for more than two decades. It uses the vector presentation of AC machine model for current VC. The current vector is controlled through its real and the imaginary components. Referring to the conventional phase variable model of AC machines, VC controls the current phase angles, in addition to the magnitude and frequency of currents, to determine the instantaneous values of the motor voltages. The main goal of a VC method is to make the two magnetic fields in AC machines independent and perpendicular as much as possible. Therefore, a separately excited DC machine may resemble the performance of an AC machine under VC, despite very different constructions of the two types of machines.

VC is implemented in different schemes, all requiring the information of rotor position or the position of the stator flux linkage vector. It also needs a modulating means to determine the switching instances of the inverter power electronic switches. The performance of an AC machine under VC is fast, accurate, and smooth. In this chapter a comprehensive presentation of VC concepts and principles in addition to a wide variety of VC schemes are presented. Also, many desirable motor operating features under VC are discussed including current and voltage limiting, flux weakening, minimum loss operation, etc.

The chapter starts with a critical review of the conventional scalar control (SC) of PMS machines followed by the analogous operation of a separately excited DC machine as an introduction to VC.

3.1 Scalar control of PMS motors

SC is still a widely applicable control method associated with induction motors. It is aimed at low dynamic applications like blowers, pumps, compressors, and valves. It is also applied to PMS machines, although the practice with these machines is less frequent with respect to SC of induction machines. SC, in general, uses steady-state mathematical models of machines in the design stage and demands low

Control of Permanent Magnet Synchronous Motors. Sadegh Vaez-Zadeh. © Sadegh Vaez-Zadeh 2018. Published in 2018 by Oxford University Press. DOI 10.1093/oso/9780198742968.001.0001

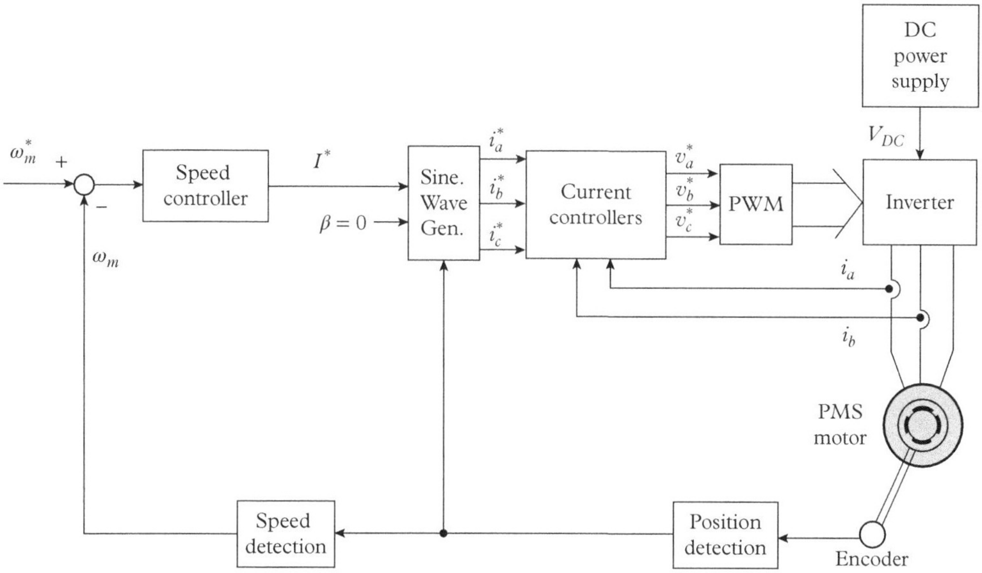

Figure 3.1 *Scalar current control system of PMS machines.*

computations when it is being implemented and run. The system is more commonly used for speed control in which a speed command is followed by the motor speed.

A schematic view of an SC system for PMS motors is shown in Fig. 3.1. The speed command is provided directly by the end user or, more commonly, is produced by an overhead system, as is the case in many industrial applications. The speed control is usually carried out through a closed loop via a feedback speed signal. The actual speed is measured by a speed sensor or a position sensor. The speed sensor is usually a tachogenerator and the position sensor is an encoder. If a position sensor is used, a detecting circuit is also employed to process the sensor output signal to determine the speed feedback signal. A resolver may also be used as a sensor in connection with a resolver to digital (RTD) circuit. The speed error as the difference between the commanded and feedback signals is continuously calculated, and fed to the speed controller. The controller is often of a proportional integral (PI) type and is implemented by a standard software routine. The controller main function is to shape the speed signal during dynamic conditions by providing a fast and smooth speed trajectory in time to reach a near zero error in steady-state.

The speed controller also provides an output stator phase current magnitude, I^*, as a command signal for the inner current loop as seen in Fig. 3.1. The output is a fixed signal at steady-state and is used as the magnitudes of all three phase currents. The phase currents are 120° apart and form a symmetrical three-phase system of signals in steady-state. The most important feature of the motor currents stems from the synchronous nature of the machine operation. The currents, in fact, produce a resultant rotating magneto-motive force (mmf) in phase with the magnet-produced flux linkage to develop a smooth constant torque in steady-state. This is achieved by injecting phase currents in accordance with a rotor pole position. That is why a rotor position signal must be available for normal operation of the machine, even without a speed control loop. The phase current commands are produced by a sine wave generation unit as seen in Fig. 3.1. The unit receives currents magnitude and position signal, I and θ_r, respectively, and generates three output signals, i_a^*, i_b^*, and i_c^*, as current commands, all having a current magnitude of I and an angle of θ_r, and being 120° apart as

$$i_a^* = I \sin\theta_r, \tag{3.1.1}$$

$$i_b^* = I \sin(\theta_r - 2\pi/3), \tag{3.1.2}$$

$$i_c^* = I \sin(\theta_r + 2\pi/3). \tag{3.1.3}$$

The current commands are compared with the actual phase currents as feedback signals obtained by current sensors. As a result, the current errors are measured all the time and fed to the current controllers. The current controllers are usually of PI type to provide fast dynamic responses, as well as zero steady-state errors. More complex types of controllers are also used to achieve optimal, adaptive, robust, and other featured performances under a wide range of operating and ambient conditions. Two current sensors on two machine phases are sufficient if the machine windings are star-connected with no neutral connection or they are delta connected. The third current is obtained from the fact that the sum of three currents is 0. The current controllers provide three-phase voltage command output signals, v_a^*, v_b^*, and v_c^*, as seen in Fig. 3.1. A schematic view of current control loop is shown in Fig. 3.2.

The phase voltage commands are used as input signals of the PWM unit. The unit compares them with a carrier wave with a fixed high frequency, in order to determine the switching gating signals of a three-phase inverter. The inverter, as a power transducer, produces motor supply voltage by processing the DC source to deliver it to the machine terminals. The DC source is usually provided by an AC/DC rectifier through a diode bridge or is a battery as mentioned in Chapter 1. The inverter input voltage, which is commonly recognized as the DC link voltage, is regarded as a constant voltage regardless of small motor load-dependent deviations caused by a little voltage drop

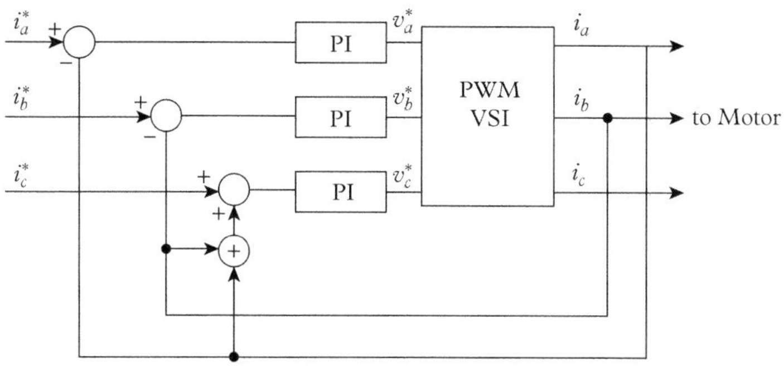

Figure 3.2 *Current control loops.*

across the DC link. The inverter output three-phase voltages are applied to the motor stator winding terminals to flow the phase currents into the winding circuits and develop the machine torque.

As previously mentioned, the rotor position signal is used in the sine wave generation unit to produce current commands that end up with a resultant rotating mmf in phase with the rotor-produced air gap flux. Although this is the most usual mode of operation, it is not always the case. In fact, at high-speed range above the rated speed, the phase-advancing mode is used under which the current-produced mmf is developed prior to reaching the rotor flux. This is achieved practically by phase-advancing the current commands. The merit of the phase-advancing mode can be fully understood by looking at the machine phase voltage equations presented by eqn (2.3.1) in the previous chapter, which is recalled here for phase "a" only as

$$v_a = R_s i_a + p\lambda_a, \tag{3.1.4}$$

where the second term on the right-hand side of eqn (3.1.4) is an induced voltage, e_a, and is recalled from eqn (2.3.2) as

$$e_a = p\lambda_a = p\begin{bmatrix} L_{aa} & M_{ab} & M_{ac} \end{bmatrix}\begin{bmatrix} i_a \\ i_b \\ i_c \end{bmatrix} - \lambda_m \omega_r \sin\theta_r. \tag{3.1.5}$$

The total induced voltage of eqn (3.1.5) consists of two components, i.e., a current-produced component, which is the first term, and a magnet-produced component, e_{ma}, which is the second. When the phase a current is in phase with the magnet-produced voltage, the total induced voltage reaches a high value, which is limited by the applied voltage, v_a. As a result, the phase current is limited and, in turn, the developed torque will be limited at high-speed range. The problem can be overcome by advancing all three-phase currents by a

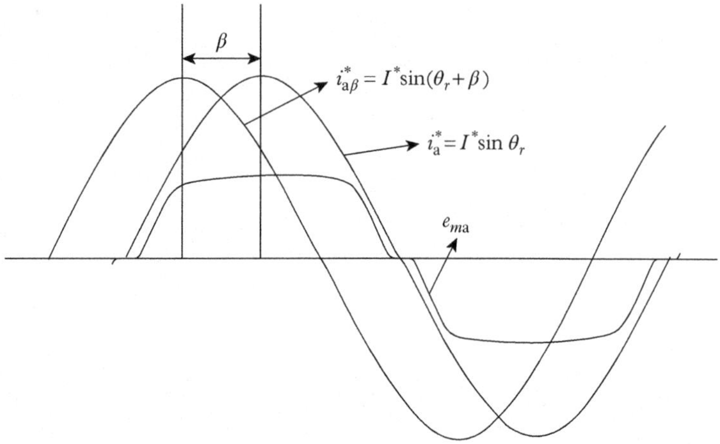

Figure 3.3 *Phase current command with and without phase-advancing control.*

predetermined angle, β. This angle is applied to the sine wave generation unit as seen in Fig. 3.1, with a non-zero value. As a result, the phase currents find an opportunity to rise sufficiently without having an excessive induced voltage, thus proving enough torque at high speed. The current command for phase "*a*," with and without phase advancement, i_a^* and $i_{a\beta}^*$, is shown in Fig. 3.3 with respect to the PM induced voltage, e_{ma}. It must be mentioned that, in practice, e_{ma} contains more or fewer harmonics, as shown in Fig. 3.3. However, it is regarded as a pure sinusoidal voltage in this chapter as it is in Chapter 2. Phase advancement is, in fact, another face of flux weakening, as will be discussed in Section 3.6.

SC of PMS machines is theoretically easy to understand and is readily designed. Its system implementation is cost effective, and uses moderately accurate and simple sensing devices. The SC system also needs limited computation. However, it suffers from sluggish transient response and low bandwidth. It is not fitted to optimal flux-weakening operation strategies as the phase variable model it uses is not suited to these kinds of operations. In fact, the machine model becomes tedious and the control system turns out to be extremely complicated under most operating strategies. Thus, the PMS motor drives cannot be effective in high performance applications that require fast and accurate dynamic control. These shortcomings open the door to great opportunities and even the vital necessity of high performance control systems, including VC to meet the challenging requirements of many current and emerging applications. Nevertheless, SC can be regarded as the first step toward high performance control as will be elaborated in Section 3.3.

3.2 Basic theory of VC

In this section, the torque production of DC machines, which is analogous to the one of the AC machines under VC, is presented first. The basics of VC of AC machines is then discussed, followed by the PMS machine model under VC.

3.2.1 Features of a DC motor control

In any electrical machine, a torque is developed due to the interaction of two magnetic fields. In a mathematical sense, the torque is proportional to the outer product of the two magnetic flux linkage vectors. In separately excited DC machines, the two magnetic fields are provided by two separate flux sources, i.e., stator or field winding, and rotor or armature winding. Furthermore, in these machines a commutation system always locates the two magnetic fields spatially perpendicular to each other, thus, providing the best space position of the two fields for the purpose of torque production. Alternatively, it is possible to regard the machine torque proportional to an outer product of a magnetic flux linkage vector and a current vector according to Lorentz's law. Thus, the torque of a DC machine can be written as

$$T_e = K_T \bar{\lambda}_f \times \bar{i}_a, \tag{3.2.1}$$

where $\bar{\lambda}_f$ is the rotor magnetic flux linkage vector and $\bar{i}_a$ is the armature current vector. As theses vectors are perpendicular, the outer product of eqn (3.2.1) becomes

$$T_e = K_T \lambda_f i_a, \tag{3.2.2}$$

where λ_f and i_a are the magnitudes of the rotor magnetic flux linkage vector and the armature current vector, respectively. For the commutation system to cause this, a full potential of torque production is used, due to the fact that the flux linkage and current vectors are 90° apart.

It is the nature of DC machines that the field winding must have many turns to provide a sufficiently strong mmf and, consequently, a strong magnetic flux linkage. It is also known that a winding inductance is proportional to its turn number squared, while its resistance is proportional to its turn number. Therefore, the field winding of a DC machine has a large time constant of $\tau_f = L_f/R_f$. This is why a torque control process based on the adjustment of λ_f would be sluggish and time-consuming. As a result, a high performance torque control may be achieved by keeping the rotor flux linkage constant, while adjusting the armature current independently. The armature current can fortunately be controlled rapidly, since the armature winding time constant is rather low due to its low turn number.

3.2.2 Basics of VC

In an AC machine, the two torque-contributing fields are neither independent nor orthogonal in general. Also, there is no commutation system to locate the fields in orthogonal position with respect to each other. The main goal of a VC method is to make the two magnetic fields independent and perpendicular as much as possible. Therefore, a separately excited DC machine may resemble the performance of an AC machine under the VC, despite totally different constructions of the two types of machines.

A VC system is a fast control system. When comparing a machine performance under VC with its performance under SC, this feature of VC must be emphasized first. A main difference between SC and VC is the fact that a steady-state model of the machine is used for determining the commanded variables of the inverter under an SC, in VC a dynamic model of the machine is used for the same purpose.When referring to the description of dynamic models in Section 2.1, we emphasize that the main feature of a dynamic model takes into account the variation of machine variables as a function of time. However, a steady-state model determines the variation of machine variables with respect to each other when the machine works at steady-state, regardless of time. Using a steady-state model in a machine control system may adjust the machine variables to their commanded values at steady-state. However, it cannot shape the variation of machine variables in transient state, when they are moving toward their commanded or desired values. Therefore, the swiftness of the control is not guaranteed. Bearing in mind that the main objective of a motor control system is to shape the motor performance at transient state, using a dynamic model of a machine in the control system is a natural choice. This is only realized by so-called high performance motor control methods including VC.

Another feature of VC is its preciseness. Using a dynamic model contributes to this feature. However, there is more in VC to provide a precise control response. In fact, a VC uses a more compact machine model in a rotating reference frame, and takes care of more aspects of a machine-controlled signal. A comparison of the treatment imposed on a machine-controlled variable under VC with respect to the way it is treated under SC is necessary to emphasize this feature of VC. A very good SC method of AC machines, i.e., a V/f control, is chosen for this comparison. In this control method, both the magnitude and frequency of the commanded voltage to the machine are determined from a steady-state machine model, according to a commanded speed. The voltage is then realized by a PWM inverter and is applied to the machine. Therefore, the instantaneous value of the motor voltage is not controlled. In other words, it may be regarded that the phase of the applied voltage is not controlled. Therefore, only two of the three aspects of the voltage (magnitude, frequency, and phase) are controlled. In VC, however, the instantaneous value of the motor voltage is controlled or it may be regarded that all three aspects of the motor voltage are controlled. By controlling the instantaneous voltage, VC tries to shape the machine response to a command signal during the transient state. Thus, a more precise control is achieved and a smoother machine performance is provided.

It must be noted that the notation of VC does not denote a type of control system or method as is understood in the control engineering discipline. VC stems from the fact that a vector quantity, instead of a scalar quantity, is being controlled. In this sense, the emphasis is placed on both the magnitude and the phase of the main controlled variable. In fact, a VC is implemented in practice by finding the space vector of machine current by having the vector components. Then both its magnitude and its angle are controlled. The angle control corresponds to the phase control of applied voltage to the machine as mentioned previously. Theoretically, VC is explained in a rotating reference frame. Therefore, a transformation of the machine model from the conventional stationary (*a–b–c*) reference frame to a rotating reference frame (as presented in Chapter 2) is an essential part of VC theory. This transformation can be regarded as a simple mapping for those familiar with the conformal mapping in engineering mathematics. Conformal mapping is used in solving heat transfer, electrostatic, and electromagnetic problems in complex bodies by simplifying border conditions of the problems. Reference frame transformation performs a similar simplifying task in machine control. More precisely, sinusoidal variables of a machine model in a stationary *a–b–c* reference frame are seen as DC variables in a rotating reference frame as seen in Chapter 2. Time-varying inductances of the machine in the former reference frame are transferred into constant inductances in the latter reference frame. The transformation simplifies the machine model and eases the current controller design. Furthermore, the machine performance under the controller is more rapid and more precise as mentioned previously.

An *a–b–c* reference frame to *d–q* reference frame transformation, as elaborated in Section 2.5, is recalled to present the principles of VC of PMS machines by using Fig. 3.4. The *d*-axis of the rotating reference

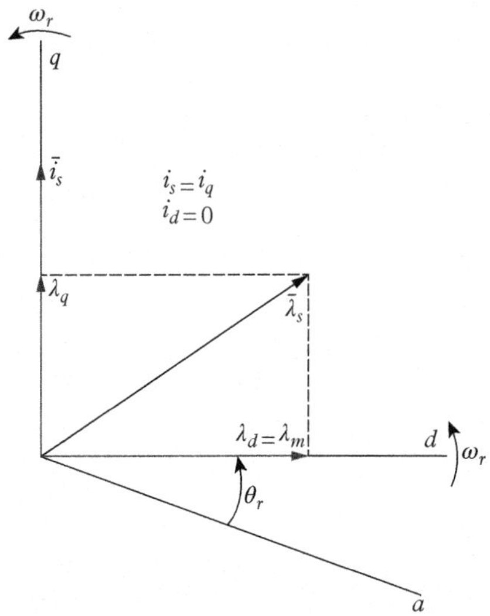

Figure 3.4 *Principles of VC in surface-mounted PMS machines.*

frame is oriented in line with the magnetic axis of the rotor magnet pole as seen in Fig. 3.4. It must be noted that the reference axes are orthogonal in electrical angle. The VC of surface-mounted PMS motors is carried out by orienting the stator current vector in line with the q-axis as seen in Fig. 3.4. This orientation provides a maximum developed torque, considering the available stator current. As a result, $i_d = 0$ under this control. Consequently, the d-axis flux linkage of the stator winding becomes constant and independent of the stator current with its value the same as the magnitude of the magnet flux linkage, i.e., $\lambda_d = \lambda_m$. This way, the torque dynamics is independent of the flux transient and follows the current transient only.

3.2.3 Motor model under VC

Transforming the machine model to a rotating reference frame simplifies the model, as seen in Section 2.5. For a PMS machine with surface-mounted magnets, the voltage equation of the machine along the q-axis is further simplified to

$$v_q = R_s i_q + L_s p i_q + \omega_r \lambda_m, \qquad (3.2.3)$$

where the d-axis current component is assumed to be 0 and L_s is the machine synchronous inductance. It is interesting to compare the voltage equation (3.2.3) with the armature voltage equation of a DC machine, which is given as

$$v_a = R_s i_a + L_a p i_a + \omega_r \lambda_f. \qquad (3.2.4)$$

It is seen that eqns (3.2.3) and (3.2.4) are essentially the same, where the q-axis quantities of PMS machines are analogues to the armature quantities of DC machines and the magnet flux linkage of the former machines is analogues to the field flux linkage of the latter machines. The same analogy exists between the torques of the two types of machines as the torque of PMS machines is presented as

$$T_e = \frac{3}{2} P \lambda_m i_q, \tag{3.2.5}$$

while, the torque of a DC machine is given as

$$T_e = K_T \lambda_f i_a. \tag{3.2.6}$$

Comparing eqns (3.2.5) and (3.2.6), the same analogy between the mentioned quantities of the PMS and the DC machines is confirmed. The dynamic equation governing the mechanical aspect of the machines is also the same. Therefore, it is concluded that a surface-mounted PMS machine, which is controlled by VC in rotor reference frame with $i_d = 0$, can be regarded as a separately excited DC machine with constant field excitation. This is justified by the fact that in such a PMS machine the alternating stator voltage is transformed into a DC voltage in the steady-state, i.e., the q-axis stator voltage. Also, the rotating magnet flux linkage is transformed into a constant flux linkage and kept orthogonal to the stator flux linkage. Therefore, they resemble the armature voltage and the field of the DC machine, respectively. Also, the speed voltage that appears in the q-axis voltage equation of PMS machines resembles the speed voltage which is induced in the armature of DC machines.

In fact, the a–b–c reference frame to d–q reference frame transformation in conjunction with $i_d = 0$ in PMS machines carries out the same function as a commutation system in DC machines. The commutation system rectifies the alternating voltage in the armature winding of DC machines at the armature winding terminals. It also keeps the flux linkage produced by the armature current perpendicular to the flux linkage produced by the field winding. The reference frame transformation together with $i_d = 0$ carries out the same tasks as mentioned previously. Therefore, there is no surprise if the models of the two types of machines are analogous.

3.3 VC in rotor reference frame with phase current controllers

VC is a method of torque control via current control. Therefore, VC is related to the most inner control loop in a motor drive control and does not matter to the outer loops including the speed control loop. The speed control in a vector-controlled motor drive is thus the same as that of SC or other control methods. Nevertheless, we consider the speed control here since it is the least inner control loop after the torque control and its output is used as an input to the current control loop, which is implemented by the VC method. However, the speed control loop is not elaborated on here.

VC is presented for surface-mounted PMS motors in this section and for interior permanent magnet (IPM) motors in the following sections. Looking back to the torque equation of PMS machines with surface-mounted poles, eqn (3.2.5), it is evident that i_d does not contribute to the torque. Therefore, the torque is a linear algebraic function of i_q only. Thus, an i_q control is, in fact, a torque control with just a constant coefficient $3/2P\lambda_m$ in between. An i_q control can be fundamentally designed based on the machine q-axis voltage equation of eqn (3.2.4). It is seen in this equation that i_q is related to v_q by a linear differential equation. It is implied that PMS machine torque can be controlled through i_q, while maintaining $i_d = 0$. Keeping $i_d = 0$ not only linearizes the q-axis voltage equation and make it possible to design a simple linear current controller, but also causes the machine to develop a maximum torque for any machine current. This happens because the whole i_s current follows along the q-axis, which contributes to the developed torque as seen in Fig. 3.4. A vector-controlled PMS motor drive system is shown in Fig. 3.5, including phase current control. As shown there, the VC system has two input signals, i.e., i_q^* and i_d^*, as two components of a current vector, where the former is an output of the speed controller. The system includes a reference frame transformation. It transforms i_d^* and $i_q{}^*$ into phase current commands i_a^*, i_b^*, and i_c^* by using a rotor position signal θ_r. The rotor position is measured by a position sensor. An optical encoder is usually used for sensing the position to provide accurate information. It is worth mentioning that position error causes a malfunction of the VC due to false orientation of the rotating reference frame. The actual phase currents, i_a, i_b, and i_c, are measured by current sensors and used as feedback signals to close the current loops. They are compared with the phase current commands to provide current errors as inputs to the three current controllers. The current controllers are usually of PI type. However, other type of controllers may be used for the sake of performance improvement in special circumstances or in the case of extra demands, such as being robust, optimal, or adaptive. They accept phase current errors and provide phase voltage commands v_a^*, v_b^*, and v_c^*. The voltage commands are applied to a PWM system to generate gating signals for the inverter. The inverter then inverts the DC input power to a three-phase AC output power

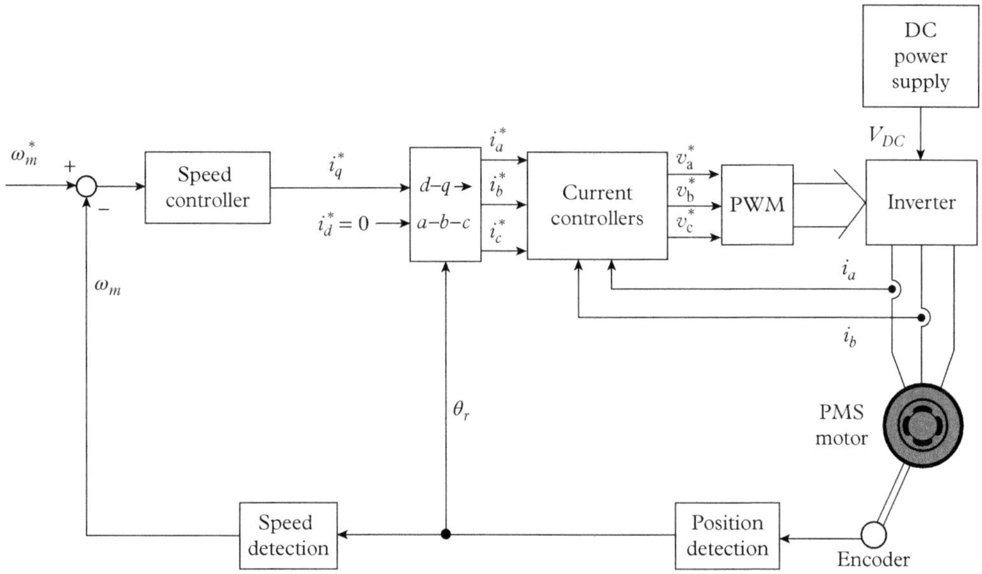

Figure 3.5 *Vector control of PMS motors in rotor reference frame with phase current controllers.*

according to the voltage commands to supply the PMS machine. The machine thus develops an accurate torque and rotates at a speed corresponding to its own inertia and load. Since a speed control is also included in the system of Fig. 3.5, the machine speed is adjusted to the speed command through the speed control loop. The actual speed signal, which is used as a feedback signal in the speed control loop, is calculated by differentiating the rotor position signal. The speed command is compared with the actual speed and the error is applied to the speed controller to produce the *q*-axis current command.

It must be mentioned that in case of $i_d = 0$ control, the machine *d*-axis flux is fixed at λ_m and no rotor flux weakening occurs. As a result, the stator flux, λ_s, is governed by i_q only. Applying a negative i_d^* to PMS machines with surface-mounted poles provides flux weakening to the machine due to $\lambda_d = L_d i_d + \lambda_m$. However, it is not usually favorable because it reduces i_q and, consequently, the developed torque under the same stator current.

It is interesting to compare the VC system of Fig. 3.5 with the SC of Fig. 3.1. It is seen that the only differences between these two control systems are two-fold. First, the sine wave generation block of SC is replaced by *d–q* to *a–b–c* reference frame transformation block in VC system. Second, I^* and $\beta = 0$ as inputs to the former block are replaced by i_q^* and $i_d^* = 0$ as inputs to the latter block, respectively, while the outputs of the two blocks are the same. This means that a duality exists between the SC system and the simplest VC system due to the similarity of the sine wave generation and the reference frame transformation. However, the VC system of Fig. 3.5 is just the beginning of a long series of VC systems with many extra features not achievable by either SC or the VC of Fig. 3.5. In fact, the system of Fig. 3.5, as the first step in VC, provides a smooth transition from SC to VC.

Vector control with phase current controllers suffers from a limited bandwidth problem. The current controllers in this system act on the machine phase currents, which are alternating by a frequency proportional to the machine speed. As the speed goes up, the actual currents must track the phase current commands, which are alternating rapidly. In this situation, the controllers' essential delay may get close to the time period of phase currents. As a result, the actual signals may not be able to follow the commanded signals and cause undesirable errors. Vector control theory has a solution to get rid of this problem as is presented in the next section.

3.4 VC in rotor reference frame with *d–q* current controllers

It is possible in a VC system to control *d–q* currents instead of *a–b–c* currents. This can happen by performing a *d–q* to *a–b–c* reference frame transformation on voltages instead of currents. This control scheme is presented in this section together with its essential decoupling circuit.

3.4.1 Basic VC scheme with *d–q* current control

A vector-controlled PMS motor drive system with *d–q* current controllers is shown in Fig. 3.6. It is seen in this figure that the system input signals, i.e., i_q^* and i_d^*, are compared with feedback currents i_q and i_d, respectively, to determine current errors. The errors are applied to two controllers to provide *q*- and *d*-axis voltage commands v_q^* and v_d^*. The controllers are shown in Fig. 3.6 together with a decoupling circuit and named as decoupling current controller. The decoupling circuit will be described later on in this section. The current controllers in this system also are usually of PI type with an option to use other type of controllers. The system includes two reference frame transformations. The first one transforms v_q^* and v_d^* to phase voltage commands, v_a^*, v_b^*, and v_c^*, by using the rotor position signal, θ_r, which is measured as in the system of Section 3.3. The second

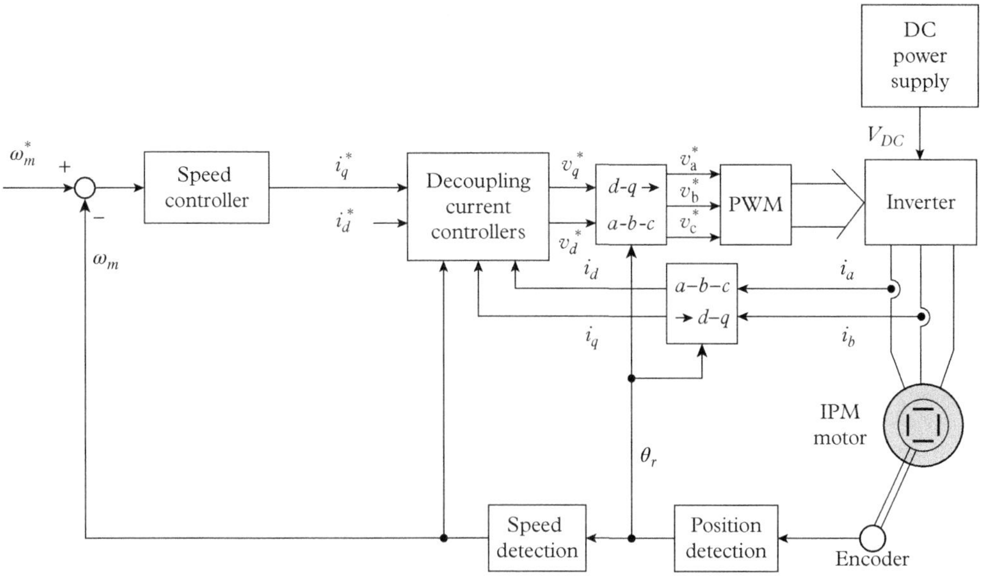

Figure 3.6 *Vector control of PMS motors in rotor reference frame with d–q current controllers.*

reference frame transformation receives actual phase currents, i_a, i_b, and i_c, from current sensors, and transforms them into i_q and i_d to be used as feedbacks in current loops. Here, the same rotor position signal is used in the transformation. Once the phase voltage commands are provided by the first reference frame transformation, the rest of system is the same as that in the system with phase current controllers.

The system is not faced with the problem of limited bandwidth as the current controllers act on the q- and d-axis currents, which are DC signals under steady-state conditions. The system, in fact, carries out algebraic calculation of reference frame transformations on alternating feedback signals, while performing the essentially delayed signal processing by current controllers on non-alternating signals. Thus, the bandwidth problem as appears in the VC system with phase current controllers is not seen in this system. This is regarded as a major advantage of the system with respect to the previous one. As a result, the d–q current control is much more popular than phase current control in VC systems.

3.4.2 Decoupling current controllers

The goal of VC is to separate torque control from flux control in AC machines like in DC machines, as mentioned in Section 3.1. This may be achieved by controlling torque via i_q and controlling flux by i_d if the

d-axis circuit is decoupled from the q-axis circuit in PMS machines. However, a problem associated with VC of interior permanent magnet motors in rotor RF is a coupling of the d- and q-axis circuits in the sense that both voltage components depend on both current components. This is seen in the voltage equations in this reference frame as presented in eqns (2.5.14) and (2.5.15), which are recalled here as

$$v_d = R_s i_d + L_d p i_d - \omega_r L_q i_q, \tag{3.4.1}$$

$$v_q = R_s i_q + L_q p i_q + \omega_r L_d i_d + \omega_r \lambda_m. \tag{3.4.2}$$

This fact deteriorates the machine performance under simple linear controllers like PI ones. In fact, only non-linear controllers may handle such non-linearity. Using non-linear current controllers in VC systems complicates the controllers design and weakens the VC merits.

The coupling problem is treated in Section 3.3 for surface-mounted PMS machine control by deciding $i_d = 0$. This solves the coupling problem in q-axis voltage equation and not in d-axis voltage equation. Furthermore, non-zero (negative) d-axis current is preferred in the IPM machine to make use of reluctance torque. Therefore, a decoupling circuit in connection with current controllers is desirable as a general solution for VC of IPM machines.

A closer look at the voltage equations (3.4.1) and (3.4.2) shows that the machine model suffers from the cross-coupling effects of speed voltage terms, i.e., $\omega_e L_d i_d$ and $\omega_e L_q i_q$. These effects are dominant in the voltage equations, especially at high speed, since L_d and L_q are relatively large in IPM motors. Thus, i_d and i_q cannot be controlled independently by simple linear controllers. Linear control theory can only be applied to the machine model to design, for example, PI controllers for both i_d and i_q if the voltage equations are linearized first. This means that the d-axis voltage must depend on i_d and its dynamics only, and q-axis voltage must depend on i_q and its dynamics only.

The voltage equations (3.4.1) and (3.4.2) may suggest a block of decoupling current controllers as presented in Fig. 3.7 in which the d- and q-axis voltage commands are each provided by a combination of two signals as

$$v_d^* = v_{d1} + v_{d0}, \tag{3.4.3}$$

$$v_q^* = v_{q1} + v_{q0}, \tag{3.4.4}$$

where v_{d1} and v_{q1} are provided by the PI current controllers $C_d(s)$ and $C_q(s)$, respectively, while non-linear terms v_{do} and v_{qo} are

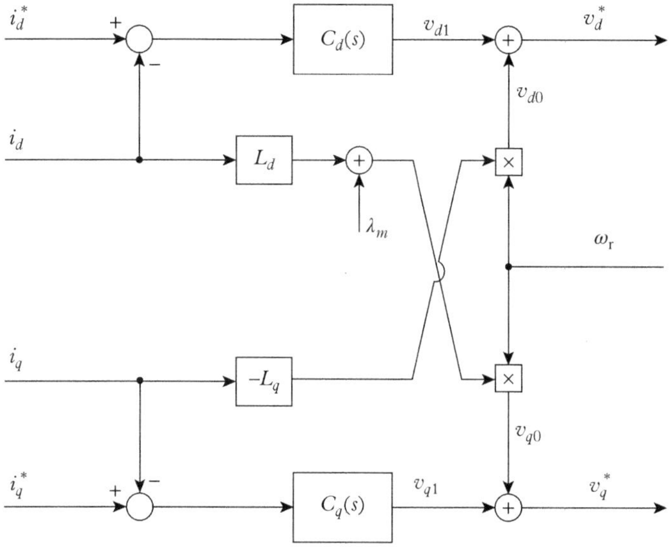

Figure 3.7 *Block diagram of decoupling current controllers.*

provided as feedforward compensation signals, which are constructed according to

$$v_{d0} = -\omega_r L_q i_q^*, \tag{3.4.5}$$

$$v_{q0} = \omega_r L_d i_d^* + \omega_r \lambda_m. \tag{3.4.6}$$

By this arrangement the PI controllers shape the dynamics of d- and q-axis voltage commands, respectively, due to their own current errors. However, the non-linear terms v_{do} and v_{qo} linearize the motor dynamics by decoupling $v_d{}^*$ from i_q and v_q^* from i_d, respectively. Therefore, the equivalent linearized system is shown along the d- and q-axes in Fig. 3.8, where the linearized transfer functions $G_d(s)$ and $G_q(s)$ are given as

$$G_d(s) = \frac{1}{L_d s + R_s}, \tag{3.4.7}$$

$$G_q(s) = \frac{1}{L_q s + R_s}. \tag{3.4.8}$$

3.5 Operating limits and limiting means

A PMS machine operating under steady-state conditions must remain in certain operating limits to fulfill the basic design specifications

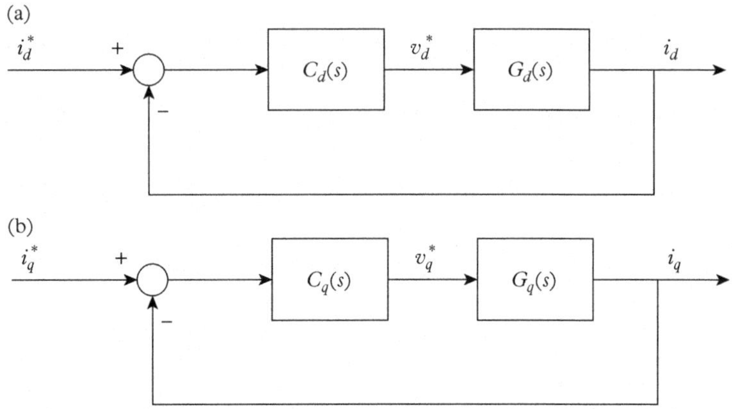

Figure 3.8 *Block diagrams of the equivalent linearized system: (a) d-axis block diagram. (b) q-axis block diagram.*

of the machine. The most important limits are for current and voltage, which define other operating limits. The machine stator current must not exceed a certain value due to some machine essential constraints. The most vulnerable machine component against overcurrent is dielectric of windings. The dielectric class determines the heat that can safely be generated in the windings and thus it decides the current limit. The cooling system also has a decisive role in setting the current limit. The limit is much more restricted in steady-state due to a relatively long time constant of the heat transfer.

3.5.1 Current limit

The steady-state current limit is around the rated current. In transient state, however, the current limit may be as high as several times the machine-rated current. The machine current limit must be observed by the control system. The limit may be imposed on the stator current vector magnitude, i_s. It is recalled here in terms of its d- and q-axis components as

$$i_s = \sqrt{i_d^2 + i_q^2} \leq i_{sL}, \tag{3.5.1}$$

where i_{sL} is the stator current limit. It is primarily shown by a circle with a radius equal to i_{sL} in an i_d–i_q coordinate system as seen in Fig. 3.9. Therefore, as the current limit is concerned, a control system must maintain the current vector inside the circle. The current limit in practice can be met by placing current limiters on the d- and q-axis current commands. A limiter passes a current command without any change if the current is less than the limit. However, it fixes

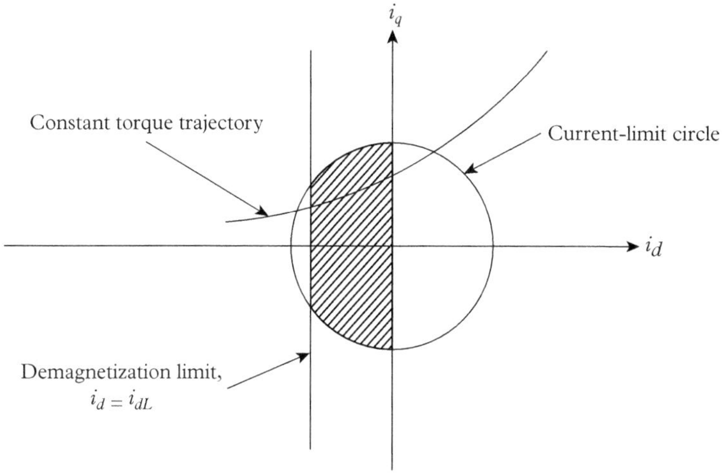

Figure 3.9 *Current limit circle and permanent magnet demagnetization limit.*

the command to the limit if it exceeds the limit. The limits for i_d^* and i_q^* are decided by the system designer according to machine basic relationships.

The limit for i_d^* is usually set by the demagnetization effect of d-axis current on the permanent magnet material of rotor poles. It is known that a negative d-axis current produces an opposing flux against the magnet flux as the d-axis is along the magnetic axis of the pole. If the current is higher than a threshold, the magnet is demagnetized permanently and the machine malfunctions, including a decreased magnet torque. The threshold is regarded as the d-axis current limit, i_{dL}, and used in the d-axis current limiter. A demagnetizing coefficient, ξ, is defined as the ratio of the d-axis armature reaction flux to the permanent magnet flux linkage as

$$\xi = -\frac{L_d i_d}{\lambda_m}. \tag{3.5.2}$$

If the coefficient is large, while the coercivity of the magnet material is not enough, then the magnet demagnetization occurs (Morimoto *et al.* 1990a). The d-axis current can increase with no risk of permanent demagnetization until the resultant flux density at the trailing edge of the magnet pole vanishes. At this operating limit, the demagnetizing coefficient reaches a limiting value of ξ_L, which is obtained in terms of machine parameters (Morimoto *et al.* 1990a). The safe operating region is therefore defined as

$$\xi \leq \xi_L. \tag{3.5.3}$$

The d-axis current limit is then obtained from eqns (3.5.2) and (3.5.3) as

$$i_{dL} = -\frac{\xi_L \lambda_m}{L_d}. \tag{3.5.4}$$

It is also known that there is no gain for a PMS machine operating with $i_d > 0$. This is because the reluctance torque becomes negative and the total torque reduces when i_d gets positive according to eqn (2.5.21) due to the fact that always $L_q \geq L_d$. Therefore, as the d-axis current is concerned, the stator current vector must be located in the dashed regions of current limit circle specified in Fig. 3.9.

The q-axis current limit is determined by eqn (3.5.1) by using the output of the i_d limiter and the maximum allowable i_s as

$$i_{qL} = \sqrt{i_{sL}^2 - i_d^{*2}}. \tag{3.5.5}$$

This is used in the i_q limiter. Therefore, an arrangement as seen in Fig. 3.10 can be used for implementing current limiters. This way both the rotor pole permanent magnet and the dielectric of windings are protected from permanent demagnetization and high temperature, respectively, due to overcurrent. Figure 3.10 shows that a lower i_d^* provides more room for a higher i_q^*, resulting in a desired utilization of torque production capability of the machine despite current limits on both current components.

Another way of limiting the motor current form exceeding the allowed value is to continuously compare the total stator current with

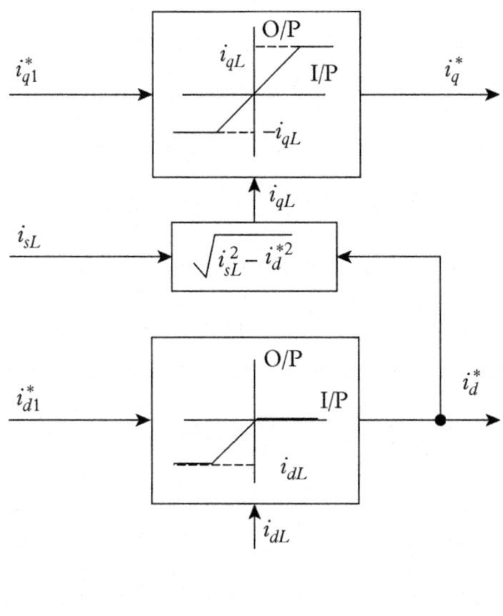

Figure 3.10 *Implementation of current limiters.*

its limit. If the current is going to exceed the limit, the reference of the less influential current component on the torque is limited. This is done by calculating the sensitivity of the developed torque with respect to i_d and i_q at each reference point as

$$S_{i_d}^{T_e} = \frac{\partial T_e}{\partial i_d} \mid_{i_q = i_q^*} = \frac{3}{2} P (L_d - L_q) i_q^* , \tag{3.5.6}$$

$$S_{i_q}^{T_e} = \frac{\partial T_e}{\partial i_q} \mid_{i_d = i_d^*} = \frac{3}{2} P \left[\lambda_m + (L_d - L_q) i_d^* \right] . \tag{3.5.7}$$

These sensitivities show how the d- and q-axis currents influence the torque at each motor operating point. The reference current corresponding to the lower sensitivity is limited. Assuming that the torque sensitivity to i_q is the higher sensitivity, the limits for the d- and q-axis current commands are presented as

$$i_{qL} = i_{sL}, \tag{3.5.8}$$

$$|i_{dL}| = \sqrt{i_{sL}^2 - i_q^{*2}}. \tag{3.5.9}$$

However, if the torque sensitivity to i_d is the higher sensitivity, the limits for the d- and q-axis current commands are presented as

$$|i_{dL}| = i_{sL}, \tag{3.5.10}$$

$$i_{qL} = \sqrt{i_{sL}^2 - i_d^{*2}}. \tag{3.5.11}$$

3.5.2 Voltage limit

The commanded voltage to the inverter must be limited, too, in order to protect the power electronic switches. It is theoretically possible to directly limit the components of voltage vector according to

$$v_s = \left(v_d^2 + v_q^2 \right)^{1/2} \leq v_{sL}, \tag{3.5.12}$$

where v_{sL} is the voltage limit, i.e., the maximum safe voltage of the switches. However, in practice it is easier to limit the voltage indirectly through the current components. This is done by considering the steady-state voltage equations as

$$v_d = -\omega_r L_q i_q , \tag{3.5.13}$$

$$v_q = \omega_r L_d i_d + \omega_r \lambda_m, \tag{3.5.14}$$

where voltage drops due to winding resistances are ignored for the sake of simplicity. Then, eqns (3.5.13) and (3.5.14) are substituted into the voltage constraint at its voltage limit, i.e.,

$$v_s = \left(v_d^2 + v_q^2\right)^{1/2} = v_{sL}. \tag{3.5.15}$$

The result would be the equation

$$i_d = -\frac{\lambda_m}{L_d} + \frac{1}{L_d}\sqrt{\frac{v_{sL}^2}{\omega_r^2} - (L_q i_q)^2}. \tag{3.5.16}$$

This is the equation of an ellipse in i_d–i_q coordinates. Therefore, the stator current vector must remain inside the ellipse to observe the voltage limit. It is seen in eqn (3.5.16) that the ellipse depends on the motor speed. Figure 3.9 shows the limit for three values of speed. It is seen that the limit is being contracted by an increasing speed. We can thus conclude that the allowed machine current is reduced at high speed. This results in a decreasing developed torque with an increasing speed. Also, it means that machine control becomes more restricted and more demanding at high speed. The voltage limit can be observed in practice by incorporating the ellipse into the VC systems at the command stage. This is not a difficult task as the control systems are implemented nowadays by software routines.

Both current and voltage limits must be observed in steady-state. In reference to Fig. 3.11 it can be said that the machine must operate inside the common area of both limits by dragging the stator current vector to the common area. This can be done by current VC.

Referring to the machine torque equation (2.5.19), torque can also be shown in the d–q reference frame of Fig. 3.11. The common area

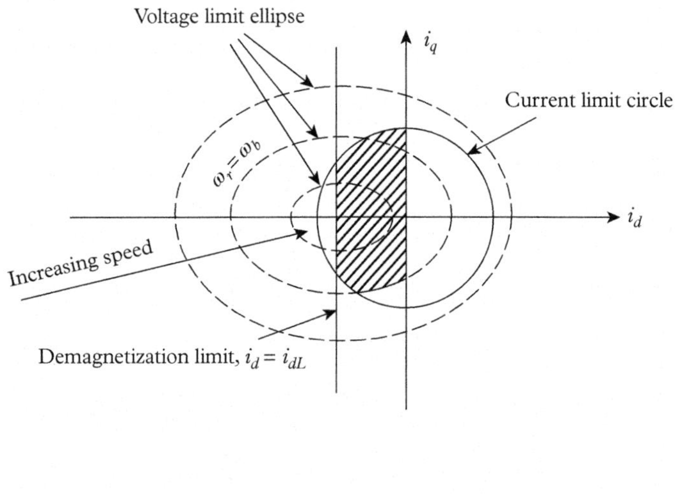

Figure 3.11 *Common area of current limit circle and voltage limit ellipse.*

as the allowed operating region imposes restriction on torque development. In fact, stator current vectors with locus on the torque trajectory located inside the allowed region are applied to the system as current reference. The corresponding implementation will be discussed in Section 3.6 in connection with the transition among the flux control modes.

3.6 Flux control

It is traditionally a common practice to divide the torque speed characteristic of an electric motor into the constant torque region below the base speed, where the motor flux is constant, and the constant power region, where the motor flux is weakened and the motor power remains fixed. The latter operating region is well justified by knowing that electric motors cannot produce rated torque above the rated speed due to the power limit. A common solution is to reduce the developed motor torque to observe the power limit. This is usually achieved by flux weakening above the base speed. In PMS machines where the rotor flux produced by magnet poles cannot be changed, flux weakening is solely carried out through the armature reaction control. This is to control the stator current such that the desired torque is developed, while the magnet flux is confronted by an armature flux component. Fortunately, VC provides an opportunity to control the torque development and flux weakening by the same control system. In a rotor reference frame in particular, a negative i_d contributes to a positive reluctance torque and at the same time produce a component of armature flux, which opposes the magnetic flux. The flux weakening phenomenon at a high speed region can also be studied from the viewpoint of voltage – current limit. As the motor input voltage (inverter output voltage) and the motor current reach their limits, it is wise to adjust the components of current vector, i_d and i_q, such that both voltage and current limits are observed, while torque development and flux weakening are controlled.

In contrast to the traditional fixed flux operation of electric motors in the constant torque region of motor torque – speed characteristic, below the rated speed where the voltage limit has not yet been reached, it is sometimes desirable to have some kind of flux weakening. This type of flux weakening is not aimed at the prevention of exceeding the motor power limit, but to achieve maximum torque for each unit of current flowing through the machine or simply maximum torque per ampere (MTPA) operation. It means that for every stator current modulus, the current components i_d and i_q are adjusted, such that the maximum torque is achieved.

A similar flux weakening may be useful in some situation to achieve unity power factor operation. In this case, current components must be adjusted and controlled such that a unity power factor is achieved. Flux weakening control systems for the above-mentioned modes of operation, i.e., MTPA, unity power factor (PF), and high speed are presented in this section. It is recalled that in all these operating modes, a negative i_d is desirable. In fact, a positive i_d leading to an increased machine flux in the d-axis direction is not common, since it reduces the developed torque and causes increased machine losses.

A challenging task in the flux weakening control of PMS motors is to observe the current and voltage limits. Another challenge is to manage the machine operation, such that the transition from one flux-weakening mode to another one happens smoothly as the motor speed changes. These challenges are also studied in this section.

3.6.1 Maximum torque per ampere control

Rapid motor acceleration is required when a motor starts from standstill or a big change in the speed is required. In these situations, motor must develop a high torque. The motor under such torque demand usually reaches its current limit. This causes a limited torque, which is not sufficient for rapid acceleration. In these situations it is desirable to have maximum possible torque for every unit of stator current to utilize the full torque production potential of the machine. This maximum torque per ampere operation is equivalent to having a specific torque with minimum possible current. Therefore, even if the machine is not needed to develop high torque, it is still desirable to operate at MTPA mode to reduce motor copper loss, and inverter and rectifier conduction loss. The MTPA control can be implemented with speed control or torque control as follows.

3.6.1.1 MTPA with speed control

The torque is presented in terms of instantaneous stator current components in rotor reference frame by eqn (2.5.19), which is recalled here as

$$T_e = \frac{3}{2} P \left[\lambda_m + (L_d - L_q) i_d \right] i_q . \qquad (3.6.1)$$

This can be depicted in an i_d–i_q coordinate system as in Fig. 3.12. The figure indicates that a specific torque can be developed by an infinite number of stator current vectors; only four of them have been depicted. The current vector with a minimum modulus fulfills the conditions of MTPA. This can be obtained graphically as a current vector belonging to a current circle tangent to the torque trajectory as seen in Fig. 3.12.

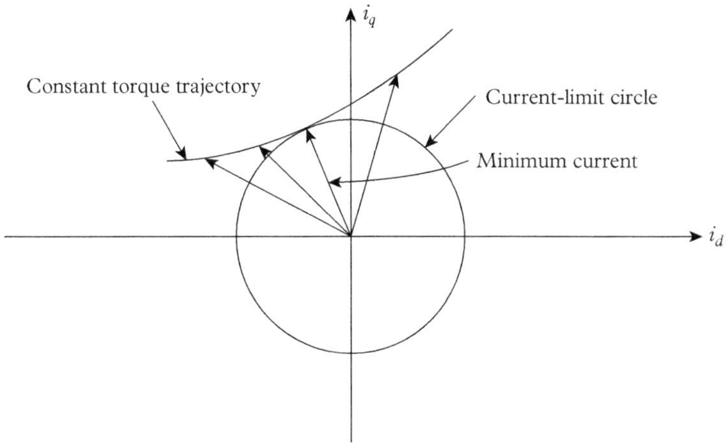

Figure 3.12 *Depicting MTPA conditions.*

In salient pole PMS machines, where $L_q > L_d$, the d-axis current under MTPA conditions is always negative due to the contribution of the reluctance torque as seen by eqn (3.6.1). The current components under MTPA conditions can be obtained mathematically by substituting for i_q from eqn (3.5.1) in terms of i_s and i_d in eqn (3.6.1), and solving the derivative of the resultant torque equation with respect to i_d, where i_s is taken as a constant (Morimoto *et al.* 1994a):

$$\frac{\partial T_e}{\partial i_d} \mid_{i_s = const.} = 0. \tag{3.6.2}$$

When in the resulting derivative of eqn (3.6.2), i_s is substituted in terms of d- and q-axis current components, the following current equation results:

$$i_d^2 - i_q^2 + \frac{\lambda_m}{L_d - L_q} i_d = 0. \tag{3.6.3}$$

Equation (3.6.3) presents a trajectory of i_s in terms of its components under MTPA conditions for every torque. This is a convex curve in the i_d–i_q coordinates with its center at the point

$$\begin{cases} i_d = \frac{\lambda_m}{(L_d - L_q)}, \\ i_q = 0. \end{cases} \tag{3.6.4}$$

The trajectory is depicted in Fig. 3.13 together with the trajectories of current limit and voltage limit (Morimoto *et al.* 1990b).

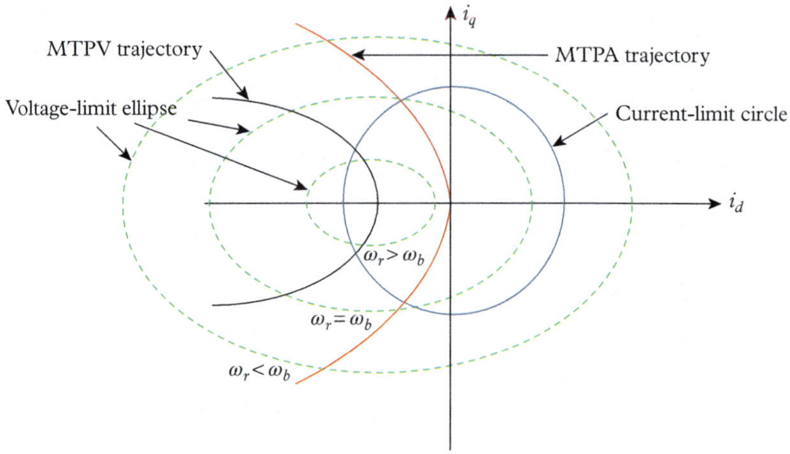

Figure 3.13 *Trajectory of stator current vector under MTPA and MTPV conditions, together with current limit circle and voltage limit ellipse.*

The d-axis current can be found from eqn (3.6.3) as a function of i_q (Morimoto *et al.* 1994a):

$$i_d = \frac{\lambda_m}{2(L_q - L_d)} - \sqrt{\frac{\lambda_m^2}{4(L_q - L_d)^2} + i_q^2}. \tag{3.6.5}$$

This equation can be used to incorporate the MTPA strategy into the VC of PMS motors by substituting the current components with their command values as depicted in Fig. 3.14.

The control system includes a block, which receives the q-axis current command and produces the d-axis current command according to eqn (3.6.5). This ensures that the current vector always remains on the MTPA trajectory. The rest of system is as the system of Fig. 3.6.

Note that in surface-mounted pole machines where $L_d = L_q$, the MTPA trajectory turns to be the q-axis as is evident from eqn (3.6.3). It means that $i_s = i_q$. Therefore, MTPA control of surface-mounted pole machines can be implemented by the control system of Fig. 3.6 with a constant $i_d = 0$. This has already been presented in Fig. 3.5 with phase current controllers, although the use of d- and q-axis current controllers are more common.

3.6.1.2 MTPA with torque control

Useful insight into the machine characteristics can be gained if the maximum torque per ampere formulation is carried out per unit or

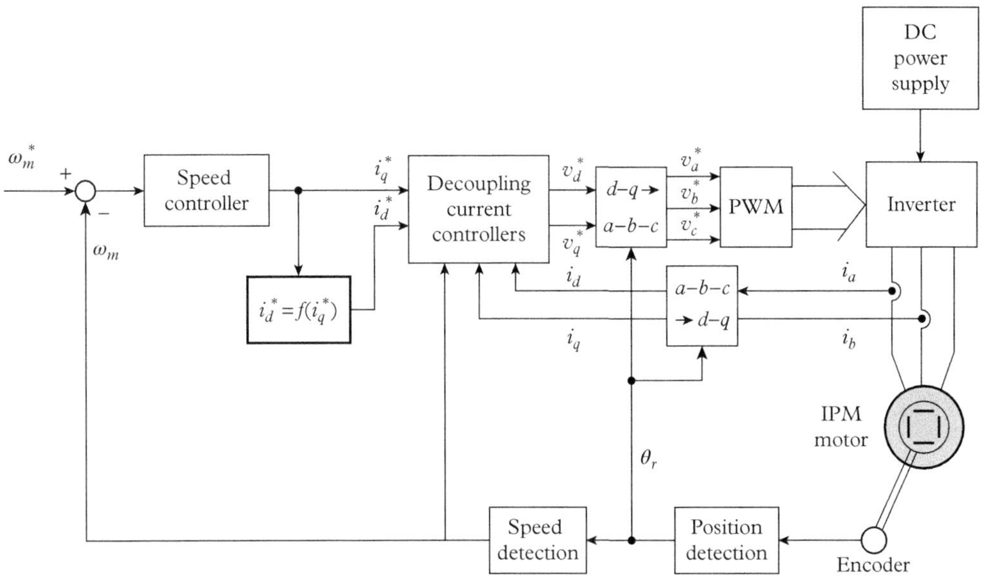

Figure 3.14 *Vector control of PMS motors in rotor reference frame with MTPA control.*

normalized basis. Normalization is done with a special choice of base values for the motor current and torque as (Jahns *et al.* 1986)

$$i_b = \frac{\lambda_m}{L_d - L_q}, \qquad T_{eb} = \frac{3}{2} P \lambda_m i_b. \tag{3.6.6}$$

The normalized values of d- and q-axis current components and developed torque are then obtained as

$$i_{dn} = \frac{i_d}{i_b}, \qquad i_{qn} = \frac{i_q}{i_b}, \qquad T_{en} = \frac{T_e}{T_{eb}}. \tag{3.6.7}$$

Using the normalized values in eqn (3.6.7), the torque equation per unit becomes

$$T_{en} = i_{qn}(1 - i_{dn}). \tag{3.6.8}$$

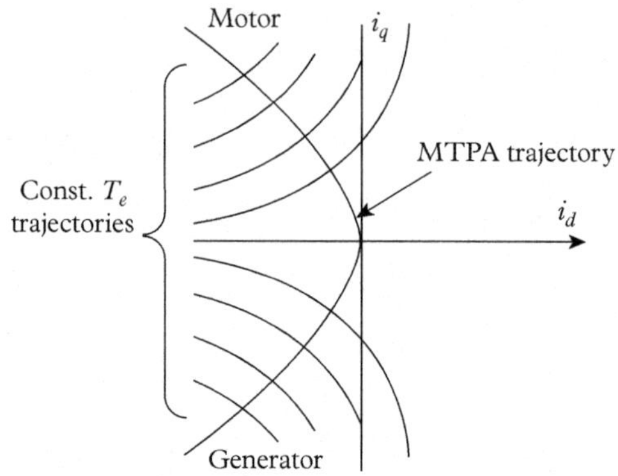

Figure 3.15 *Trajectory of the stator current vector under MTPA conditions together with trajectories of different constant torques at motoring and generating modes.*

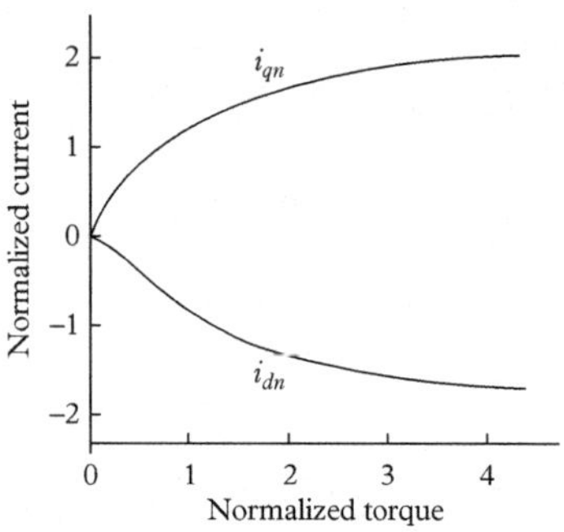

Figure 3.16 *Normalized current components versus normalized torque for MTPA control.*

It is seen that all machine parameters are eliminated from the torque equation (3.6.8). This provides an insight into the MTPA mode of operation as seen in Fig. 3.15 (Jahns *et al.* 1986).

Also in Fig. 3.15 the MTPA trajectory is tangent to the q-axis at the origin, which is the trajectory of the magnet torque alone and tends to asymptotes to 45°, which is the trajectory of reluctance torque alone (Jahns *et al.* 1986). This is due to the fact that in a PMS machine in general, the reluctance torque gets more dominant as the torque increases. Now it is possible to get i_{dn} and i_{qn} in terms of the normalized torque from the equations

$$T_{en} = \sqrt{i_{dn}(i_{dn} - 1)^3}, \tag{3.6.9}$$

$$T_{en} = \frac{i_{qn}}{2}\left[1 + \sqrt{1 + 4i_{qn}^2}\right]. \tag{3.6.10}$$

The normalized currents are plotted versus normalized torque in Fig. 3.16 (Jahns *et al.* 1986).

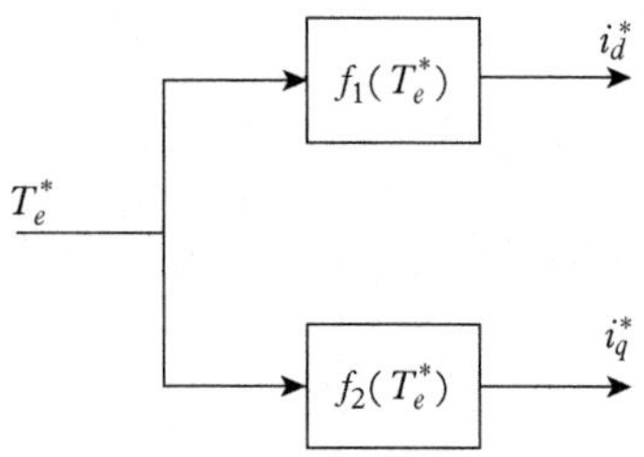

Figure 3.17 *Feedforward torque control system in rotor reference frame.*

These current trajectories suggest an alternative VC system, i.e., feedforward torque control for PMS machines under MTPA strategy. This is shown in Fig. 3.17 (Jahns *et al.* 1986). It is seen that the current commands are generated by functions f_1 and f_2, which are implemented by eqns (3.6.9) and (3.9.10), respectively, by using only the torque command. The torque command itself may directly be applied by the end user or it may be provided as an output of an overhead system, e.g., speed controller. Having the d- and q-axis current commands, the VC can be implemented with a–b–c current controllers or d–q current controllers as was elaborated before.

3.6.2 Maximum torque per voltage control

When the inverter output voltage reaches its maximum available value, it is desirable for the machine to use maximum potential of available voltage. This is to develop maximum torque for every unit of voltage. Therefore, an MTPV control is desired. To get the operating conditions under this control strategy, it is necessary to present the torque equation in terms of voltage components instead of the conventional presentation of torque in terms of current components. A procedure is presented as the following.

Using the steady-state voltage equations (3.5.13) and (3.5.14), one gets the d- and q-axis currents in terms of d- and q-axis voltages as

$$i_d = \frac{v_q - \omega_r \lambda_m}{\omega_r L_d}, \tag{3.6.11}$$

$$i_q = -\frac{v_d}{\omega_r L_q}. \tag{3.6.12}$$

These equations in connection with the torque equation (3.6.1) give the torque equation in terms of v_d and v_q. Also it is known that

$$v_s = \left(v_d^2 + v_q^2\right)^{1/2}, \tag{3.6.13}$$

which can be used to give the torque as a function of v_d and v_s. Now the following derivative equation provides a MTPV condition as

$$\frac{\partial T_e}{\partial v_d}\Big|_{V_s = const.} = 0. \tag{3.6.14}$$

Equation (3.6.14) in connection with (3.6.13) gives

$$v_q^2 - v_d^2 + \frac{\omega_r L_q \lambda_m}{L_d - L_q} v_q = 0. \tag{3.6.15}$$

This is the trajectory of MTPV in d–q voltage coordinates. When the d- and q-axes voltages from eqns (3.5.13) and (3.5.14) are substituted for in the above-mentioned trajectory, the MTPV trajectory in d–q current coordinates is obtained as

$$L_d^2 i_d^2 - L_q^2 i_q^2 + \frac{2L_d - L_q}{L_d - L_q} L_d \lambda_m i_d + \frac{L_d}{L_d - L_q} \lambda_m^2 = 0. \tag{3.6.16}$$

This equation gives the current vector trajectory for MTPV in terms of i_d and i_q as depicted in Fig. 3.13. The figure reveals that the trajectory does not pass through the coordinate system origin. The MTPV control strategy can be incorporated into the rotor-oriented VC by finding i_d from eqn (3.6.16) as a function of i_q and substituting this function into the d-axis current command-producing block in the control system of Fig. 3.14. The block then gives an i_d for every i_q, which ensures MTPV operation.

3.6.3 Unity power factor control

It is well known that the unity power factor operation of a synchronous motor drive leads to a reduced rating of a power electronic inverter, and, thus, saves system capital costs. The power factor is an important measure of motor drive performance at steady-state. Therefore, steady-state models are used for mathematically formulating its condition.

Power factor control can be incorporated into a VC system in any reference frame. It is presented in this section in rotor reference frame. It is implemented by i_d control in this RF. The condition for unity power factor operation is derived first. This is obtained by considering the fact that under this control strategy, the power factor is always kept constant at unity. Accordingly, the stator current vector and the stator voltage vector must be aligned under this control strategy. This is translated, in rotor reference frame, into the relationship

$$\frac{v_d}{v_q} = \frac{i_d}{i_q}. \tag{3.6.17}$$

By substituting for the voltage components in (3.6.17) from the steady-state voltage equations and ignoring voltage drops due to winding resistance, the following stator current vector trajectory can be obtained in terms of its d- and q-axis current components:

$$L_d i_d^2 + L_q i_q^2 + \lambda_m i_d = 0. \tag{3.6.18}$$

When eqn (3.6.18) is solved for i_d, the following d-axis current is obtained in terms of i_q:

$$i_d = -\frac{\lambda_m}{2L_d} - \sqrt{\frac{\lambda_m^2}{4L_d^2} - \frac{L_q}{L_d} i_q^2}. \tag{3.6.19}$$

Again, replacing the current components in eqn (3.6.19) with their command values, it can be used in the d-axis current command-producing block of Fig. 3.14 to fulfill unity power factor control.

3.6.4 Transition among flux control modes

It is desirable to incorporate different flux control modes of operation into a VC system and facilitate a transition between them to always achieve an optimum motor performance corresponding to the motor operating point and at the same time to take into account the current and voltage limits. A transition between the MTPA control and the MTPV control as the most common transition, which ensures maximum motor power, will be presented here (Morimoto *et al.* 1990a,b). Another transition of control modes will be presented in hybrid loss minimization and MTPA control in Section 3.9.4.

The stator current vector trajectories under the MTPA and the MTPV, together with the current limit circle and the voltage limit ellipse at two speeds, are depicted in i_d–i_q coordinates in Fig. 3.18. The stator current vector under MTPA conditions is on OA1 trajectory, where point "O" on the trajectory corresponds to zero torque. The stator current can increase on the trajectory up to A1, which is the cross point of the trajectory and the current circle, corresponding to the maximum torque with allowable current. The motor may work with this current on the so-called constant torque region with the maximum torque. The maximum motor speed at A1 is ω_{r1}. At this operating point the motor provides maximum power, observing both the current and the voltage limits. The voltage limit ellipse at this speed is depicted by a dash line. As the motor speed increases beyond ω_{r1}, the stator current vector cannot remain on the MTPA trajectory anymore if the maximum power is to be delivered by the motor. In fact, the current vector approaches A2 on the current circle, and the voltage ellipse shrinks. Therefore, A1–A2 on the circle corresponds to the so-called constant power operation of the machine. The speed ω_{r2}

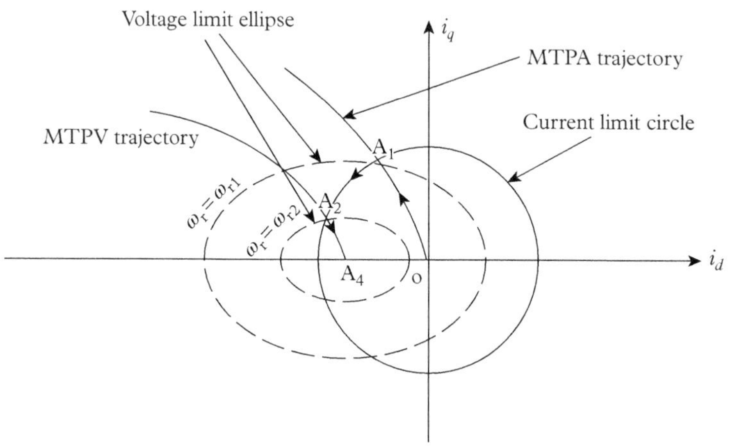

Figure 3.18 *Transition among flux control modes for machines with* $\lambda_m < L_d i_{sL}$.

is the maximum speed for the maximum motor power, considering voltage limit. Above this speed, the voltage-limited maximum motor power cannot be obtained, since the MTPV trajectory intersects the voltage ellipse beyond the current circle.

Considering the motor performances presented previously, the motor can produce maximum power over the entire speed range, while observing the current and voltage limits, if the stator current vector in Fig. 3.19 is forced to follow the three modes of operation as follows (Morimoto *et al.* 1990b):

1. $\omega_r \leq \omega_{r1}$: The current vector is fixed at A1. The d- and q-axis stator current components are given by solving the MTPA conditions and the current limit, i.e., eqns (3.6.3) and (3.5.1). Therefore, $i_s = i_{sL}$ and $v_s < v_{sL}$.
2. $\omega_{r1} < \omega_r \leq \omega_{r2}$: The current vector moves along the current limit circle from A1 to A2, while the speed increases from ω_{r1} toward ω_{r2}. The d- and q-axis stator current components are given by solving the current limit circle and the voltage limit ellipse, i.e., eqns (3.5.1) and (3.5.15). Therefore, $i_s = i_{sL}$ and $v_s = v_{sL}$.
3. $\omega_r \geq \omega_{r2}$: The current vector moves along the MTPV trajectory from A2 to A4, while the speed increases beyond ω_{r2}. The d- and q-axis stator current components are given by solving the MTPA conditions and the voltage ellipse, i.e., eqns (3.6.3) and (3.5.16). Therefore, $i_s < i_{sL}$ and $v_s = v_{sL}$.

Figure 3.18, as explained previously, is associated with the operating modes of PMS machines in which $\lambda_m < L_d\, i_{sL}$. If the machine design

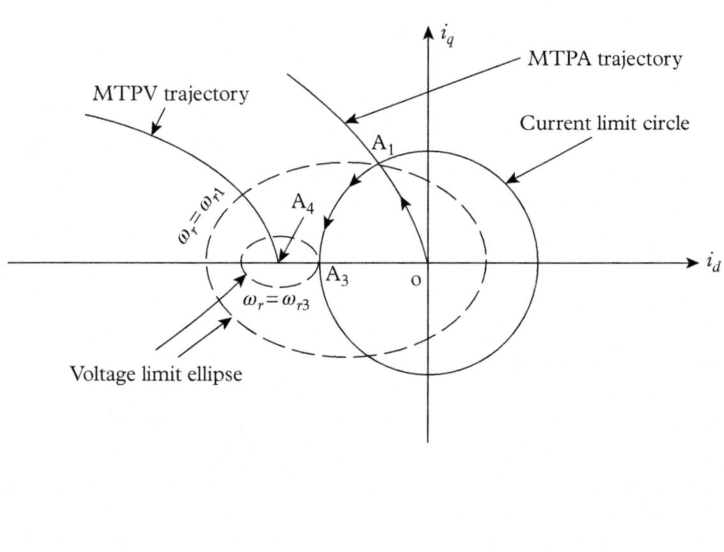

Figure 3.19 *Transition among flux control modes for machines with* $\lambda_m > L_d i_{sL}$.

is such that $\lambda_m > L_d\, i_{sL}$, then the MTPV trajectory falls outside the current limit circle as shown in Fig. 3.19. Therefore, the stator current vector at $\omega_r \geq \omega_{r2}$ cannot move along the MTPV trajectory. As a result, the stator current vector moves on the current limit circle toward point A3, as seen in Fig. 3.19. At this point $i_q = 0$ and, therefore, the machine developed torque and subsequently the machine power becomes 0, while the motor speed reaches its maximum value at $\omega_r = \omega_{r3}$, which is given by

$$\omega_{r3} = \frac{v_{sL}}{\lambda_m - L_d\, i_{sL}}. \tag{3.6.20}$$

At a special machine design that $\lambda_m = L_d\, i_{sL}$, the maximum machine speed theoretically approaches infinity and the voltage limit ellipse vanishes as seen from eqn (3.6.20).

In surface-mounted PMS machines, the same modes of operation as described previously are held. However, Figs 3.18 and 3.19 must be redrawn such that the voltage limit becomes a circle, the MTPA trajectory lays on the q-axis, and the MTPV trajectory becomes a vertical line at $i_d = -\lambda_m / L_d$.

The transition of machine operation among the previously mentioned three modes of operation when the motor speed varies over its entire range can be carried out in practice by a transition routine. The routine is implementation as a part of the overall machine control system and determines the d-axis current command in response to the q-axis current command. It replaces the flux weakening block (i_d^* determination block) in Fig. 3.14.

3.6.5 Saturation of current controllers

Saturation of current controllers refers to a malfunction of current controllers during which the current control is partially lost. The malfunction is due to neither the reaching of the winding current capacity to its limit nor a design problem in the current controller. It is, in fact, caused by the inability of a motor current to follow its command signal due to a voltage limit at high speed. Considering the machine d-axis voltage equations is helpful in clarifying the cause of the problem. The equation is recalled here as

$$v_d + \omega_r L_q i_q = R_s\, i_d + L_d p i_d. \tag{3.6.21}$$

It is seen in eqn (3.6.21) that when the machine speed increases, the speed voltage along the d-axis increases. As a result, the total voltage on the right-hand side of eqn (3.6.21), which acts as a driving potential to flow the d-axis current through the machine, decreases. This

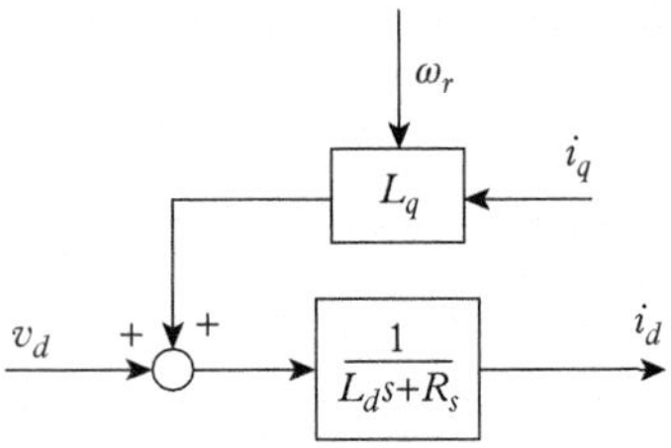

Figure 3.20 *Speed voltage as an internal feedback in the machine, causing saturation of the d-axis current controller.*

is because the d-axis terminal voltage is negative due to the negative d-axis current in IPM machines. Therefore, the d-axis current control is partially lost, which is denoted as the saturation of the current controller. The speed voltage acts as an internal feedback for the machine as seen in the machine block diagram of Fig. 3.20 corresponding to eqn (3.6.21). When the feedback increases at high speed, the error to the current dynamic block and thus the block output, i.e., i_d, reduces. This, in turn, causes a drift in the developed torque. The problem can be avoided if extra terminal voltage is applied to the machine. However, since the machine operation beyond the base speed is maintained by applying full voltage to the machine corresponding to the DC link voltage of the inverter, there is no extra voltage available to avoid saturation of the current controller at high speeds.

One solution is to lower the speed voltage $\omega_r L_q i_q$ in eqn (3.6.21) by limiting more severely the q-axis current to provide enough voltage to the dynamic block in Fig. 3.20. However, a reduction of i_q will change the machine developed torque, unless i_d changes accordingly. Fortunately, a decrease in i_q increases the total negative voltage on the left-hand side of eqn (3.6.21), which acts as a driving potential to flow i_d, and, therefore, increases the negative i_d and increases the developed torque. For a negative i_q, which occurs during the deceleration, a higher limit must be enforced.

The saturation of the current controller and its solution can also be explained graphically by recalling that the current space vector must always remain inside the voltage limit ellipse. The saturation of the current controller occurs when a current vector command, i_{s1}, outside the voltage ellipse is applied to the machine to develop a desired torque, as seen in Fig. 3.21. The command cannot be met due to the voltage limit. The solution is to move back the current vector command inside the ellipse, whereas it is retained on the desired torque locus, as seen in Fig. 3.21 by i_{s2}.

The solution can be implemented by a system as depicted in Fig. 3.22 (Jahns 1987). The saturation of d-axis current is detected by the error between the commanded and feedback values of d-axis current, Δi_d. This error is high when the saturation does occur. It is applied to a PI current regulator to provide a current component i_{df} as a flux weakening current. This current will adjust the q-axis current limit. As the q-axis current command is limited and the system wants to develop the same torque as desired by the torque command, i_d gets more negative and its error reduces. Thus, the current control is saturated. The values of i_0 and i_{qmax} are constant and are determined such that the normal operation and the flux weakening operation are

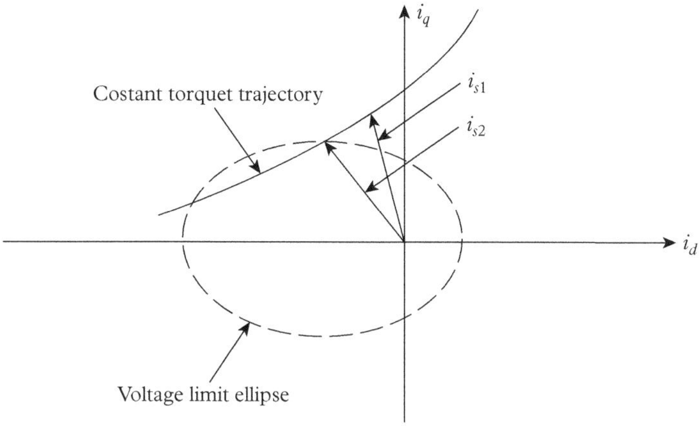

Figure 3.21 *Solution to the saturation of the current controller by moving the stator current vector into the voltage limit ellipse.*

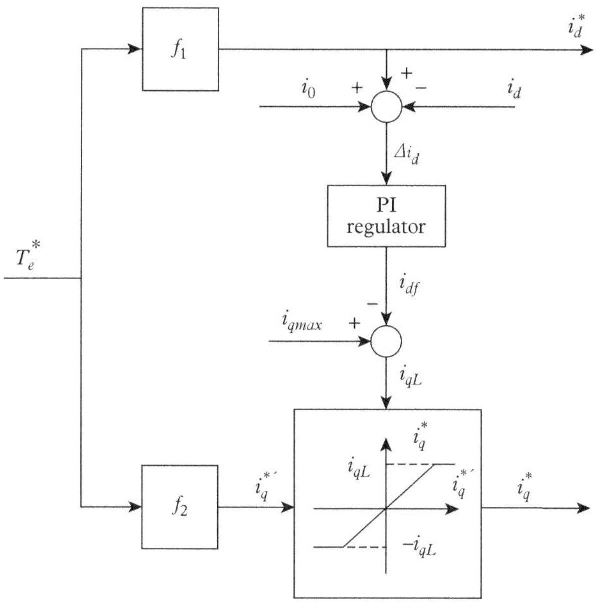

Figure 3.22 *Block diagram of modified IPM motor control to prevent current controller saturation.*

adequately decoupled and a desirable dynamics during the transition from one operating mode to the other is ensured.

3.7 VC in stator flux reference frame

A PMS machine model in the stator flux (x–y) reference frame shows a developed torque proportional to $\lambda_x i_y$ as presented by eqn (2.6.13). This compact torque equation is a salient feature of the model in this reference frame with respect to the machine model in rotor reference frame in which the machine torque is more complex and proportional to $\lambda_d\, i_q - \lambda_q\, i_d$ as presented by eqn (2.5.18). A compact torque equation in *d–q* RF proportional to $\lambda_m\, i_q$ can only be obtained when either the machine is of surface-mounted magnet type or it is controlled under $i_d = 0$ as presented in Section 3.3. It is sometimes mentioned that the machine dynamics is faster under VC in *x–y* RF than in *d–q* RF due to the one-term nature of the general torque equation in the former RF. However, it must be considered with caution as will be elaborated later in sub-section 3.7.2.

VC in *x–y* RF is essentially a stator current VC in *x–y* reference frame. The VC in this RF may be realized in two versions, i.e., with *x–y* current controllers and with *a–b–c* current controllers as is the case with VC in *d–q* RF presented in Sections 3.3 and 3.4, respectively. Here, both versions are presented.

3.7.1 VC in stator flux reference frame with phase current controllers

A vector-controlled PMS motor drive system is shown in Fig. 3.23 including phase current controllers. It is seen that the torque command may be provided as the output of the speed controller. Alternatively, it may be decided directly by the applicant. The flux linkage

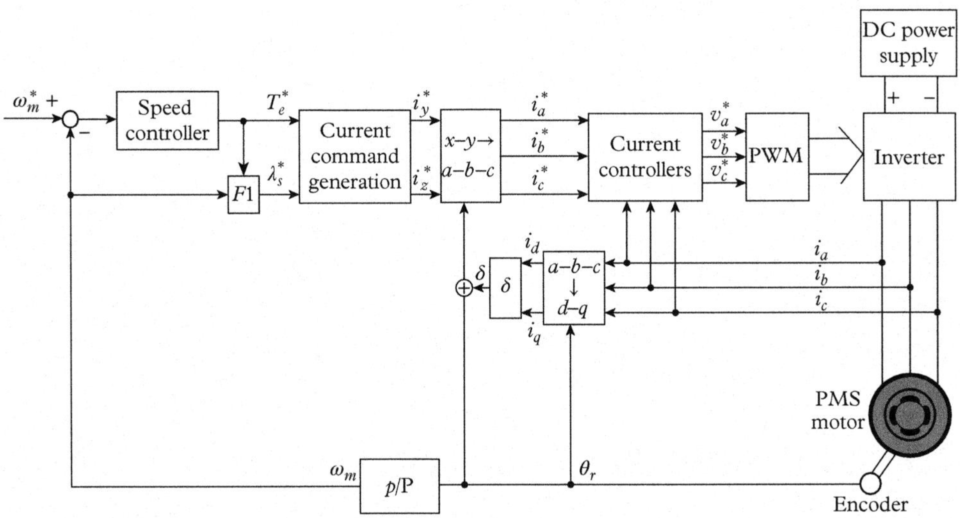

Figure 3.23 *Vector control of PMS motors in stator flux reference frame with phase current controllers*

command is also provided according to the type of desired flux regulation. A conventional flux command can be generated by function $F1$ of Fig. 3.23. The function provides the rated flux linkage in the constant torque region from standstill up to the rated speed. However, it weakens the flux linkage command above the rated speed. The flux weakening can be done such that the motor output power remains constant at rated power. The current commands are then generated by the reference current generating block if the torque and the stator flux linkage commands are available, as seen in Fig. 3.23. The block is elaborated in Fig. 3.24. It is seen in this figure that the y-axis current command, i_y^*, is produced from the torque commands by eqn (2.6.13).

It is also possible to determine i_y^* as an output of the speed controller, especially if an explicit torque control is not needed. The x-axis current command can be determined as a function of i_y^* and λ_s^* by using eqns (2.6.22) and (2.6.23). The function is shown by $F2$ in Fig. 3.24. In case of surface mounted PMS motors, it can be calculated online if a fast processor is used. It is also possible to calculate it offline for a range of i_y^* and λ_s^*, and save the results as a look-up table. In this case, more memory is needed. With the advanced digital signal processors available today for implementing motor control systems, either of the current-producing methods is readily practical. In case of IPM motors, the calculation is too involved for online computation.

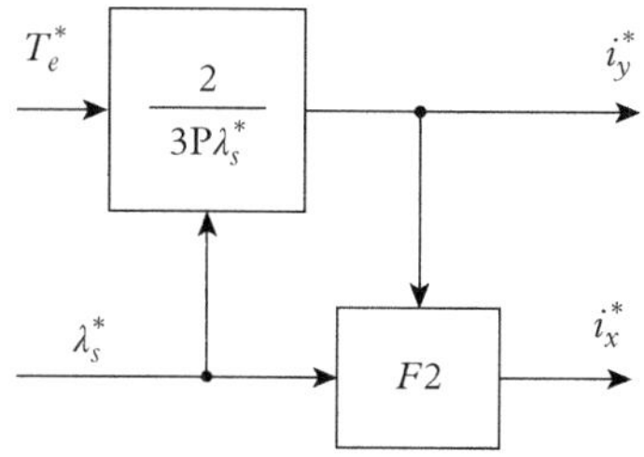

Figure 3.24 *Reference current generation.*

Going back to Fig. 3.23, it is seen that the system includes a reference frame transformation from i_x^* and i_y^* to phase current commands i_a^*, i_b^*, and i_c^* by using a stator flux position signal δ_s. The position is determined by eqn (2.6.9) as recalled next:

$$\delta_s = \delta + \theta_r. \tag{3.7.1}$$

In eqn (3.7.1), the rotor position θ_r is measured by a position sensor and the power angle δ is calculated from eqns (2.5.10), (2.5.11), and (2.6.2), if i_d and i_q are available as seen in Fig. 3.23. The latter current components are provided through an RF transformation using the actual phase currents as is carried out by block δ calculation in Fig. 3.23. The actual phase currents, i_a, i_b, and i_c, are measured by current sensors, and used as feedback signals to close the current loops. They are compared with the phase current commands to provide current errors as inputs to three current controllers. The current controllers are usually of PI type. However, other type of controllers may be used for the sake of performance improvement in special circumstances or in the case of extra demands like being robust, optimal, or adaptive. They accept phase current errors and provide phase voltage commands v_a^*, v_b^*, and v_c^*. The voltage commands are applied to a PWM system to generate gating signals for the inverter. The inverter then inverts the DC input power to a three-phase AC output power according to the voltage commands to supply the PMS machine. The machine thus develops an accurate torque and rotates at a speed corresponding to its own inertia and load. Since a speed control is also included in the system of Fig. 3.23, the machine speed is adjusted to the speed command through the speed control loop. The actual speed signal that is used as the feedback signal in the speed control loop is calculated by differentiating the rotor position signal.

3.7.2 VC in stator flux reference frame with *x–y* current controllers

VC with phase current controllers suffers from a limited bandwidth problem. The current controllers in this system act on the machine phase currents, which alternate at a frequency proportional to the machine speed. As the speed increases, the actual currents must track the phase current commands, which alternate rapidly. In this situation, the controllers' essential delay may get close to the time period of phase currents. As a result, the actual signals may not be able to follow the commanded signals and cause undesirable errors. VC in stator reference frame with *x–y* current controllers may provide a solution to get rid of this problem as shown in Fig. 3.25.

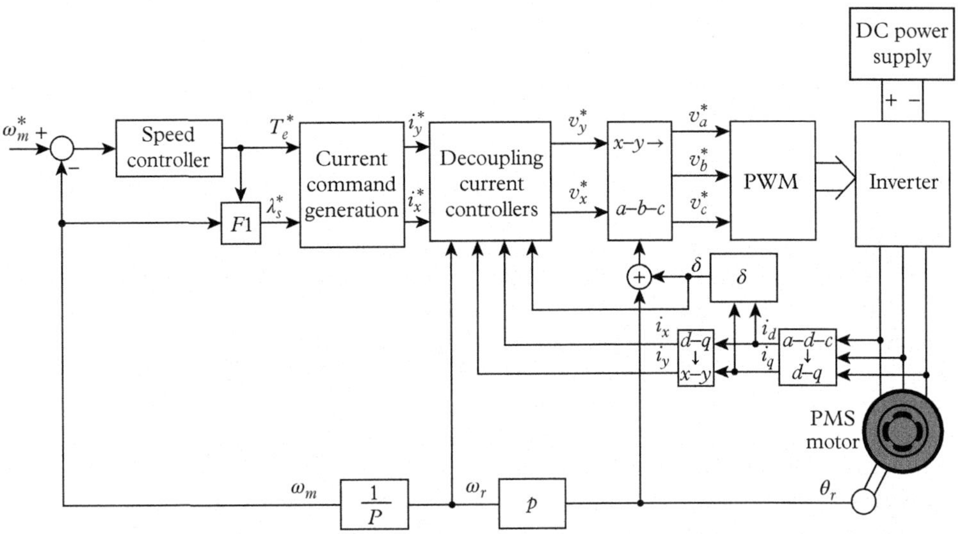

Figure 3.25 *Vector control of PMS motors in stator flux RF with x–y current controllers.*

It is seen in Fig. 3.25 that the current commands, i_x^* and i_y^*, can be provided as in the previous control system. It is also possible to determine i_y^* as an output of the speed controller especially if an explicit torque control is not needed, as is the case with the previous control system. The commanded current components are then applied to the decupling current controllers to provide voltage commands as they are elaborated later. The current commands are compared with the actual machine currents in the stator flux RF to determine current errors. Therefore, the actual *a–b–c* phase currents must be sensed and then transformed to *x–y* currents by eqns (2.5.4) and (2.6.5) using the rotor position and load angle, respectively. The current errors are applied to two current controllers to control machine torque and flux indirectly through *y*- and *x*-axis current controllers. The current controllers produce *y*- and *x*-axis voltage command components. The controllers are often of PI type. However, non-linear controllers may be used to enhance the machine dynamic performance. The voltage commands are then transformed into *a–b–c* voltage commands by an *x–y* to *a–b–c* RF transformation similar to that carried out with currents in Fig. 3.24. Once the voltage commands in *a–b–c* RF is provided, they are applied to PWM. The rest of the control system is the same as that for the VC system in Fig. 3.24.

The control system described so far has focused on independent *x*- and *y*-axis current control loops in order to achieve independent flux control and torque control, as they are the main objectives of VC. However, the *x*- and *y*-axis current control loops are, in fact, not independent. It can be elaborated by referring to the torque equation of eqn (2.6.13), which denotes a torque control by controlling i_y. However, eqn (2.6.21) shows that i_x depends on i_y. Therefore, during an i_y transient period, i_x and therefore stator flux linkage, λ_s, change. As a result, the torque dynamics is influenced by the transient behaviors of both i_y and λ_s as denoted by eqn (2.6.13). Bearing in mind that the flux dynamics is inherently slow, a non-linear and thus sluggish torque dynamics is caused despite the main objective of VC. The problem can be tackled by designing a decoupling circuit in *x–y* RF as done in the case of VC in *d–q* RF presented in Section 3.5. The decoupling circuit here, however, is different from that in *d–q* RF due to the differences in the voltage equations in the two RFs. It is designed as follows by recalling the voltage equations in *x–y* RF from Chapter 2 as

$$v_x = R_s\, i_x + p\lambda_s, \tag{3.7.2}$$

$$v_y = R_s\, i_y + \lambda_s p\delta + \omega_r\, \lambda_s. \tag{3.7.3}$$

A decoupling method regards the right-hand side of eqn (3.7.2) as a combination of two parts, a fast dynamic part consisting of the first two terms and a slow dynamic part consisting of the third term. This partition may suggest that the y-axis voltage command is provided by a combination of two signals as

$$v_y^* = v_{y1} + v_{y0}, \tag{3.7.4}$$

where the first term is a fast dynamic signal provided by the y-axis current controller and the second term is a feedforward compensating signal, which is defined as

$$v_{y0} = \omega_r \lambda_s^*. \tag{3.7.5}$$

This compensating signal is added to the output of the y-axis current controller during the implementation of the control system to partially decouple the x- and y-axis dynamics.

3.7.3 Maximum torque per ampere control in *x–y* RF

The MTPA conditions can be obtained by determining an optimal reference flux linkage as a function of commanded torque. This can be achieved by finding reference values of i_d and i_q as functions of the commanded torque according to a procedure similar to that followed in feedforward MTPA control in rotor reference frame in sub-section 3.6.1. Then, the reference flux linkage is obtained by substituting the above-mentioned current components into

$$\lambda_s = \sqrt{(\lambda_m + L_d i_d)^2 + L_q^2 i_q^2}. \tag{3.7.6}$$

As a result, the reference stator flux linkage is given as a function of commanded torque regardless of reference frame as

$$\lambda_s^* = f\left(T_e^*\right). \tag{3.7.7}$$

Then, in the control system of Fig. 3.25, the reference flux linkage is produced by eqn (3.7.7) instead of being provided by $F1$.

3.7.4 Maximum torque per voltage control in *x–y* RF

It is possible to control PMS machines under maximum torque per voltage control strategy. The objective and merits of such control is discussed in d–q reference frame in sub-section 3.6.2. The conditions of MTPV in x–y reference frame can be derived by obtaining the reference stator flux linkage in terms of commanded torque. This is achieved by substituting for *d*- and *q*-current components in terms of *d*- and *q*-voltage components in steady-state in torque equation to have the torque as a function of v_s and v_q only. Solving the derivative of the torque with respect to v_q then will give an optimal v_q and, in turn, λ_q. Subsequently, the optimal value of λ_d and, finally, λ_s can also be obtained. The control system is the same as that of the conventional system of Fig. 3.25 except for the reference flux linkage, which is given by a flux linkage versus torque block with the reference torque as its input.

3.7.5 Unity power factor control in *x–y* RF

Unity power factor control can also be implemented in stator flux reference frame. The stator current vector and the stator voltage vector must be aligned under this control strategy. This results in

$$\frac{v_x}{v_y} = \frac{i_x}{i_y}. \tag{3.7.8}$$

On the other hand, the steady-state machine voltage equations in stator flux reference frame are reduced to

$$\begin{aligned} v_x &= 0, \\ v_y &= \omega_s \lambda_s, \end{aligned} \tag{3.7.9}$$

where the resistance voltage drops are ignored. These equations in connection with eqn (3.7.8) result in

$$\begin{aligned} i_x &= 0, \\ i_y &= i_s. \end{aligned} \tag{3.7.10}$$

Therefore, a VC system with a zero *x*-axis current command fulfills unity power factor control in stator flux reference frame. This is, in fact, an easier condition than the unity power factor condition in rotor reference frame obtained as eqn (3.6.18). Subsequently, the conventional VC system in this RF as presented in Fig. 3.25 evolves into a less involved system.

3.8 VC in polar coordinates

In this chapter, VC has been presented in Cartesian coordinates so far. VC can be implemented in polar coordinates too. In polar coordinates the modulus and the angle of the machine vector variables, instead of the direct- and the quadrature-axis components, are used to model the machine, and therefore are controlled. As a consequence, if the machine torque and flux linkage commands are set as the inputs to the VC system, the modulus and the angle of current vector must be produced by some means.

VC in polar coordinates can be implemented in different reference frames according to the RF orientations, including rotor RF and stator flux linkage RF as the most popular ones. As mentioned in Section 2.7, the current vector module is the same in all conventional reference frames used for modeling and control of AC machines. Therefore, the control of current modulus is invariant with respect to the RF. However, the angle of current vector depends on the RF in which the machine model is presented. The angle can be controlled in rotor, stator flux linkage, or other reference frames. Also, the modulus current controller can be implemented in rotating or stationary RFs. Thus, one or two RF transformations are needed to produce voltage commands for a PWM block and also provide the feedback current to the current controller. As also presented in Section 2.7, the transformation of a space vector among reference frames in space vector domain is carried out by vector rotation. This is done in a space VC by adding or subtracting the angle between the two reference frames to or from the angle of the current vector. This simplifies the control system implementation and reduces the execution time of the control software. It is not necessary to present all possible configurations of VC in polar coordinates. However, it is essential to present a sample of such control to emphasize the features of this type of control mentioned previously in general form.

3.8.1 Basic control system in polar coordinates

In this sub-section a rotor-oriented VC in polar coordinates, since it's the most popular, will be elaborated. The system is aimed at controlling the torque of a surface-mounted motor. However, the speed and position control loops can also be added to the core system of torque control. The control system is presented in Fig. 3.26, where a torque command is provided by the end user or by a speed controller (Vas 1998). The torque command is compared with the feedback torque to determine the torque error during the transient mode of operation. The error is applied to a torque controller, which is usually for a PI controller with the option of a more complex controller for achieving extra performance specifications. The controller outputs a positive or negative current vector modulus, depending on the accelerating or decelerating mode of operation that the machine seeks

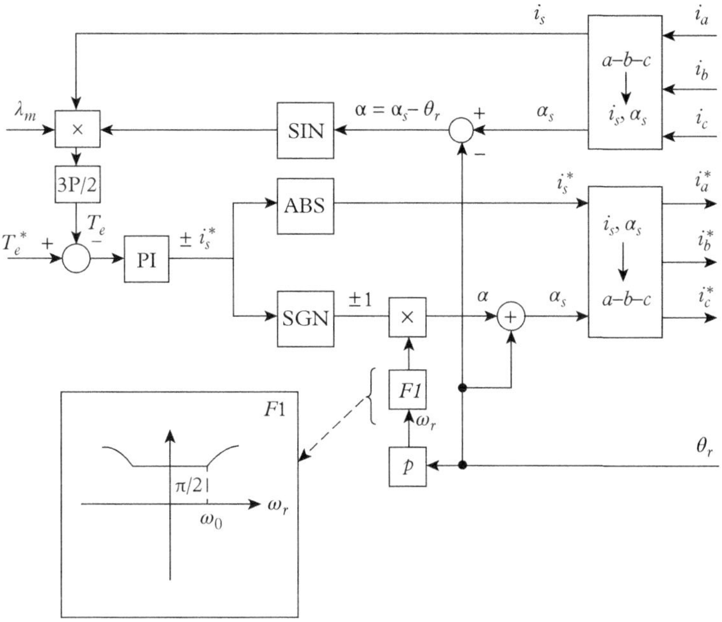

Figure 3.26 *Vector control of PMS motors in polar coordinates.*

to undergo. The absolute value of the controller output is extracted by a simple routine determined by the ABS block in Fig. 3.26.

The machine mode of operation can be determined by a sign block which provides a unity output of positive (accelerating mode) or negative (decelerating mode). This is multiplied by the output of the flux weakening block to provide the current vector angle in a rotor reference frame. The flux weakening block may have different forms, depending on the purpose of the flux weakening. A traditional type flux weakening suitable for surface-mounted PMS motors with surface-mounted magnets is used here to provide a current VC perpendicular to the rotor flux linkage vector up to the motor base speed and an increased angle beyond the base speed. The current angle can also be determined to fulfill the objectives of flux weakening methods presented in Section 3.6 for IPM motors. Now having determined the current vector angle in the rotor RF, it is transformed into the current vector angle in the steady-state RF by adding the rotor position to it. The rotor position can be obtained by an encoder or through an estimator. The current vector is then transformed from the stationary polar coordinate into a phase variable RF by the inverse of eqn (2.7.10) to provide phase current commands. The commands are applied to a current-controlled PWM inverter to produce the required voltages to the motor. The feedback torque can be prepared by the actual motor current. It is done by transforming the measured phase currents into the stationary polar coordinates. Subtracting the rotor position from the angle of the current vector in the stationary RF gives the angle of the current vector in the rotor RF. Now the feedback torque is calculated by multiplying the current modulus by the sine of the angle and then by $3P\lambda_m/2$.

3.8.2 Maximum torque per ampere control in polar coordinates

The maximum torque per ampere trajectory in i_d–i_q coordinates has already been given by eqn (3.6.3) and is recalled here as

$$i_d^2 - i_q^2 + \frac{\lambda_m}{(L_d - L_q)} i_d = 0. \tag{3.8.1}$$

The d- and q-axis components of stator current vector in polar coordinates are also given as

$$i_d = i_s \cos\alpha, \tag{3.8.2}$$

$$i_q = i_s \sin\alpha, \tag{3.8.3}$$

where α is the angle of stator current space vector in rotor RF, as seen in Fig. 2.19. Substituting the current components into the trajectory of eqn (3.8.1) yields the MTPA trajectory in polar coordinates as

$$i_s \cos 2\alpha + \frac{\lambda_m}{L_d - L_q} \cos\alpha = 0. \tag{3.8.4}$$

Also,

$$\cos\alpha = -\frac{\lambda_m}{4(L_d - L_q)i_s} - \sqrt{\frac{\lambda_m^2}{16(L_d - L_q)^2 i_s^2} + \frac{1}{2}}. \tag{3.8.5}$$

Then, the control system under MTPA in polar coordinates is given by the control system of Fig. 3.26, except for the reference current angle which is provided from eqn (3.8.5) with the stator current vector magnitude command as the input.

3.8.3 Unity power factor control in polar coordinates

Under the unity power factor operation, the stator current vector and the stator voltage vector must be aligned. On the other hand, the steady-state voltage vector, ignoring resistance voltage drop, is π/2 rad. out of phase with respect to the stator flux linkage vector, i.e.,

$$\bar{v}_s = j\omega_s \bar{\lambda}_s. \tag{3.8.6}$$

Now, it is evident that a unity power factor is only achieved if the stator current vector is also $\pi/2$ rad. out of phase with respect to the stator flux linkage vector, i.e.,

$$\chi = \alpha - \delta = \frac{\pi}{2}. \tag{3.8.7}$$

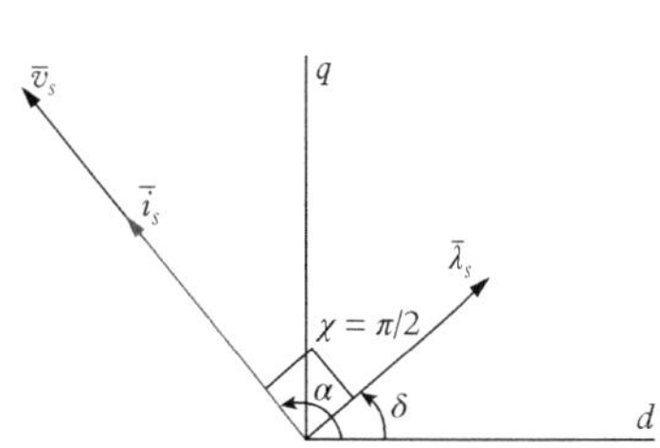

Figure 3.27 *Vector diagram of PMS machine at unity power factor.*

Figure 3.27 depicts the corresponding vector diagram where the resistance voltage drop is ignored. The torque equation, therefore, reduces to its simplest form as

$$T_e = \frac{3}{2} P \lambda_s i_s. \tag{3.8.8}$$

This compact equation facilitates a less involved control system than the more common VC system in rotor reference frame in Cartesian coordinates, which was presented earlier. It is depicted in Fig. 3.28.

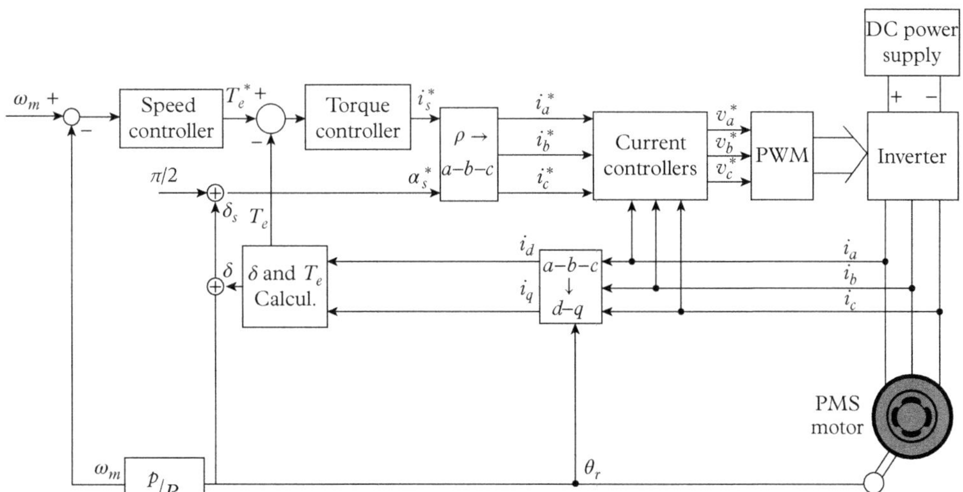

Figure 3.28 *Vector control of PMS motors in polar coordinates at unity power factor.*

It is seen that a torque controller as the basic control of subsection 3.8.1 provides the magnitude of the stator current vector command. The current angle command is obtained by adding a $\pi/2$ rad. to $\delta_s = \delta + \theta_r$. Then, the current vector is transformed into the *a–b–c* RF to be applied to the PWM inverter. The angle δ is provided by the ratio of q- and d-axis flux linkage components, which in turn are calculated according to eqns (2.5.10) and (2.5.11). Having these flux linkage components together with the d- and q-axis currents, the torque feedback is also calculated by either eqn (2.5.18) or (2.5.19).

3.9 Loss minimization control

As mentioned in Chapter 1, PMS motors in general enjoy high efficiency, mainly due to the lack of excitation winding in rotor and slip frequency. However, the potential of high efficiency can be exploited in these motors under a loss minimization control (LMC) scheme. The LMC is a control scheme under which the total electrical loss of the machine reduces to a minimum, regardless of the machine operating point. Assuming no iron loss in the machine, there is only the stator copper loss that must be minimized. This can be achieved by minimizing the stator current. Therefore, the LMC with this assumption reduces to MTPA control. However, in many PMS machines, particularly the modern ones, which use rare earth PM materials, the machine iron loss cannot be ignored due to high flux density provided by high energy PM material of rotor poles. Therefore, the LMC is usually useful in saving energy under steady-state operation. LMC is a type of flux control similar to MTPA control, MTPV control, and unity power factor control, as presented in Section 3.6. However, due to the increasing demand for environmental friendly means and methods, LMC is given higher priority in this book and is presented in full detail in this section.

3.9.1 Methods of loss reduction

The reduction of losses in electrical motor drives can be achieved by different methods including the following:

1. Selecting the motor by matching closely the application requirements and the motor specifications. For an electric vehicle, for instance, it is important to consider the single or multiple motor drive system, fixed gear or shifting gears,

maximum torque needed, maximum voltage available from the battery, and so on.

2. Designing the motor for low loss (efficiency motor). This is achieved by design optimization methods, and by using more and higher quality materials. For instance, the use of more copper to reduce the copper loss and the use of low hysteresis laminated steel cores to reduce the iron losses.
3. Improving voltage and current waveforms of the motor power supply to reduce harmonic losses. Waveform shaping techniques are used to produce desirable voltage and current by inverters.
4. Controlling motor electrical losses at a possible minimum value at all operating conditions. It is referred to as LMC or efficiency optimization control in the literature. The LMC is based on the fact that each operating point (speed and torque) can be obtained by many combinations of independent motor variables like voltage and current, and giving different levels of total losses. The one that results in the minimum losses is chosen. This method of loss minimization has received increasing attention for DC and induction motors. F. Nola in particular received overwhelming response for the loss minimization control of induction motors via power factor control. He adjusted the applied voltage as a function of load (Nola 1980). The LMC is presented for permanent magnet motors, too, in many references.

Loss minimization control so far has been researched in two ways. One is the optimal control that minimizes the energy losses for a closed cycle or over a prespecified speed profile (transient loss minimization). The other is loss minimization at every operating point of the torque-speed characteristic (steady-state loss minimization). The latter includes offline LMC in which the optimal stator flux linkage or stator current vector is obtained in terms of motor operating point (load and speed), from the steady-state mathematical model of PMS machines, and stored in a look-up table. Then they are applied as reference signals at each operating point. The other method is the online method in which the input power to the inverter is monitored regularly and the machine flux linkage or a stator current component is intentionally varied in search for a minimum input power, while the output power is kept constant by some means. The online LMC comes in different schemes including stepwise and continuous changes of the control variable. Figure 3.29 summarizes the LMC

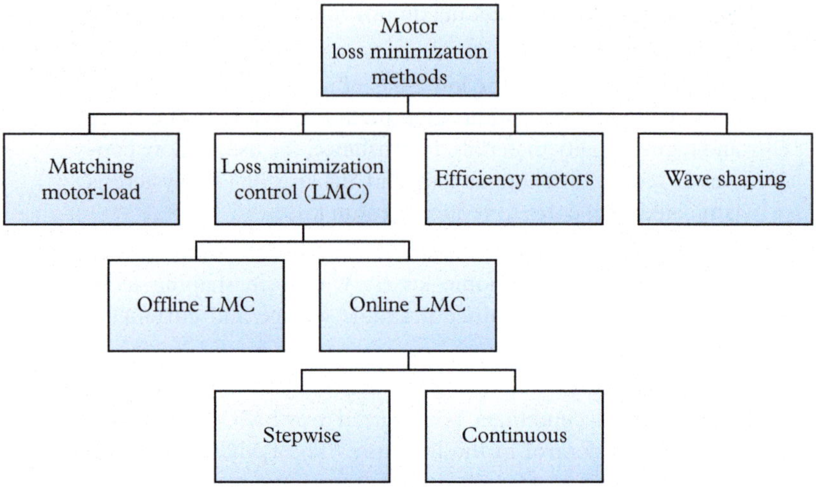

Figure 3.29 *Classification of loss minimization control methods.*

methods as described above. Only the steady-state loss minimization control is elaborated in this section.

3.9.2 Modeling of electrical loss

Motor loss consists of several components: among them the electrical loss plays the most important role. The motor efficiency can be improved if the electrical loss reduces. Referring to the steady-state machine model in rotor reference frame presented in Section 2.8, the motor electrical loss, P_L, as a combination of copper loss and iron loss can be found as

$$P_L = \frac{3}{2}R_s\left(i_d^2 + i_q^2\right) + \frac{3}{2}R_c\left(i_{dc}^2 + i_{qc}^2\right), \qquad (3.9.1)$$

where the first term is the motor copper loss and the second term is the motor iron loss. At a specific value of the motor speed and load, and assuming constant motor parameters, it can be shown that the electrical loss is a convex function of the d-axis component of stator current, i_d. This is done by substituting all the current components in eqn (3.9.1) in terms of i_d and then plotting P_L versus i_d. An alternative approach is to represent the electrical loss in terms of i_{dT} and i_{qT} by using the current equations obtained from the equivalent circuit of Fig. 2.20 as

$$i_{dT} = i_d - i_{dc}, \quad i_{qT} = i_q - i_{qc},$$

$$i_{dc} = -\frac{\omega_r \rho L_d i_{qT}}{R_c}, \quad i_{qc} = \frac{\omega_r(\lambda_m + L_d i_{dT})}{R_c}. \tag{3.9.2}$$

The result is given as

$$P_L = R_s\left\{\left(i_{dT} - \frac{\omega_r \rho L_d i_{qT}}{R_c}\right)^2 + \left(i_{qT} + \frac{\omega_r(\lambda_m + L_d i_{dT})}{R_c}\right)^2\right\}$$

$$+ \frac{\omega_r^2(\rho L_d i_{qT})^2}{R_c} + \frac{\omega_r^2(\lambda_m + L_d i_{dT})^2}{R_c}. \tag{3.9.3}$$

Also, the torque is presented in terms of these current components as

$$T_e = \frac{3}{2} P i_{qT}\left[\lambda_m + \left(L_d - L_q\right) i_{dT}\right]. \tag{3.9.4}$$

Now, substituting for i_{qT} from eqn (3.9.4) into eqn (3.9.3), P_L for every torque and speed is obtained as a function of one variable only, i.e., i_{dT}. This function is depicted in Fig. 3.30 by a solid line, in addition to the motor copper loss, P_{Cu}, and iron loss, P_{Fe} with a dot-dash and dash lines, respectively, for a motor with the specifications presented in Table 3.1 (Vaez and John 1995). The shape of P_L versus i_{dT} stems from the fact that the stator current versus i_{dT} in PMS motors is a V-shaped curve.

The optimal value of i_{dT} corresponding to a minimum P_L can be found by solving the derivative of P_L with respect to i_{dT}, which is (Morimoto *et al.* 1994b)

$$AB = T_e^2 C, \tag{3.9.5}$$

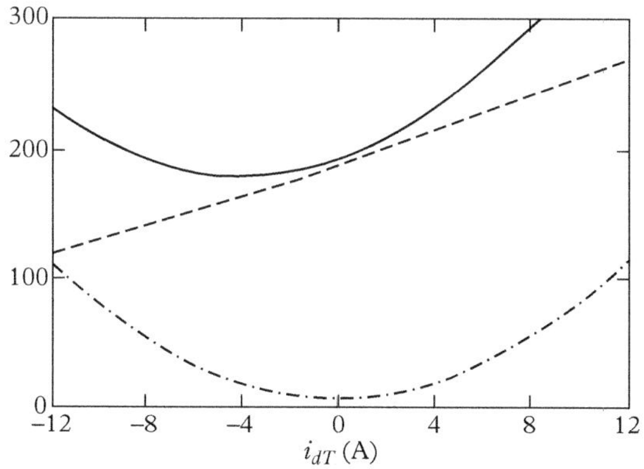

Figure 3.30 *Motor losses (Watt) versus i_{dT} for a typical motor; P_L solid, P_{Fe} dash, P_{Cu} dot-dash (Vaez and John 1995).*

Table 3.1 *Motor specifications (Morimoto* et al. *1994b).*

Rated speed, rpm	1800	$R_s, R_c\ \Omega$	1.93, 460
Rated torque, Nm	3.96	L_d, *mH*	42.44
Rated current, A	3	L_q, *mH*	79.57
P, No. of pole pairs	2	λ_m, *Wb*	0.314
B, Viscous coefficient, Nm / rad / sec.	0.0008	*J, rotor inertia , Kg. m*^{2}	0.003

where

$$A = 4P^2\left[R_sR_c^2 i_{dT} + \omega_r^2 L_d\,(R_s + R_c)\,(\lambda_m + L_d i_{dT})\right],$$
$$B = [\lambda_m + (1-\rho)L_d i_{dT}]^3, \tag{3.9.6}$$
$$C = \left[R_sR_c^2 + (R_s + R_c)\,(\omega_r \rho L_d)^2\right]\,(1-\rho)L_d.$$

The optimal value of i_{dT} can now be found from eqn (3.9.6) in terms of torque and speed by numerical calculations. If the motor is of non-salient type, where $L_d = L_q = L_s$ and thus $\rho = 1$, then $C = 0$ and, subsequently, $A = 0$. The optimal i_{dT} is now obtained from the condition of $A = 0$ as

$$i_{dT} = \frac{\omega_r^2 L_s(R_s + R_c)\lambda_m}{R_sR_c^2 + \omega_r L_s^2(R_s + R_c)}\,. \tag{3.9.7}$$

It is seen that the optimal i_{dT} is independent of torque for this type of machine. The current i_{qT} corresponding to the optimal i_{dT} is found by substituting eqn (3.9.7) in the torque equation of eqn (3.9.4). The optimal values of i_d and i_q, which are needed in the control system as the reference currents, are then decided by using eqn (3.9.2) in connection with eqn (3.9.4).

In general the motor parameters vary over wide ranges, depending on the operating conditions and the ambient temperature. In many PMS motors these variations are dominant due to the motor construction. In particular L_q, L_d, and magnet flux are affected by saturation; R_s changes due to a temperature rise; and R_c varies depending on the motor speed, and also due to the saturation in the iron bridges between rotor magnets. These variations affect the motor characteristics and performance including the minimum loss operation. A way of accounting for parameter variations is to model motor parameters in terms of motor variables, e.g., stator current components. The relevant relations expressing L_q, L_d, and λ_m as functions of i_{dT} and i_{qT}

can be found by experimental measurements as presented by eqns (2.9.1)–(2.9.6). R_c can also be presented as a function of motor frequency by eqn (2.8.4) to reflect both components of iron loss, i.e., hysteresis loss and eddy current loss.

Using variable parameters in the equivalent circuit of Fig. 2.20, it is still possible to model P_L as a function of i_{dT} only and find out the optimal value of i_{dT}. It is evident that the combination of changes in different motor parameters adds to the difference between the values of optimal i_{dT}, with and without parameter variations. The important implication of these results is the fact that the optimal value of i_{dT} is greatly parameter dependent. A further consideration confirms, however, that the parameter dependency is minimal when the motor works under the rated values of speed and torque (Vaez and John 1995).

3.9.3 Offline loss minimization control

An offline LMC by current VC can be applied to PMS motors to achieve energy saving. A schematic view of the control system is depicted in Fig. 3.31 (Morimoto *et al.* 1994b). The system is the same as flux control systems described in Section 3.6, except for the method of determining i_d^*. In this system, the stator current components are controlled in such a way that the d-axis current is at its optimal value

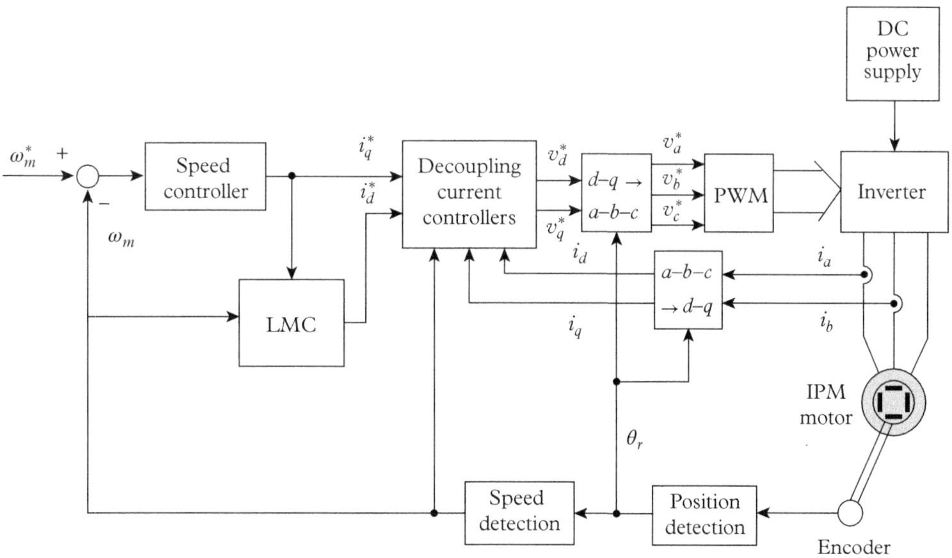

Figure 3.31 *Offline loss minimization control (LMC) system of PMS motors.*

all the time. The optimal value of i_d and the corresponding value of i_q are calculated offline for many operating points according to the above procedure. In practice, the optimal value of i_d and the corresponding value of i_q for many values of motor speed are stored in a memory as a look-up table. The look-up table gives an output (optimal i_d corresponding to a minimum motor loss) in response to a pair of inputs, i.e., motor speed and i_q command. The latter signal in turn is the output of the speed controller. The output from the look-up table in conjunction with the speed controller output, i.e., the d- and q-axis current commands, respectively, are applied to the motor current controllers. In this way, the energy saving operation of the motor is ensured at every operating point.

3.9.4 Hybrid LMC and MTPA control

The optimal operating conditions as presented above results in minimum motor electrical loss, but may not ensure desirable motor dynamics when a new speed command is applied to the machine. This is due to the development of non-optimal torque under LMC. A solution is to switch to an optimal torque development control like a maximum torque per ampere control during the transient period. This is done by always monitoring the motor speed error. When the error increases beyond a prespecified threshold, it is regarded as a drift from the steady-state operation and the need for MTPA control instead of LMC. Therefore, the LMC is halted and the MTPA control is triggered. The motor operates under this control mode until the speed error reduces below the threshold and the LMC mode is reactivated. A hybrid LMC and MTPA control system is presented in Fig. 3.32 to ensure both maximum efficiency at steady-state and maximum dynamics at transient state (Vaez-Zadeh *et al.* 2006a). It is seen that the d-axis current command can be switched between the two control modes. The sensitivity current limiting method, as presented in sub-section 3.5.1, is also implemented in the system as seen in Fig. 3.32.

The performance of the motor with data as presented in Table 3.1 is presented in Figs 3.33–3.35 (Vaez-Zadeh *et al.* 2006a). The no-load motor start-up in response to a speed step of 2000 rpm is shown in Fig. 3.33. The motor starts rapidly under MTPA control, as seen in Fig. 3.33(a). This is due to a high i_q generated by the speed controller as seen in Fig. 3.33(b). It is also due to a very negative i_d provided by the MTPA block as seen in Fig. 3.33(c). The current limiter, as seen in Fig. 3.32, takes effect during the transient state and limits the less effective current component such that the total current does not exceed the current limit, i_{sL}, which is set to three times the rated current.

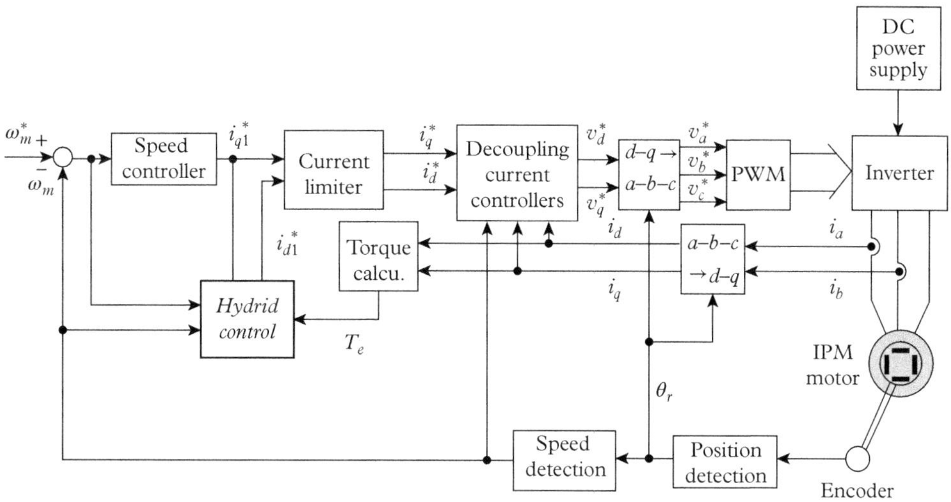

Figure 3.32 *Hybrid LMC and MTPA control.*

The action of the current limiter can be seen on the i_d plot under transient conditions, where the current frequently goes under limit. However, as the machine reaches steady-state, not only does i_q reduce to a low value corresponding to no-load conditions, but also i_d is adjusted by LMC to a new value. Figure 3.34 shows the motor performance under a load disturbance test. It must be mentioned that usually LMC provides more negative i_d than MTPA control at the same operating point. However, here the torque is not the same under MTPA control and LMC. The potential energy saving under the LMC is depicted in Fig. 3.35 in terms of motor efficiency and motor electrical loss, over a speed range of 0–3000 rpm and a constant load of 1.65 Nm. The efficiency increases with the speed, reaching a high value of about 95% as seen in Fig. 3.35(a), corresponding to the low electrical loss as seen in Fig. 3.35(b).

3.9.5 Online loss minimization control

Despite advantages of offline LMC, its implementation faces some difficulties in practice. In fact developing a variable parameter model for the calculation of optimal i_d values is complicated due to the extensive measurement and modeling needed. This task becomes more difficult in the situation where some motor parameters have multiple dependencies. For instance, R_c may depend on both motor frequency and i_d. Therefore, it is concluded that an offline loss minimization control is not suitable for motors with parameter variations,

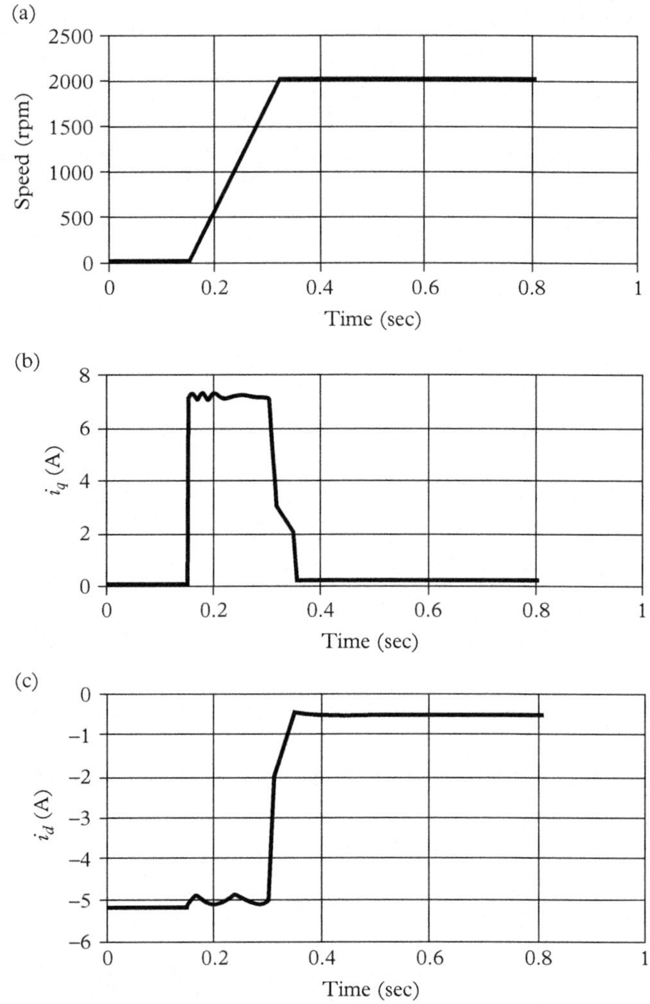

Figure 3.33 *Motor performance under hybrid LMC and MTPA control at step response: 0–2000 rpm: (a) speed, (b)* q*-axis current, and (c)* d*-axis current state (Vaez-Zadeh* et al. *2006a).*

working under a wide range of operating conditions as in an electric vehicle.

In online loss minimization control, in contrast to the offline approach, the optimum value of control variable corresponding to a minimum loss is found by an online search. A common practice in this control strategy is to activate the control algorithm right after the motor speed reaches steady-state value. Then apply a step change to a control variable, e.g., motor flux linkage or the d-axis component of stator current as seen in Fig. 3.36(a), and wait for some time, long enough for the motor to pass the subsequent transient and

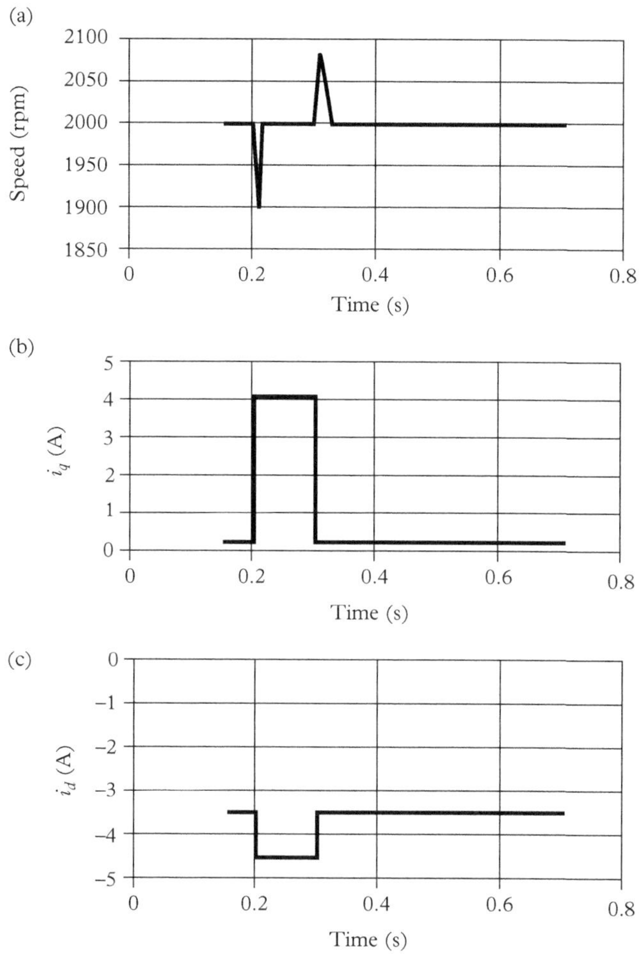

Figure 3.34 *Motor performance under hybrid LMC and MTPA control, load disturbance test at 2000 rpm: (a) speed, (b)* q*-axis current, and (c)* d*-axis current state (Vaez-Zadeh* et al. *2006a).*

come to a fairly steady-state situation. Motor input power values are measured before and after the change made in the control variable. If the power reduces, as in Fig. 3.36(b), another step is applied to the control variable in the same direction. Otherwise, the second step is applied in the opposite direction. This procedure continues until a step change in the control variable does not change the input power. This means that a minimum input power has been obtained, as seen in Fig. 3.36(b). The motor output power is kept constant during the search by a regulator, as seen in Fig. 3.36(c). If a change in the motor commanded speed or torque occurs, the LMC algorithm is deactivated in response to a noticeable speed error and the

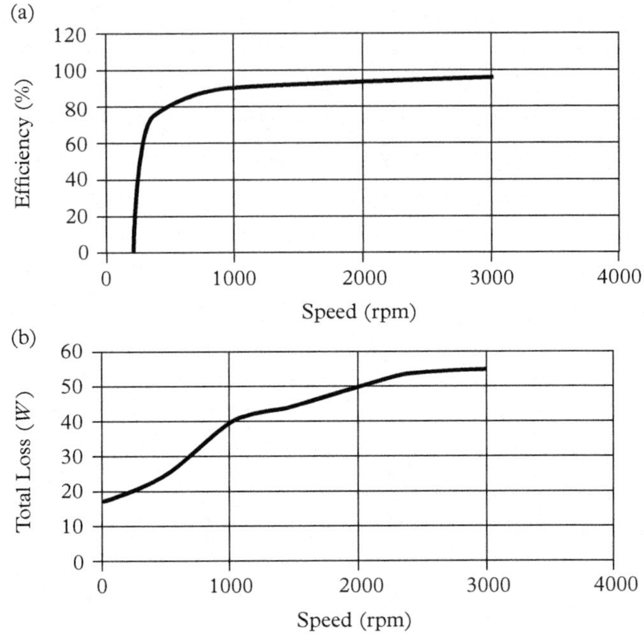

Figure 3.35 *Energy saving under LMC at a load of 1.65 Nm over a speed range of 0–3000 rpm: (a) motor efficiency, and (b) motor total electrical loss state (Vaez-Zadeh* et al. *2006a).*

motor control returns to usual control to produce enough torque and provides desirable motor dynamics.

A functional block diagram of the online loss minimization control system is shown in Fig. 3.37 (Vaez-Zadeh and Hendi 2005). It consists of a steady-state condition detector, a loss minimization core, and a torque compensator. All three parts are software-based. However, voltage, current, and speed sensors provide them with the required input signals.

The online loss minimization controller works at steady state, i.e., when the motor speed reaches the commanded speed signal. Therefore, the steady-state detector enables both the loss minimization core and the torque compensator when the speed error signal (the commanded speed minus the actual speed) becomes sufficiently small. The detector receives motor speed signal from the speed sensor and compares it repeatedly with the commanded speed signal in short intervals. If the error is less than a small predetermined value, the steady-state is detected. Otherwise, the detector keeps working in a loop until the steady-state is detected at some later time. When the steady-state is detected, the enabling signals are sent to other parts of the controller as seen in Fig. 3.37. As a result, the loss minimization core changes i_d stepwise by applying successive values of Δi_d toward its optimum value corresponding to a minimum input power to the

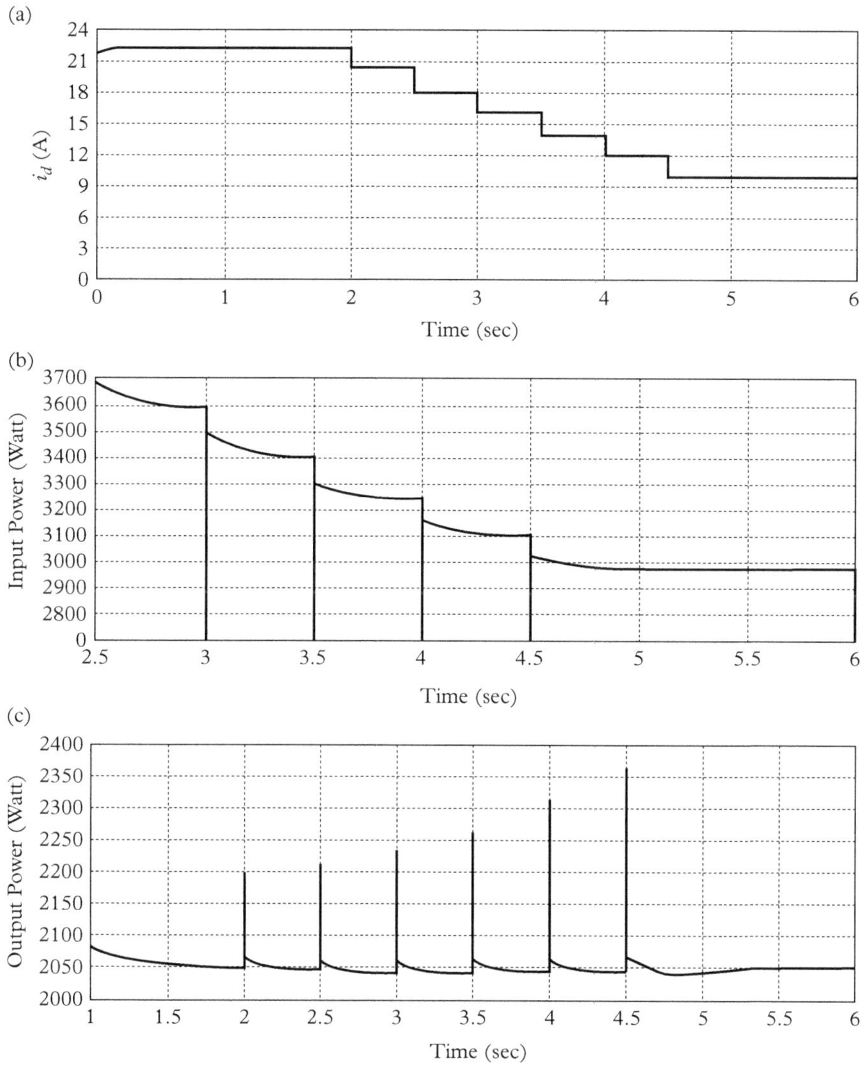

Figure 3.36 *Stepwise online loss minimization control: (a)* d-*axis current, (b) drive input power, and (c) motor output power (Vaez-Zadeh and Hendi 2005).*

drive. The correct direction of change in i_d (sign of Δi_d) must be determined first, to ensure a reduction of the drive input power. The direction of change in i_d must be reversed throughout the loss minimization period. The loss minimization core keeps changing i_d and reducing the input power until the optimum value of i_d is reached. At this current the input power is at its minimum value and further

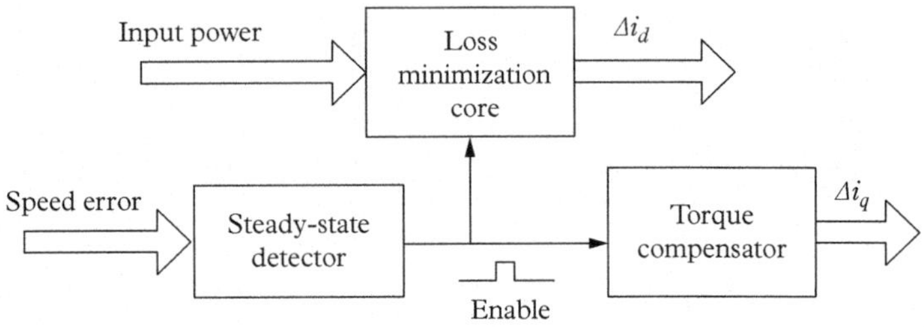

Figure 3.37 *Elements of an online LMC (Vaez-Zadeh and Hendi 2005).*

change of i_d results in the input power to increase. Therefore, the direction of step change alters repeatedly and the input power swings in the vicinity of its minimum value, until the steady-state detector determines a transient state and disables the loss minimization core. A transient state may occur due to applying a load or a command signal to the motor drive system.

A change in i_d during the loss minimization period changes the motor flux linkage. Therefore, the developed torque and ultimately the motor speed change. As a result, the output power does not remain constant. On the other hand, it is known that a reduction of the input power in response to change in i_d results in a reduction of the system losses only if the motor output power remains unchanged. Since the loss minimization core works under a constant motor speed, the constant output power can be ensured by retaining the motor torque unchanged, despite the change in i_d. A torque compensator is employed for this purpose. The compensator adjusts i_q in response to the change in i_d, such that the torque remains unchanged. This can be done in different ways as will be described later.

Fast speed dynamics of the speed controller can maintain the torque and reduce the transient period after a step change is made into i_d (Kirschen *et al.* 1985). This method works if the subsequent transient change in the motor speed is permissible and a slow loss minimization controller is acceptable for the application. Otherwise, an additional torque compensation method can be used (Vaez *et al.* 1999; Vaez-Zadeh 2001). It calculates the change in the q-axis current command, Δi_q, required for the compensation of the change in the torque, which results from the change in the LMC control variable (d-axis current command) according to

$$\Delta i_q \approx -\frac{(L_d - L_q)i_q}{\lambda_m + (L_d - L_q)i_q}\Delta i_d, \qquad \forall\ \Delta i_d << i_d. \tag{3.9.8}$$

It must be mentioned that the torque compensator is parameter dependent (see eqn (3.9.8)). Therefore, it rather jeopardizes the non-parameter dependency of online LMC core. An alternative compensating method is presented in the literature, which is not parameter dependent at the cost of more complication, as will be mentioned later (Vaez-Zadeh 2001).

3.9.6 Continuous online LMC

Online LMC seems very attractive since it does not need a system loss model. Furthermore, the motor parameter variations are taken into account naturally, resulting in more energy saving. Despite these features, an important deficiency of online approach, when applied to high performance applications, is the long search time that it takes to reach the minimum input power. In many applications with frequent changes in operating conditions (speed and load), e.g., electric vehicles, the steady-state period is short and there is not enough time to reach minimum input power conditions. This is because the search for an optimum value of the control variable starts only after the machine reaches a steady-state. During transient states, when the operating point is changing, the reduction in input power cannot be used as a criterion for energy saving. Even if the steady-state period permits finding the minimum input power, some energy is lost during a long search time. If the transient state is repeated frequently as in an electric vehicle, the total energy saving may not be significant.

The long search time is mainly caused by the stepwise change of the control variable and the sluggish response of the compensator responsible for maintaining the output power. By avoiding the stepwise change of the control variable, the relatively long transient after each step and subsequent wait period between steps are avoided. As a result, a faster and smoother energy saving operation occurs.

An online LMC can be used in which the stepwise change of the control variable, the d-axis current, is replaced by its continuous adjustment toward its optimal value (Vaez *et al.* 1999). A linear pattern of change for i_d with respect to time is considered. As a result, the drive input power contiguously reduces toward its minimum value. The principles of LMC are described in connection with Fig. 3.38.

During the motor speed transient, when the speed is changing toward its commanded value and the speed error is large, the LMC is off and the motor works under an original i_d, resulting in a desirable dynamics, as seen in Figs 3.38(a)–(c). Once the steady-state speed condition is detected by the steady-state detector, as described in connection with Fig. 3.37, the loss minimization core is activated by the enabling signal. The loss minimization core determines the direction of change in i_d by which the input power reduces. This is done by examining the input power variation in response to continuous

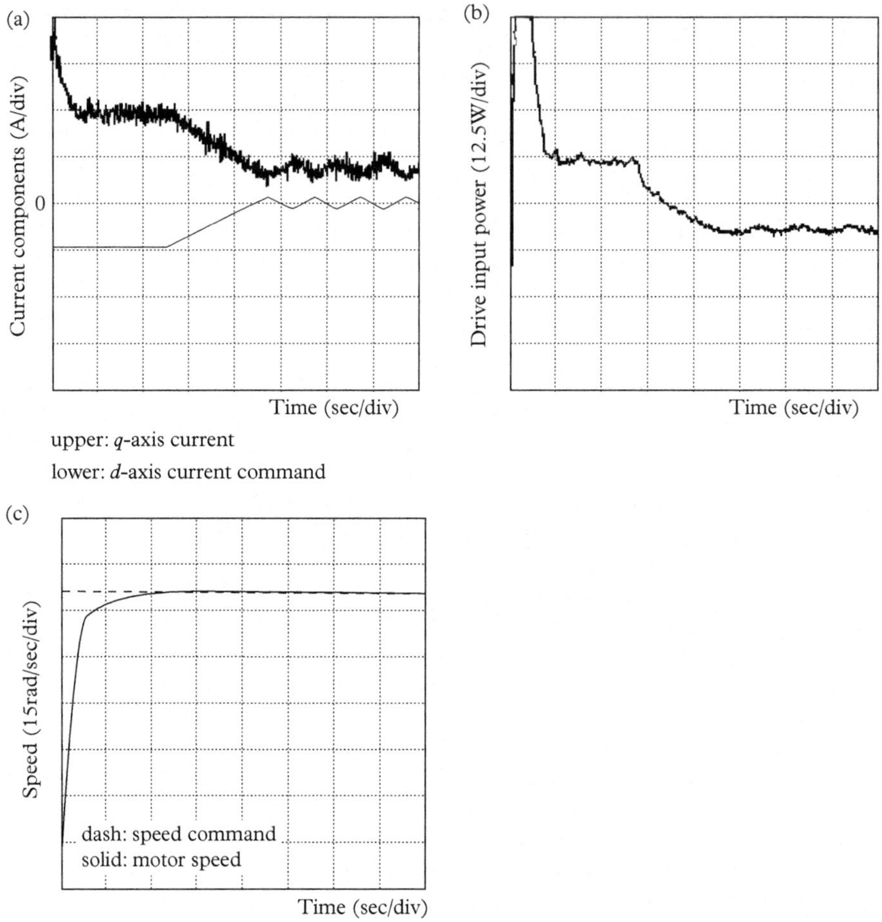

Figure 3.38 *Principles of continuous online loss minimization control: (a)* d- *and* q-*axis stator current components, (b) drive input power, and (c) motor speed (Vaez* et al. *1999).*

change of i_d^* for a short period of time. Then i_d varies in the determined direction toward its optimal value, as in Fig. 3.38(a), resulting in a reduction in the input power, P_{DC}, toward its minimum value, as in Fig. 3.38(b). The minimum input power is determined when the change in P_{DC} during a short interval falls below a limit. At this point the input power becomes flat, which means that a further change in i_d^* does not have a noticeable effect on the input power. A new mode of operation starts next in which i_d varies in a triangular manner by changing i_d for short periods in opposite directions repeatedly. This ensures an energy-saving operation if the motor operating point changes slowly.

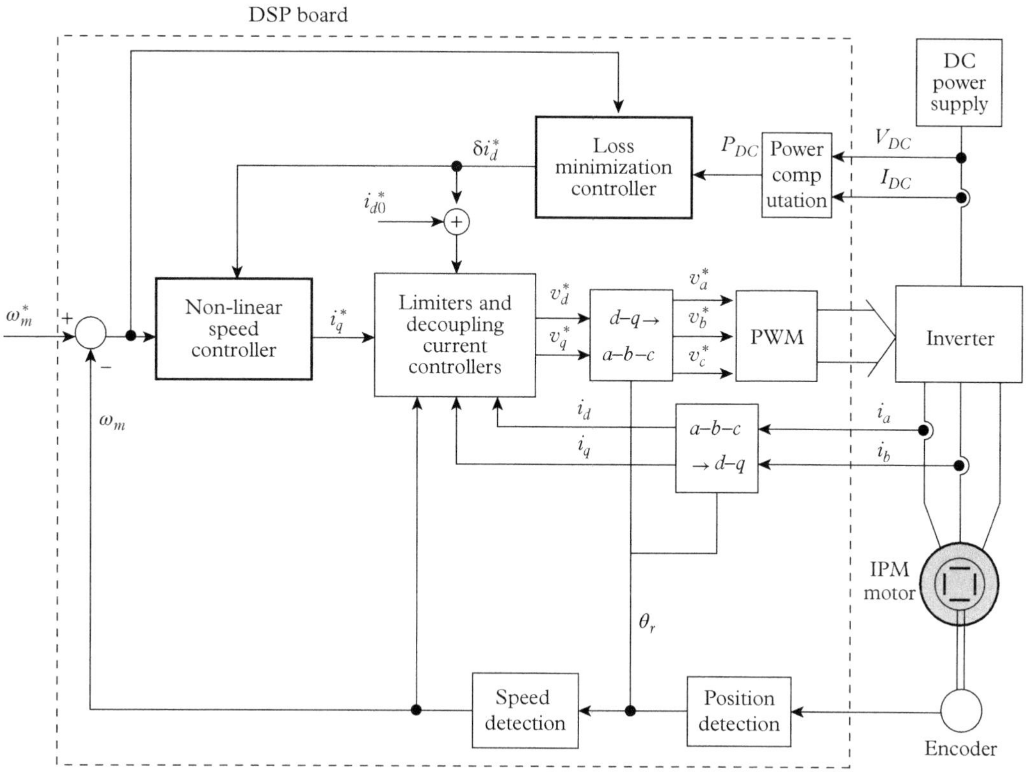

Figure 3.39 *Continuous online loss minimization control system of PMS motors in rotor reference frame.*

A VC system, including the continuous online LMC and a non-linear speed controller/compensator, is presented schematically in Fig. 3.39 (Vaez-Zadeh 1999). The original d-axis current command, i_{d0}^*, can be determined such that a desirable motor dynamic performance is achieved. It can be obtained from a maximum torque per ampere (MTPA) scheme, for instance, as presented in Section 3.6. The output of LMC continuously changes this signal toward the optimal i_d^*. As i_d changes to reduce the drive input power, the developed torque tends to vary. Since the load is assumed to be constant, the change in torque results in a change in the motor speed. However, the speed controller tries to maintain the speed at its commanded value. Therefore, it modifies the q-axis current command in response to a change in i_d. This natural compensation process depends on the motor torque speed dynamics, which is slow in nature. As a result, once i_d is changed by the LMC, the speed undergoes a transient period before the natural compensation process restores the commanded speed. The motor speed transient is compensated by strengthening the natural compensation process when i_d is changed by the LMC. This is done by a non-linear speed controller. The controller augments the input to the speed controller, i.e., the speed error signal, when the LMC is active.

The higher the change in i_d^*, the higher the effective speed error signal. This results in an output of the speed controller, i_q^*, which is affected by changes in i_d^*. Therefore, the change in the reluctance torque caused by i_d^* is compensated for by the change in the magnet torque caused by i_q^*. As a result, the total motor torque and, thus, the motor speed remain constant. However, the speed controller works normally when the motor loss reaches its minimum and the speed error is not augmented (Vaez-Zadeh 2001).

Figure 3.40 shows the experimental results of the drive performance under the LMC. The motor is started under a low load condition in response to a medium step speed command as in Fig. 3.40(a). The original d-axis current command, i_{d0}^*, is set at some negative value to provide fast dynamics. At about $t = 2.7$ seconds, a steady-state speed is detected. Then, the continuous adjustment of i_d starts toward its optimal value with the corresponding modification in i_q to keep the torque unchanged, as shown in Fig. 3.40(b). After about 2 seconds, the input power reaches its minimum value, as shown in Fig. 3.40(c). This is about a 30% reduction in the input power. Then, the triangular mode of operation is initiated. A very small variation in the motor speed can be seen when the d-axis current starts to change. However, it dies out quickly by the action of the speed controller.

The motor dynamics under the LMC is validated by applying a speed command change at $t = 11$ seconds, As a result, the loss minimization core is deactivated and the d-axis current command returns to its original value, i_{d0}^*. The motor undergoes the transient period with this current until a new steady-state speed is obtained when the loss minimization core becomes active again, and another minimum input power is obtained at the new operating point, as seen in Figs 3.40(a)–(c).

3.10 Summary

The chapter begins with a description of an SC of PMS motors with sine wave current command generation, phase advancement, and phase current controllers. It is regarded as a first step, non-matured control method in terms of corresponding motor modeling, and control design and performances, which justify the attempts toward high performance VC of PMS motors. Fundamentals of PMS

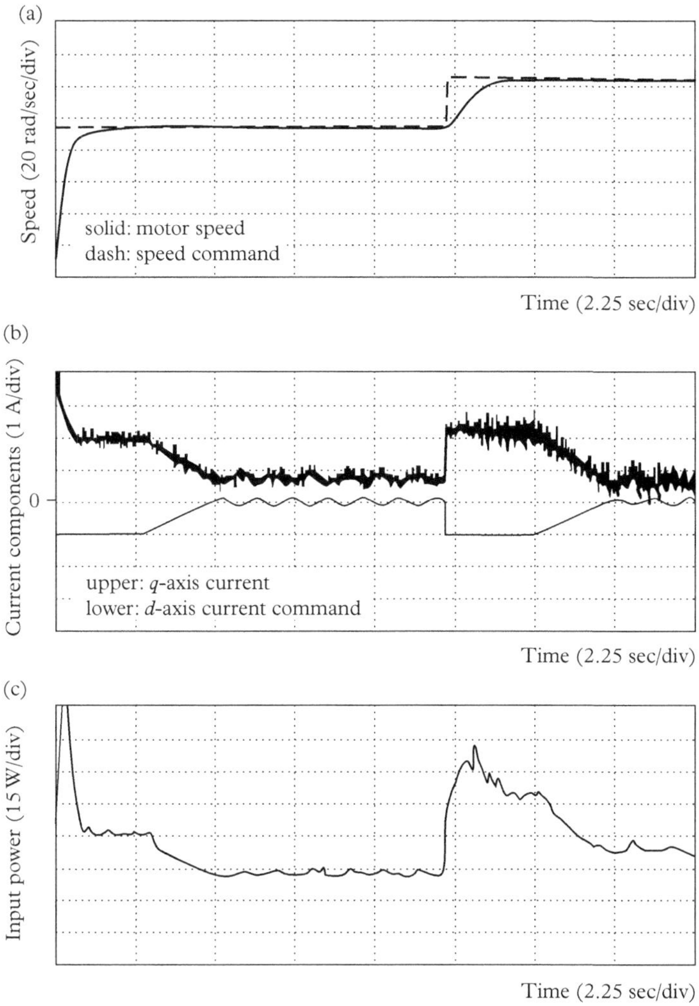

Figure 3.40 *Experimental results of continuous online loss minimization control of PMS motor: (a) motor speed, (b)* d- *and* q-*axis stator current components, and (c) drive input power (Vaez-Zadeh 2001).*

motor VC are then presented with an eye on the analogy with DC motor operating principles. The VC of surface-mounted permanent magnet pole motors is elaborated in a rotor reference frame with phase current controllers and *d–q* current controllers. It is followed by VC of interior PMS motors in which a type of flux weakening is normally carried out. Current and voltage operating limits of PMS motors are shown mathematically and graphically in *d–q* current coordinates system, and the means for incorporating them into the control system are discussed. Flux control modes of operation of PMS motors along with the corresponding control means are also presented in detail as a particular feature of this book. These include maximum torque per ampere control, maximum torque per volt control, and unity power factor control. After that, the basics of VC in stator flux linkage reference frame are presented, in addition to different flux weakening control systems in this reference frame. Also, VC in polar coordinates is discussed by means of space phasors in different reference frames. The saturation of current controllers and its control remedy are presented next. Finally, loss minimization control by offline and online strategies is elaborated after the method of motors loss reduction and the PMS motor loss modeling are presented.

PROBLEMS

P.3.1. Find the focal points of voltage limit ellipse for PMS machines in terms of i_d and i_q and draw their locus for a wide range of ω_r, and discuss the zero speed and speed limit.

P.3.2. In a PMS machine with the specifications given as in Table 3.1, the demagnetization limit is where $\lambda_d = 0$. Determine the d- and q-axis currents at maximum developed torque at the verge of demagnetization.

P.3.3. Calculate modules of the stator current vector and the stator flux linkage vector in terms of load angle (δ) under the unity power factor mode of operation.

P.3.5. Figure 3.12 depicts MTPA operation as a minimum possible stator current with a specific motor torque. Alternatively, MTPA operation can be regarded more explicitly as a maximum possible motor torque with a specific stator current. Show that the two interpretations of MTPA are the same by depicting the second one and comparing it with the one in Fig. 3.12.

P.3.6. Calculate the d- and q-axis current command under MTPA control, where the motor current command exceeds the current limit i_{sL}.

P.3.7. Determine f_1 and f_2 in the feedforward control of Fig. 3.17 as functions of T_{en}.

P.3.8. Determine the MTPA trajectory in terms of normalized current components presented in eqn (3.6.7).

P.3.9. Is it possible for an IPM motor to operate under the conditions of MTPA and PF = 1 simultaneously? If so, determine the mathematical conditions. Otherwise, prove that it is impossible to do so.

P.3.10. Develop a flowchart for the software routine which implements the transition among flux control modes.

P.3.11. Determine the flux linkage vector trajectory under unity power factor operation in rotor-oriented polar coordinates.

P.3.12. Suppose it is desirable for an IPM machine to operate under constant flux. Determine the maximum torque trajectory in terms of load angle under the constant flux conditions.

P.3.13. The offline loss minimization control as presented in Section 3.9 is based on the machine model with a fixed iron loss resistance, R_c. Consider the influence of supply frequency (motor speed) on this resistance, and the machine iron loss and LMC of PMS motors.

BIBLIOGRAPHY

Adnanes, A., Sulkowski, W., Aga, L., and Norum, L. (1991). A fully digital permanent magnet synchronous motor drive with flux weakening. In: *5th International Conference on Electrical Machines and Drives*, pp. 227–231. IET, London.

Adnanes, A.K., Nilsen, R., Loken, R., and Norum, L. (1993). Efficiency analysis of electric vehicles, with emphasis on efficiency optimized excitation. In: *Conference Record of the IEEE Annual Meeting of the Industry Applications Society*, pp. 455–462. IEEE, Piscataway, NJ.

Andersen, H.R. and Pedersen, J.K. (1996). Low cost energy optimized control strategy for a variable speed three-phase induction motor. In: *27th Annual Meeting of the IEEE, Power Electronics Specialists Conference*, pp. 920–924. IEEE, Piscataway, NJ.

Babb, M. (1995). Premium efficiency motors promise to save billions. *Control Eng.* 42(6), 1–4.

Bates, B. (1992). *Electric Vehicles, a Decade of Transition, Selected Papers Through 1990.* Society of Automotive Engineers, Warrendale, PA.

Bilewski, M., Fratta, A., Giordano, L., Vagati, A., and Villata, F. (1993). Control of high-performance interior permanent magnet synchronous drives. *IEEE Trans. Ind. Appl.* 29(2), 328–337.

Binns, K. and Wong, T. (1984). Analysis and performance of a high-field permanent-magnet synchronous machine. *IEE Proc. B-Electric Power Appl.* 131(6), 252–258.

Bonnett, A.H. (1992). Understanding the changing requirements and opportunities for improvement of operating efficiency of AC motors. In: *Conference Record of Annual Pulp and Paper Industry Technical Conference*, pp. 600–610. IEEE, Piscataway, NJ.

Bose, B.K. (1986). *Power Electronics and AC Drives*. Prentice-Hall, Upper Saddle River, NJ.

Bose, B.K. (1988). A high-performance inverter-fed drive system of an interior permanent magnet synchronous machine. *IEEE Trans. Ind. Appl.* 24(6), 987–997.

Cavallaro, C., Di Tommaso, A.O., Miceli, R., Raciti, A., Galluzzo, G.R., and Trapanese, M. (2005). Efficiency enhancement of permanent-magnet synchronous motor drives by online loss minimization approaches. *IEEE Trans. Ind. Electron.* 52(4), 1153–1160.

Cerruto, E., Consoli, A., Raciti, A., and Testa, A. (1995). A robust adaptive controller for PM motor drives in robotic applications. *IEEE Trans. Power Electron.* 10(1), 62–71.

Chaaban, F., Chedid, R., and Mellor, P. (1996). Steady state and sensitivity analysis of high-field permanent magnet machines. *Electric Machines Power Syst.* 24(6), 639–651.

Chalmers, B. (1992). Influence of saturation in brushless permanent-magnet motor drives. *IEE Proc. B-Electric Power Appl.* 139(1), 51–52.

Chan, C. (1993). An overview of electric vehicle technology. *Proc. IEEE* 81(9), 1202–1213.

Chang, L. (1993). Recent developments of electric vehicles and their propulsion systems. *IEEE Aerospace Electron. Syst. Mag.* 8(12), 3–6.

Chang, L. (1994). Comparison of AC drives for electric vehicles-a report on experts opinion survey. *IEEE Aerospace Electron. Syst. Mag.* 9(8), 7–11.

Chapman, P., Sudhoff, S., and Whitcomb, C. (1999). Optimal current control strategies for surface-mounted permanent-magnet synchronous machine drives. *IEEE Trans. Energy Convers.* 14(4), 1043–1050.

Chen, S. and Yeh, S-N. (1992). Optimal efficiency analysis of induction motors fed by variable-voltage and variable-frequency source. *IEEE Trans. Energy Convers.* 7(3), 537–543.

Cleland, J. G., McCormick, V. E., and Turner, M. W. (1995). Design of an efficiency optimization controller for inverter-fed AC induction motors. In: *Conference Record of the IEEE 30th IAS Annual Meeting, Industry Applications Conference*, pp. 16–21. IEEE, Piscataway, NJ.

Colby, R.S. and Novotny, D.W. (1987). Efficient operation of surface-mounted PM synchronous motors. *IEEE Trans. Ind. Appl.* 23(6), 1048–1054.

Colby, R.S. and Novotny, D.W. (1988). An efficiency-optimizing permanent-magnet synchronous motor drive. *IEEE Trans. Ind. Appl.* 24(3), 462–469.

Consoli, A., Raciti, A., Dobrucky, B., and Hrabovcova, V. (1993). Static and dynamic behaviour of disk synchronous motors. In: *Proceedings of IEEE International Symposium on Industrial Electronics*, pp. 646–651. IEEE, Piscataway, NJ.

Consoli, A. and Renna, G. (1989). Interior type permanent magnet synchronous motor analysis by equivalent circuits. *IEEE Trans. Energy Convers.* 4(4), 681–689.

Doncker, R.D., Pulle, D.W., and Veltman, A. (2010). *Advanced Electrical Drives, Analysis, Modeling, Control.* Springer Science & Business Media, Berlin.

Famouri, P. and Cathey, J.J. (1989). Loss minimization control of an induction motor drive. In: *Conference Record of the IEEE Industry Applications Society Annual Meeting* , pp. 226–231. IEEE, Piscataway, NJ.

Famouri, P. and Cooley, W. L. (1994). Design of DC traction motor drives for high efficiency under accelerating conditions. *IEEE Trans. Ind. Appl.* 30(4), 1134–1138.

Favre, E., Cardoletti, L., and Jufer, M. (1993). Permanent-magnet synchronous motors: A comprehensive approach to cogging torque suppression. *IEEE Trans. Ind. Appl.* 29(6), 1141–1149.

Fukino, M., Irie, N., and Ito, H. (1992). *Development of an Electric Concept Vehicle with a Super Quick Charging System*, SAE Technical Paper Series 920422, pp. 25–27. SAE, Oxford.

Funabiki, S. and Fukushima, T. (1991). Current commands for high-efficiency torque control of DC shunt motor. *IEE Proc. B-Electric Power Appl.* 138(5), 227–232.

Holtz, J. and Springob, L. (1996). Identification and compensation of torque ripple in high-precision permanent magnet motor drives. *IEEE Trans. Ind. Electron.* 43(2), 309–20.

Honsinger, V. (1980). Performance of polyphase permanent magnet machines. *IEEE Trans. Power Apparat. Syst.* 4(PAS-99), 1510–1518.

Jahns, T.M. (1987). Flux-weakening regime operation of an interior permanent-magnet synchronous motor drive. *IEEE Trans. Ind. Appl.* 23(4), 681–689.

Jahns, T.M. (1994). Motion control with permanent-magnet AC machines. *Proc. IEEE* 82(8), 1241–1252.

Jahns, T.M., Kliman, G.B., and Neumann, T.W. (1986). Interior permanent-magnet synchronous motors for adjustable-speed drives. *IEEE Trans. Ind. Appl.* 22(4), 738–747.

Jahns, T.M. and Soong, W.L. (1996). Pulsating torque minimization techniques for permanent magnet AC motor drives-a review. *IEEE Trans. Ind. Electron.* 43(2), 321–330.

Kadjoudj, M., Benbouzid, M.E.H., Ghennai, C., and Diallo, D. (2004). A robust hybrid current control for permanent-magnet synchronous motor drive. *IEEE Trans. Energy Convers.* 19(1), 109–115.

Kim, J-M., Park, K., Kang, S-J., Sul, S-K., and Kwon, J-L. (1996). Improved dynamic performance of interior permanent magnet synchronous motor drive in flux-weakening operation. In: *27th Annual Meeting of the IEEE, Power Electronics Specialists Conference*, pp. 1562–1567. IEEE, Piscataway, NJ.

Kim, J-M. and Sul, S-K. (1997). Speed control of interior permanent magnet synchronous motor drive for the flux weakening operation. *IEEE Trans. Ind. Appl.* 33(1), 43–48.

King, R., Turnbull, F., Delgado, E., and Szczesny, P. (1988). *High-performance ETX-11 70-hp Permanent Magnet Motor Electric Drive System.* Electric Vehicle Association of Canada, Vancouver.

Kirschen, D. S., Novotny, D. W., and Lipo, T. A. (1985). On-line efficiency optimization of a variable frequency induction motor drive. *IEEE Trans. Ind. Appl.* 21(3), 610–616.

Krause, P., Nucera, R., Krefta, R., and Wasynczuk, O. (1987). Analysis of a permanent magnet synchronous machine supplied from a 180° inverter with phase control. *IEEE Trans. Energy Convers.* 2(3), 423–431.

Krause, P.C. (1986). *Analysis of Electric Machinery.* McGraw-Hill, New York, NY.

Krishnan, R. (2010). *Permanent Magnet Synchronous and Brushless DC Motor Drives.* CRC Press, Boca Raton, FL.

Kusko, A. and Galler, D. (1983). Control means for minimization of losses in AC and DC motor drives. *IEEE Trans. Ind. Appl.* 19(4), 561–570.

Leonhard, W. (2001). *Control of Electrical Drives.* Springer Science & Business Media, Berlin.

Macminn, S.R. and Jahns, T.M. (1988). Control techniques for improved high-speed performance of interior PM synchronous motor drives. In: *Conference Record of the IEEE Industry Applications Society Annual Meeting*, pp. 272–280. IEEE, Piscataway, NJ.

Mademlis, C. and Agelidis, V.G. (2001). On considering magnetic saturation with maximum torque to current control in interior permanent magnet synchronous motor drives. *IEEE Trans. Energy Convers.* 16(3), 246–252.

Mademlis, C., Kioskeridis, I., and Margaris, N. (2004). Optimal efficiency control strategy for interior permanent-magnet synchronous motor drives. *IEEE Trans. Energy Convers.* 19(4), 715–723.

Miller, T. J. E. (1989). *Brushless Permanent-magnet and Reluctance Motor Drives.* Oxford University Press, Oxford.

Morimoto, S., Takeda, Y., and Hirasa, T. (1990a). Current phase control methods for permanent magnet synchronous motors. *IEEE Trans. Power Electron.* 5(2), 133–139.

Morimoto, S., Takeda, Y., Hirasa, T., and Taniguchi, K. (1990b). Expansion of operating limits for permanent magnet motor by current vector control considering inverter capacity. *IEEE Trans. Ind. Appl.* 26(5), 866–871.

Morimoto, S., Hatanaka, K., Tong, Y., Takeda, Y., and Hirasa, T. (1993a). Servo drive system and control characteristics of salient pole permanent magnet synchronous motor. *IEEE Trans. Ind. Appl.* 29(2), 338–343.

Morimoto, S., Takeda, Y., Hatanaka, K., Tong, Y., and Hirasa, T. (1993b). Design and control system of inverter-driven permanent magnet synchronous motors for high torque operation. *IEEE Trans. Ind. Appl.* 29(6), 1150–1155.

Morimoto, S., Sanada, M., and Takeda, Y. (1994a). Wide-speed operation of interior permanent magnet synchronous motors with high-performance current regulator. *IEEE Trans. Ind. Appl.* 30(4), 920–926.

Morimoto, S., Tong, Y., Takeda, Y., and Hirasa, T. (1994b). Loss minimization control of permanent magnet synchronous motor drives. *IEEE Trans. Ind. Electron.* 41(5), 511–517.

Muni, B., Pillai, S., and Saxena, S. (1996). A PC based internal power factor angle controlled interior permanent magnet synchronous motor drive. In: *27th Annual Meeting of the IEEE Power Electronics Specialists Conference*, pp. 931–937. IEEE, Piscataway, NJ.

Myers, C.J., Christian, J.M., and Reibsamen, G.G. (1980). *World Guide to Battery-Powered Road Transportation, Comparative Technical and Performance Specifications.* McGraw-Hill, New York, NY.

Nakai, H., Ohtani, H., Satoh, E., and Inaguma, Y. (2005). Development and testing of the torque control for the permanent-magnet synchronous motor. *IEEE Trans. Ind. Electron.* 52(3), 800–806.

Nakamura, Y., Kudo, T., Ishibashi, F., and Hibino, S. (1995). High-efficiency drive due to power factor control of a permanent magnet synchronous motor. *IEEE Trans. Power Electron.* 10(2), 247–253.

Nola, F. (1980). Power factor controller-An energy saver. In: *IEEE Industry Application Society Annual Meeting*, pp. 194–198. IEEE, Piscataway, NJ

Ojo, O. and Cox, J. (1996). Investigation into the performance characteristics of an interior permanent magnet generator including saturation effects. In: *Conference Record of the IEEE 31st IAS Annual Meeting, Industry Applications Conference*, pp. 533–540. IEEE, Piscataway, NJ.

Ojo, O., Osaloni, F., Wu, Z., and Omoigui, M. (2003). A control strategy for optimum efficiency operation of high performance interior permanent magnet motor drives. In: *Conference Record of the IEEE 38th IAS Annual Meeting, Industry Applications Conference*, pp. 604–610. IEEE, Piscataway, NJ.

Parasiliti, F. and Poffet, P. (1989). A model for saturation effects in high-field permanent magnet synchronous motors. *IEEE Trans. Energy Convers.* 4(3), 487–494.

Pillay, P. and Krishnan, R. (1989). Modeling, simulation, and analysis of permanent-magnet motor drives. I. The permanent-magnet synchronous motor drive. *IEEE Trans. Ind. Appl.* 25(2), 265–273.

Pillay, P. and Krishnan, R. (1990). Control characteristics and speed controller design for a high performance permanent magnet synchronous motor drive. *IEEE Trans. Power Electron.* 5(2), 151–159.

Pillay, P. and Krishnan, R. (1991). Application characteristics of permanent magnet synchronous and brushless DC motors for servo drives. *IEEE Trans. Ind. Appl.* 27(5), 986–996.

Rahman, M., Zhone, L., and Lim, K. (1996). A DSP based instantaneous torque control strategy for interior permanent magnet synchronous motor drive with wide speed range and reduced torque ripples. In: *Conference Record of the IEEE 31st IAS Annual Meeting, Industry Applications Conference*, pp. 518–524. IEEE, Piscataway, NJ.

Schiferl, R. and Lipo, T. A. (1988). Power capability of salient pole permanent magnet synchronous motors in variable speed drive applications. In: *Conference Record of the IEEE Industry Applications Society Annual Meeting*, pp. 23–31. IEEE, Piscataway, NJ.

Sneyers, B., Novotny, D.W., and Lipo, T.A. (1985). Field weakening in buried permanent magnet ac motor drives. *IEEE Trans. Ind. Appl.* 21(2), 398–407.

Soong, W.L. and Miller, T. (1994). Field-weakening performance of brushless synchronous AC motor drives. *IEE Proc. Electric Power Appl.* 141(6), 331–340.

Sousa, G.C., Bose, B.K., and Cleland, J.G. (1995). Fuzzy logic based on-line efficiency optimization control of an indirect vector-controlled induction motor drive. *IEEE Trans. Ind. Electron.* 42(2), 192–198.

Sun, T., Wang, J., and Chen, X. (2015). Maximum torque per ampere (MTPA) control for interior permanent magnet synchronous machine drives based on virtual signal injection. *IEEE Trans. Power Electron.* 30(9), 5036–5045.

Sun, T., Wang, J., Koc, M., and Chen, X. (2016). Self-Learning MTPA Control of interior permanent-magnet synchronous machine drives based on virtual signal injection. *IEEE Trans. Ind. Appl.* 52(4), 3062–3070.

Toliyat, H. A., Campbell, S. G. (2003). *DSP-based Electromechanical Motion Control.* CRC Press, Boca Raton, FL.

Vaez, S. and John, V.I. (1995). Minimum loss operation of PM motor drives. In: *Canadian Conference on Electrical and Computer Engineering*, pp. 284–287. IEEE, Piscataway, NJ.

Vaez, S., John, V., and Rahman, M. (1997). Energy saving vector control strategies for electric vehicle motor drives. In: *Proceedings of the* IEEE, *Power Conversion Conference*, pp. 13–18. IEEE, Piscataway, NJ.

Vaez, S., John, V., and Rahman, M. (1999). An on-line loss minimization controller for interior permanent magnet motor drives. *IEEE Trans. Energy Convers.* 14(4), 1435–1440.

Vaez-Zadeh, S. (2001). Variable flux control of permanent magnet synchronous motor drives for constant torque operation. *IEEE Trans. Power Electron.* 16(4), 527–534.

Vaez-Zadeh, S. and Hendi, F. (2005). A continuous efficiency optimization controller for induction motor drives. *Energy Convers. Manag.* 46(5), 701–713.

Vaez-Zadeh, S., Nasiri-Gheidari, Z., and Tootoonchian, F. (2006a). Vector control of IPM motor with maximum torque and efficiency. In: *Proceedings of 14th Iranian Conference on Electrical Engineering.* Tehran, Iran [In Persian.]

Vaez-Zadeh, S., Zamanifar, M., and Soltani, J. (2006b). Nonlinear efficiency optimization control of IPM synchronous motor drives with online parameter estimation. In: *37th IEEE Power Electronics Specialists Conference*, pp. 1–6. IEEE, Piscataway, NJ.

Vas, P. (1998). *Sensorless Vector and Direct Torque Control.* Oxford University Press, Oxford.

Williamson, S. and Cann, R.G. (1984). A comparison of PWM switching strategies on the basis of drive system efficiency. *IEEE Trans. Ind. Appl.* 20(6), 1460–1472.

Zeng, Z., Zhou, E., and Liang, D. (1996). A new flux weakening control algorithm for interior permanent magnet synchronous motors. In: *Proceedings of the IEEE IECON 22nd International Conference on Industrial Electronics, Control, and Instrumentation*, pp. 1183–1186. IEEE, Piscataway, NJ.

Direct Torque Control

4

DTC as the main competing control method against VC has attracted huge interest among researchers since its inception in the mid-1980s (Depenbrock 1986; Takahashi and Noguchi 1986). DTC was applied to PMS motors in the 1990s and, since then, has gained great attention in the literature, although its market share in the motor drive sector is a different story. Nevertheless, the method has found full recognition and some applications in the industry.

DTC at the time of inception took into account already known facts regarding AC machines to form effectively a viable control system of high performance. The facts are presented in the first section of this chapter as the principles of the method. A major shift in the motor control field, initiated by DTC, is to consider the indispensable discrete behavior of the inverter in the motor control system to avoid extra burden in the system implementation. As a result, in contrast to VC, DTC determines specifically the desired voltage vector in each instance. This vector is applied to the machine through the inverter and thus controls the machine during a switching interval of the inverter. In this way, a closer control to the machine behavior under the inverter is provided.

After presenting the principles of DTC, the chapter continues with the basic DTC system. The PMS operating conditions, including current, voltage, and flux limits, are then discussed followed by the flux control schemes under DTC. Alternative DTC systems, including optimal DTC with space vector modulation (SVM), are also studied. DTC in connection with loss minimization control is presented next. Finally, a comparison between DTC and VC is attempted before the ending summary.

4.1 Principles of DTC

In this section, fundamental principles of DTC are reviewed. Systematic understanding of these principles is essential for deep insight into DTC. The principles include the relationship of motor torque deviation with load angle deviation; the relationship of flux linkage

Control of Permanent Magnet Synchronous Motors. Sadegh Vaez-Zadeh. © Sadegh Vaez-Zadeh 2018.
Published in 2018 by Oxford University Press. DOI 10.1093/oso/9780198742968.001.0001

deviation with voltage space vector; the generation of voltage vectors by inverter; the rule for selecting the most appropriate voltage vector; and finally, the means for limiting the variation of flux linkage magnitude during the motor operation.

4.1.1 Torque deviation versus torque angle deviation

Electromagnetic torque of a PMS machine with surface-mounted magnets, in the stator flux RF, can be presented by

$$T_e = \frac{3}{2} P \lambda_s i_y, \tag{4.1.1}$$

where λ_s and i_y stand for the x- and y-components of the stator flux linkage vector and stator current vector, respectively, as presented in Chapter 2 by eqn (2.6.13). Referring to Chapter 2, the i_y for surface-mounted magnet machines is given by

$$i_y = \frac{1}{L_s} \lambda_m \sin \delta, \tag{4.1.2}$$

where λ_m is the magnitude of rotor flux linkages vector and δ is the angle between these two flux linkage vectors, i.e., the load angle as seen in Fig. 4.1. Also, L_s is the stator winding inductance. Substituting eqn (4.1.2) into eqn (4.1.1) yields

$$T_e = \frac{3}{2} \frac{P}{L_s} \lambda_m \lambda_s \sin \delta. \tag{4.1.3}$$

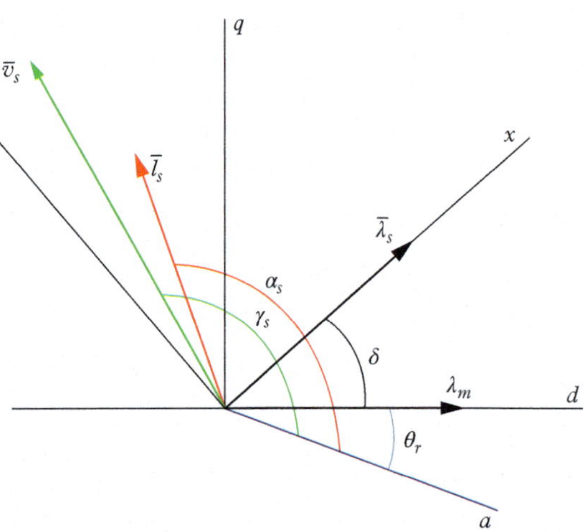

Figure 4.1 *A vector diagram of PMS machines showing current, voltage, and flux linkage vectors in rotor and stator flux linkage RFs.*

The magnitude of rotor flux linkage depends on the type of the magnet material, the pole dimensions, and motor configuration. Therefore, it is fixed for a specific motor if other marginal effects are ignored. The stator flux linkage magnitude is also assumed to be constant as a constant flux control is a common practice in many motor control methods for achieving fast motor dynamics. A constant stator flux linkage magnitude is realized in DTC by a particular means, as will be presented shortly. Therefore, the torque depends on a single variable, δ, only, as seen by eqn (4.1.3). A time variation of torque in a very short duration can be represented by a differentiation of eqn (4.1.3) as given by

$$\frac{dT_e}{dt} = \frac{3}{2}\frac{P}{L_s}\lambda_m\,\lambda_s\,\dot{\delta}\,\cos\delta, \qquad \text{at } t = 0, \tag{4.1.4}$$

where a dot over a variable denotes a derivative. During the tiny interval of differentiation, δ experiences a variation, which is influential on torque deviation, but negligible with respect to δ itself. Thus, cos δ can be regarded constant in the interval. Therefore, eqn (4.1.4) shows that a small deviation of torque depends almost solely on the time derivative of δ. This result has an important implication that lays down the foundation of DTC, which is that the torque dynamics is almost proportional to the "rate of change" in the angle between the stator and rotor flux linkage vectors as shown in Fig. 4.1. It is worth mentioning that

$$\delta = (\omega_s - \omega_r)t + \delta_0. \tag{4.1.5}$$

At steady-state, the stator and rotor flux linkage vectors rotate with the same speed. Thus, the derivative of eqn (4.1.5) is zero, i.e.,

$$\delta = \delta_0 \Rightarrow \dot{\delta} = 0. \tag{4.1.6}$$

Therefore, the torque deviation of eqn (4.1.5) vanishes and the torque is constant. However, at transient state, the stator flux linkage and the rotor flux linkage vectors rotate with different speeds. Thus, under transient conditions

$$\dot{\delta} = \omega_s - \omega_r. \tag{4.1.7}$$

Referring to Fig. 4.1, eqn (4.1.4) means that for a fast torque response, the speed difference in eqn (4.1.5) must be high. Therefore, a desirable motor control can be provided if the speed difference becomes as high as possible when the motor torque command changes. Bearing in mind that the rotor flux linkage rotates with the rotor speed and the rotor as a mechanical body has a large time constant, the rotor speed cannot change rapidly. As a result, this is the speed of stator flux linkage that must change rapidly. This is, in fact, the essence of the DTC method to change the speed of rotation of the stator flux linkage vector in response to a motor command in order to change δ.

For IPM machines, where $L_d \neq L_q$, the torque equation (4.1.3) turns into eqn (2.6.17), which is recalled here as

$$T_e = \frac{3P\lambda_s}{4L_dL_q}[2\lambda_m L_q \sin\delta + \lambda_s(L_d - L_q)\sin 2\delta]. \qquad (4.1.8)$$

Then the derivative of eqn (4.1.8) becomes

$$\frac{dT_e}{dt} = \frac{3P\lambda_s}{2L_dL_q}[\lambda_m L_q \cos\delta + \lambda_s(L_d - L_q)\cos 2\delta]\dot{\delta}, \qquad at\ t = 0. \quad (4.1.9)$$

It is seen again that, in the case of IPM machines, the torque deviation almost depends on the load angle deviation, $\dot{\delta}$. Therefore, the same principle is valid for IPM motors, too.

4.1.2 Alignment of flux linkage deviation with voltage space vector

The voltage equation of PMS motors in space vector domain can be written as

$$\overline{v}_s = R_s \overline{i}_s + p\overline{\lambda}_s, \qquad (4.1.10)$$

where $\overline{v}_s$, $\overline{i}_s$, and $\overline{\lambda}_s$ are the stator voltage, current, and flux linkage vectors, respectively, in a stationary RF and p is a derivative operator. The equation has already been elaborated on in Chapter 2. The first term on the right-hand side of the equation corresponds to the voltage drops on stator windings and the second term corresponds to the induced voltage in stator windings. The former is usually small with respect to the stator voltage under normal machine operation due to the small value of R_s. The only exception is in very low speed operations of inverter-fed machines, when the applied voltage is low and the relative importance of the term with respect to the applied voltage cannot be neglected. Setting aside the very low speed operations, one may ignore the term when efficiency analysis is not a concern. Using this reasonable simplification, the stator voltage space vector is the same as the induce voltage space vector:

$$\overline{v}_s \approx p\overline{\lambda}_s. \qquad (4.1.11)$$

Linearizing eqn (4.1.11) yields

$$\Delta\overline{\lambda}_s \approx \overline{v}_s\,\Delta t, \qquad (4.1.12)$$

where Δ in front of a variable stands for a linear increment in the variable. This simple equation is of prime importance when it comes to principles of DTC. It simply says that a certain flux linkage deviation vector can be obtained by applying a specific voltage vector to the stator windings for a short interval. Bearing in mind that the term Δt in eqn (4.1.12) is, in fact, a scalar, the equation further proves that the two space vectors, $\Delta\overline{\lambda}_s$ and $\overline{v}_s$, are in the same direction. On the other hand, knowing a desired $\Delta\overline{\lambda}_s$ and Δt, it is possible to find out the right voltage vector that must be applied to the machine.

Now let us look at the facts just extracted from eqn (4.1.12) in connection with the essence of DTC discussed previously. Figure 4.2 is helpful for visualizing the facts. It depicts the flux linkage vector at two instances before and after an incremental deviation with respect to the rotor flux linkage vector, i.e., $\Delta\delta$. The space vector flux linkage deviation of eqn (4.1.12) is shown in Fig. 4.2 as the subtraction of the two flux linkage vectors. As mentioned previously, the voltage vector that causes $\Delta\delta$, is parallel to $\Delta\overline{\lambda}_s$ as shown in Fig. 4.2 by $\overline{v}_s$.

Altogether, a desired torque deviation demands an incremental rotation of the stator flux linkage vector and this, in turn, determines a specific voltage vector to be applied to the machine. This is shown by the cause and effect relationships

$$\Delta T_e \Rightarrow \Delta\delta \Rightarrow \Delta\overline{\lambda}_s \Rightarrow \overline{v}_s. \tag{4.1.13}$$

In order to find $\overline{v}_s$ by the relationships given in eqn (4.1.13), it must be produced by an inverter and applied to the machine windings for a duration of Δt. Basically, this is the way DTC works.

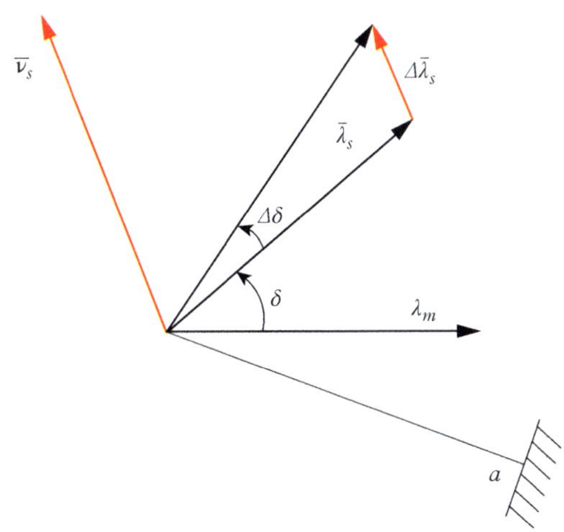

Figure 4.2 *Load angle and torque adjustment by voltage vectors.*

4.1.3 Space vector voltages of an inverter

Having found the desired voltage vector, this must be produced by an inverter. The available voltage vectors provided by a two-level, three-phase voltage source inverter has already been presented briefly in Section 1.2 and will be further explained here. Figure 1.2 shows such an inverter with two power electronic switches on each of its three legs. The switches can have two operating states—either ON, i.e., conducting, or OFF, i.e., non-conducting state. However, the two switches of a leg must have opposite states to prevent a short circuit through the switches, which is hazardous to the switches. The switching states of the inverter can be seen in Table 1.1. There are six non-zero voltage vectors and two zero voltage vectors for a two-level inverter, depending on the on/off state of switches. They can be represented by

$$\bar{v}_s = \frac{2}{3} V_{DC} \left(S_a + a\, S_b + a^2\, S_c\right), \tag{4.1.14}$$

where a and a^2 are unit vectors of 120° and 240° as

$$a = e^{j\frac{2\pi}{3}}, \qquad a^2 = e^{j\frac{4\pi}{3}}, \tag{4.1.15}$$

and S_a, S_b, and S_c are used to show the state of each leg in the inverter, which is either 0 or 1. The state of a leg is 0 when the leg is connected to zero voltage; i.e., the lower switch is on and the upper switch is off. In contrast, the state of a leg is 1 when the leg is connected to the DC bus voltage, V_{DC}; i.e., the upper switch is on and the lower switch is off. The voltage vector is zero when all legs have the same state, either 0 or 1. This occurs when the three upper switches have the same state, either on or off, while all the lower switches have the opposite state, i.e., either off or on, respectively. The voltage vectors are depicted in Fig. 4.3, where the state of switches is shown by a three-digit number, e.g., $\bar{v}_2$ (110). It is seen that the vectors have the same magnitude, except for the zero voltages. The vectors are 60° apart in their directions, however.

4.1.4 Golden rule of DTC

It was proved that the torque dynamics depends on the rotation speed of the stator flux linkage vector with regard to the magnet flux linkage vector. In other words, to get a fast torque response, one should arrange to change the load angle, δ, as quickly as possible. Referring to Fig. 4.2 and bearing in mind that $\Delta\bar{\lambda}_s$ is aligned with $\bar{v}_s$, it is ideal

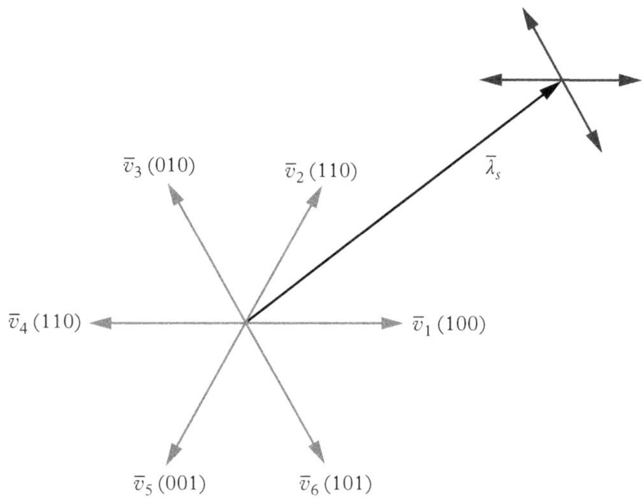

Figure 4.3 *Voltage vector options to cause fast dynamics for either increasing or decreasing the developed machine torque.*

to keep the voltage vector always perpendicular to $\overline{\lambda}_s$ to get the fastest rotation of $\overline{\lambda}_s$. This is the golden rule of DTC. The inverter is capable of producing six non-zero voltages only, however, which are not perpendicular to $\overline{\lambda}_s$ in general. Therefore, a reasonable solution is to compromise to the best possible voltage vector among the inverter voltages in each inverter switching period. Such a voltage vector must have an angle to $\overline{\lambda}_s$ as close as possible to 90° as a condition for a fast torque response. There are always two pairs of voltage vectors that better provide this condition, depending on the position of the flux linkage vector. Figure 4.3 shows these voltages, i.e., $\bar{v}_3$ and $\bar{v}_4$, as well as $\bar{v}_1$ and $\bar{v}_6$ as the best options for a specific stator flux linkage position. It can be seen that applying $\bar{v}_3$ or $\bar{v}_4$ to the machine will rotate $\overline{\lambda}_s$ to increase δ in order to increase torque. Similarly, applying either of the two voltages, $\bar{v}_1$ or $\bar{v}_6$, to the machine will rotate $\overline{\lambda}_s$ to decrease δ in order to reduce torque. A systematic approach is followed to determine all such voltage vectors for different positions of flux linkage vector. This is started by dividing the plane of inverter voltage vectors into six equal regions or sectors, each covering 60° of the plane as seen in Fig. 4.4. A voltage vector is located in the middle of each region. The regions are numbered by the number of their voltage vectors. A distinct set of four voltage vectors corresponding to the flux linkage vector in each region can be determined as explained earlier. The set can be presented by the following logics:

$$\text{for getting } \Delta\delta > 0, \quad \text{the applied voltages must be } \bar{v}_{k+1} \text{ or } \bar{v}_{k+2}, \quad \forall\, k = 1-6,$$

$$\text{for getting } \Delta\delta < 0, \quad \text{the applied voltages must be } \bar{v}_{k-1} \text{ or } \bar{v}_{k-2}, \quad \forall\, k = 1-6, \tag{4.1.16}$$

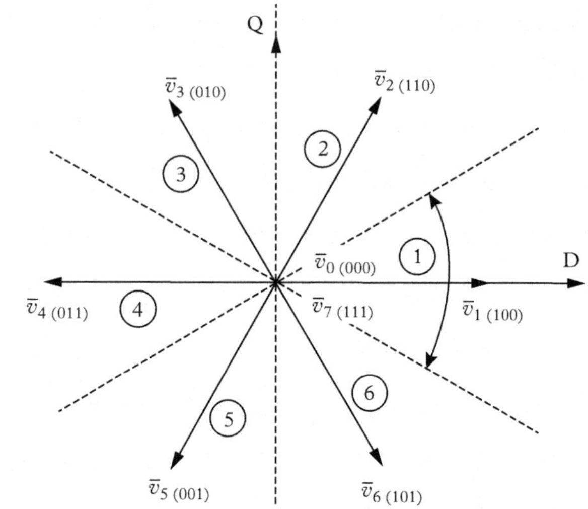

Figure 4.4 *Flux regions (sectors) of the machine based on voltage vectors of the inverter.*

where k is the region number in which the flux linkage vector is located. The first logic increases the torque, while the second logic decreases it according to eqn (4.1.4) or (4.1.9). It must be mentioned that the logic corresponds to the counter-clockwise rotations of the rotor.

4.1.5 Limiting the variation of flux linkage magnitude

It is desired to keep the magnitude of $\overline{\lambda}_s$ constant under the machine transient state to improve machine dynamics. A constant flux linkage magnitude is also assumed in the derivation of eqns (4.1.4) and (4.1.9) as the basis of DTC method. This can only be achieved during the period of a rapidly rotating flux linkage vector if an inverter can provide infinite number of voltage vectors to the motor, thus, keeping the applied voltage perpendicular to the flux linkage vector in all instances. However, a two-level voltage source inverter can only produce six non-zero voltage vectors. When these voltages are applied to a PMS motor, they usually rotate $\overline{\lambda}_s$, as well as change its magnitude. Looking back at Fig. 4.3, if $\bar{v}_3$ and $\bar{v}_4$ are applied to the motor, they will rotate $\overline{\lambda}_s$ to increase δ in order to increase torque, while they change the magnitude of $\overline{\lambda}_s$, by increasing or decreasing it, respectively. Similarly, by applying either $\bar{v}_1$ or $\bar{v}_6$ to the machine, they rotate $\overline{\lambda}_s$ to decrease δ in order to reduce torque, whereas $\bar{v}_1$ increases and $\bar{v}_6$ decreases the magnitude of $\overline{\lambda}_s$. Therefore, it is wise to retain the magnitude of $\overline{\lambda}_s$ with less possible change when a voltage vector is

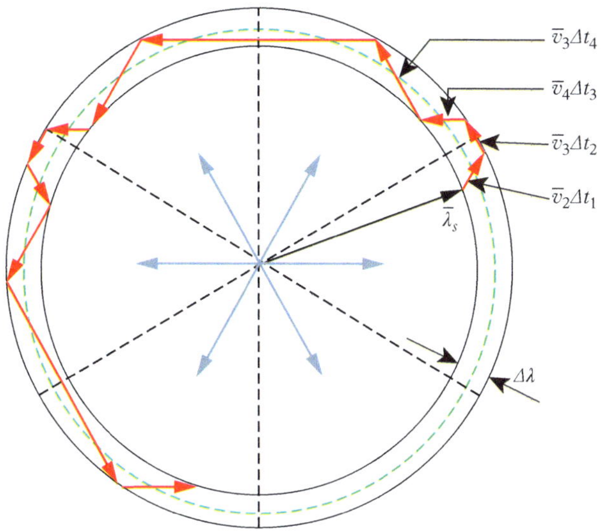

Figure 4.5 *Flux linkage band and consecutive voltage vectors in DTC.*

applied to the machine during an inverter switching period. A DTC method includes a means to do the task by limiting the unwanted changes of flux linkage magnitude within a pre-specified amount, during an inverter switching period, while increasing the rotation of the flux linkage vector. To define the limits, a flux linkage band with a lower and an upper limits is defined. This is shown in Fig. 4.5 by its width, $\Delta\lambda$. The band allows a flux linkage variation about a commanded value of the flux linkage signal. The commanded flux linkage trajectory is shown by a dashed circle in Fig. 4.5. When a voltage vector is applied to the machine, the flux linkage changes inside the band, until it reaches a limit. The inverter automatically switches the applied voltage, when the flux linkage reaches a band limit. Figure 4.5 shows that the voltage vectors consecutively are applied to the machine in different switching instances according to eqn (4.1.16). By this way, the flux linkage vector rotates fast, while its magnitude is limited within the flux linkage band. It must be noted that the flux linkage band in Fig. 4.5 is shown much larger with respect to the commanded flux linkage for the sake of clearness.

4.2 Basic DTC system

A DTC realizes the principles stated previously by the system shown in Fig. 4.6. A torque and a flux linkage magnitude command, T_e^* and λ_s^*, respectively, are compared with their estimated values, and the errors are applied to the corresponding hysteresis controllers to

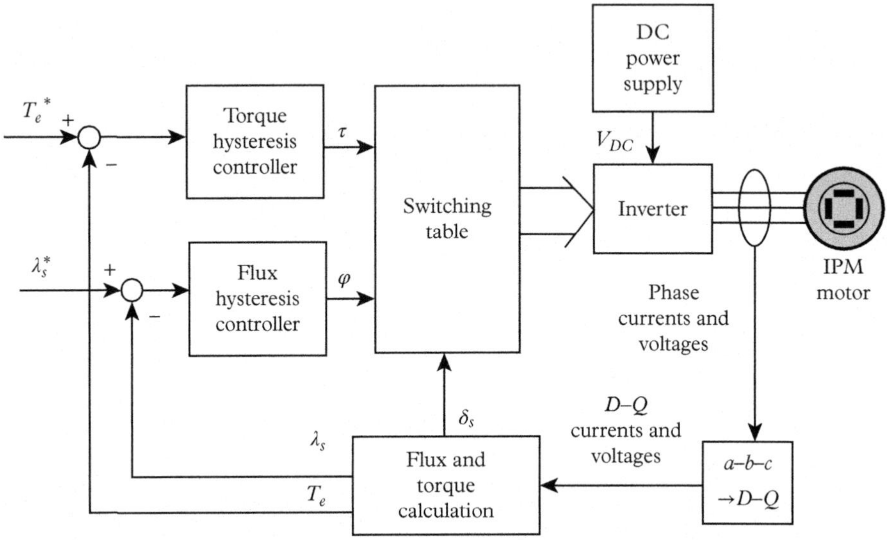

Figure 4.6 *Basic DTC system of PMS motors.*

produce torque and flux linkage flags, τ and φ, respectively. The flags, together with the stator flux linkage vector position, go to a switching table to determine a desired voltage vector command. An inverter realizes the commanded voltage and applies it to the machine. The machine then responds swiftly to reduce torque and flux linkage errors. The estimated torque and flux linkage magnitude are given by processing measured phase voltages and currents. The system components of Fig. 4.6 are described here.

4.2.1 Hysteresis controllers

In every motor control system, the torque command is of prime concern and is decided by the applicant, directly or indirectly. It is decided directly, such as in a traction application, or indirectly, as in an output of a speed controller, such as in an elevator. In industry applications, it may be decided by an overhead control system. The commended torque is the same as the machine-estimated developed torque at steady-state within an acceptable tiny deviation. However, commanded and developed torques are different at transient state, after the applicant changes the command to a new value. The commanded torque is compared with the estimated developed torque to produce the torque error.

The hysteresis torque controller accepts the torque error and provides a torque flag, τ, of 0 or 1 at its output, in each inverter

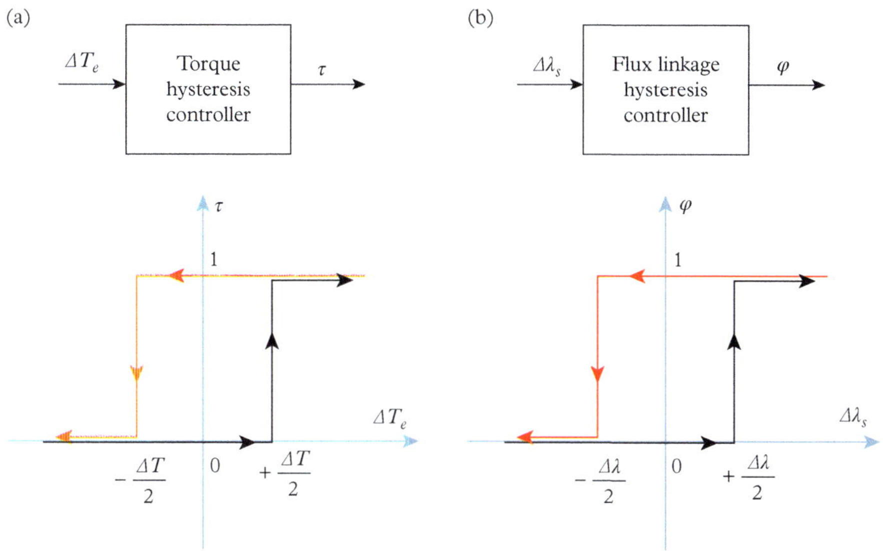

Figure 4.7 *Hysteresis controllers: (a) torque controller and (b) flux linkage controller.*

switching instance depending on the present and previous error values. The fundamental behavior of the torque hysteresis controller can be seen in Fig. 4.7(a), where the error and the flag are shown on the horizontal axis and vertical axis, respectively. A tiny torque hysteresis band, ΔT, is defined by the lower and higher limits as the acceptable deviation of the estimated torque from the commanded torque at steady-state, as seen in Fig. 4.7(a). If the torque error is within the band, the controller output remains as it is. However, the output changes if the input goes beyond the band in the direction depicted in the figure. That is, the flag changes from 0 to 1 if the error exceeds $+\Delta T/2$. Also, the flag changes from 1 to 0 if the error reduces beyond $-\Delta T/2$. Therefore, the flag observes the following rules, outside the band:

$$\begin{aligned} \tau &= 1, \quad \text{if} \quad \Delta T_e > \Delta T/2, \\ \tau &= 0, \quad \text{if} \quad \Delta T_e < -\Delta T/2. \end{aligned} \tag{4.2.1}$$

The DTC system provides an opportunity to directly apply a desired stator flux linkage magnitude to the system. It is usually produced by a flux weakening procedure, as will be described later. The commended flux linkage is the same as the estimated flux linkage at steady-state within an acceptable tiny deviation. However, the commanded and the estimated flux linkages are different at transient state, after the command changes to a new value. The commanded flux linkage is compared with the estimated flux linkage to produce the flux linkage error.

The hysteresis flux linkage controller accepts the flux linkage error and produces a flux linkage flag, φ, of 0 or 1, as its output, in each inverter switching instance, depending on the present and previous error values. The fundamental behavior of the flux hysteresis controller can be seen in Fig. 4.7(b), where the error and the flag are shown on the horizontal and vertical axis, respectively. The tiny flux linkage band, $\Delta\lambda$, is defined as the acceptable deviation of the estimated flux linkage from the commanded one at steady state. It is determined by the lower and higher limits as seen in Fig. 4.7(b). This band has already been shown in Fig. 4.5. If the flux linkage error is within the band, the controller output remains as it is. However, the output changes if the input goes beyond the band in the direction depicted in the figure. That is, the flag changes from 0 to 1 if the error exceeds $+\Delta\lambda/2$. Also, the flag changes from 1 to 0 if the error reduces beyond $-\Delta\lambda/2$. Therefore, the flag observes the following rules, outside the band:

$$\begin{aligned} \varphi &= 1, \quad \text{if} \quad \Delta\lambda_s > \Delta\lambda/2, \\ \varphi &= 0, \quad \text{if} \quad \Delta\lambda_s < -\Delta\lambda/2. \end{aligned} \tag{4.2.2}$$

4.2.2 Switching table

Table 4.1 *Switching table.*

τ	φ	*Voltage Vector*
1	1	$\bar{v}_{k+1}$
	0	$\bar{v}_{k+2}$
0	1	$\bar{v}_{k-1}$
	0	$\bar{v}_{k-2}$

The switching look-up table, as presented in Table 4.1, is an easy means of realizing the logics of eqn (4.1.16) in a systematic way. The table takes three inputs, i.e., the two flags and the region number of the stator flux linkage vector, k, and determines a desired voltage vector command as its output.

The table determines whether the machine torque and flux linkage magnitudes are to be increased or decreased by their corresponding flags in the first and second columns. The six regions in the first row are the flux linkage regions of Fig. 4.4, which are determined by $k = 1$–6. The third column of the switching table lists four voltage choices for a flux region. These are the same voltage options presented in Fig. 4.3 for a specific flux linkage vector located in the second region where $k = 2$. In this case,

$$\begin{aligned} \bar{v}_{k+1} &= \bar{v}_3, \\ \bar{v}_{k+2} &= \bar{v}_4, \\ \bar{v}_{k-1} &= \bar{v}_1, \\ \bar{v}_{k-2} &= \bar{v}_6. \end{aligned} \tag{4.2.3}$$

The torque flag, having a value of 1 or 0, selects two voltage vectors among the four choices and the flux linkage flag, having a value of 1 or 0, determines the desired voltage vector among the recent two options. In this way, having values of two flags and a flux region number, a unique voltage is determined by the switching table and is applied to the inverter.

4.2.3 Flux linkage and torque estimation

Flux linkage and torque estimation is necessary to determine the corresponding errors. This requires the means to sense the phase currents and voltages of the machine by corresponding sensors. The synthesis of Fig. 4.8 can then be used to carry out the estimations. The phase currents and voltages are sensed and used in a stationary three- or two-axis RF transformation to provide the current and voltage components in *D–Q* RF, as elaborated in Section 2.4. They are then employed in calculating the *D*- and *Q*-axis flux linkages as

$$\lambda_D = \int_{T_s} (v_D - R_s\, i_D)\, dt, \tag{4.2.4}$$

$$\lambda_Q = \int_{T_s} (v_Q - R_s\, i_Q)\, dt. \tag{4.2.5}$$

The estimated stator flux linkage magnitude, λ_s, and angle, δ_s, are then calculated as

$$\lambda_s = \sqrt{\lambda_D^2 + \lambda_Q^2}, \tag{4.2.6}$$

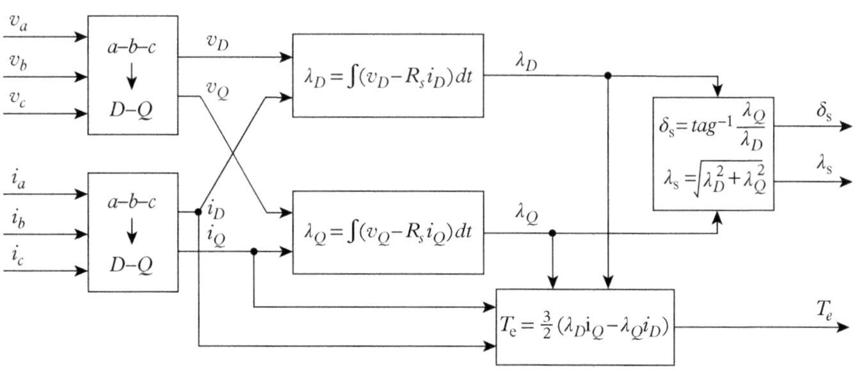

Figure 4.8 *Synthesis of stator flux linkage and torque.*

$$\delta_s = \tan \frac{\lambda_Q}{\lambda_D}. \tag{4.2.7}$$

Finally, the estimated electromagnetic torque is obtained by

$$T_e = \frac{3}{2} P \left[\lambda_D \, i_Q - \lambda_Q \, i_D \right]. \tag{4.2.8}$$

The estimated value of flux linkage angle, δ_s, determines the region in which the flux linkage vector is located. Thus, the value of k as an input to the switching table is obtained. The estimated values of flux linkage magnitude, λ_s, and torque, T_e, are used as feedback signals and compared with their corresponding commanded values as mentioned previously.

4.3 Operating limits and limiting means in DTC

PMS motors operation with current and voltage limits has already been discussed under VC in Section 3.5. The same reasons as presented in the section for VC are valid for observing PMS motor operating limits under DTC.

4.3.1 Current limit

In contrast to VC, in DTC motor current commands are not used. Therefore, preventing the motor overcurrent operation is not possible by limiting current commands. However, the actual motor currents are measured in DTC to estimate flux linkage and torque. These measured currents may thus be monitored for overcurrent prevention. A possible procedure is to transform motor phase currents into current components in the stationary two-axis RF, i_D and i_Q, and calculate the magnitude of stator current vector, i_s, to be compared with the current limit, i_{sL}. If i_s is getting close to i_{sL}, then the torque and/or flux linkage commands are to be reduced to prevent overcurrent. In this method, a certain strategy needs to be decided on and followed for optimal adjustment of torque and flux linkage, taking into account the relationship among motor torque, flux, and current.

An alternative offline method for overcurrent prevention is to limit the torque and flux linkage commands according to i_{sL}. In this method, torque and flux linkage are calculated in terms of i_s, which in turn must be limited by i_{sL}. The calculations are initiated in the d–q RF. The d- and q-axis components of the stator current, however, are later omitted from the equations as follows.

Consider the stator current, torque and flux linkage in terms of i_d and i_q as

$$i_s = \sqrt{i_d^2 + i_q^2}, \tag{4.3.1}$$

$$T_e = \frac{3}{2}P\left[\lambda_m + \left(L_d - L_q\right)i_d\right]i_q, \tag{4.3.2}$$

$$\lambda_s = \sqrt{\lambda_d^2 + \lambda_q^2} = \sqrt{[\lambda_m + L_d i_d]^2 + L_q^2 i_q^2}. \tag{4.3.3}$$

Using eqns (4.3.1)–(4.3.3), it is possible to derive T_e as a function of i_s and λ_s by omitting i_d and i_q as

$$T_e = f(i_s, \lambda_s)\,. \tag{4.3.4}$$

Now, if the reference signal for stator flux linkage magnitude is fixed or is given as a function of the developed torque, then the torque limit is obtained as a function of current limit:

$$T_{eL} = f(i_{sL})\,. \tag{4.3.5}$$

Then, the torque command is limited by the torque limit:

$$T_e^* \leq T_{eL}. \tag{4.3.6}$$

The implementation of a torque limiter in practice needs eqn (4.3.6), which is complicated and not suitable for online computation in a real system. However, having i_{sl}, eqn (4.3.6) can be calculated offline.

4.3.2 Voltage limit

The prevention of overvoltage in DTC is needed as it is needed in VC. This is implemented in terms of motor flux linkage command. The voltage equation of PMS motors in stationary RF is presented as

$$\overline{v}_s = R_s \overline{i}_s + p\overline{\lambda}_s. \tag{4.3.7}$$

Neglecting the voltage drop across winding resistance, eqn (4.3.7) implies that the stator voltage is equal to the derivative of stator flux linkage. This is presented at steady-state as

$$v_s = \omega_r \lambda_s. \tag{4.3.8}$$

Knowing the voltage limit of the machine as v_{sL}, the flux linkage command must, therefore, be limited by

$$\lambda_{sL} = \frac{v_{sL}}{\omega_r}. \tag{4.3.9}$$

It can be seen that the flux linkage limit depends on the motor speed. Then, the flux linkage command is limited by the flux linkage limit:

$$\lambda_s^* \leq \lambda_{sL}. \tag{4.3.10}$$

4.3.3 Flux linkage limit

Proper selection of stator flux linkage command in DTC of PMS motors is very important as it directly affects the machine performance. It can help to decrease stator current or machine losses at steady-state. Also, one needs to adjust both stator flux linkage angle and magnitude in order to get a fast torque response especially at motor start-up. Figure 4.9 shows the electromagnetic torque with respect to λ_s and δ for a typical PMS machine with magnetic saliency. It can be seen that smaller values of flux linkage magnitude result in near linear torque curve, while the bigger values result in non-linear curves. This means that a fast torque response with respect to δ is achieved with lower values of flux linkage magnitude. In particular, the slope of torque curve at vicinity of $\delta = 0$ becomes negative for a high flux linkage magnitude. This means that an increase in δ at this flux linkage decreases the motor torque instead of increasing it as the DTC principles states in Section 4.1.1. The argument can be summarized as that the proper dynamics of a motor under DTC depends on the flux linkage magnitude. On the other hand, the maximum achievable electromagnetic

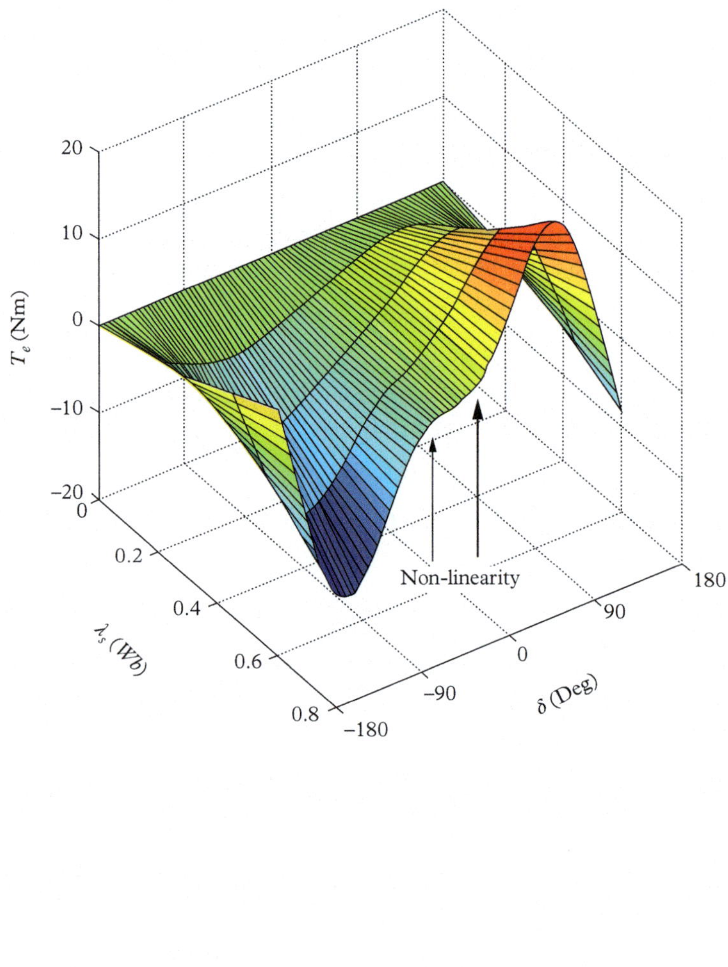

Figure 4.9 *Motor developed torque as a function of stator flux linkage magnitude and angle.*

torque is smaller with the smaller values of the flux linkage. This is a disadvantage, especially in the case where the machine is heavily loaded and needs a bigger electromagnetic torque to start up rapidly. So, it is important to calculate a maximum flux linkage value as a limit for having an increasing torque with increasing δ for flux linkage values lower than the limit.

Recall eqn (4.1.9) at $\delta = 0$ as

$$\frac{dT_e}{dt} = \frac{3P\lambda_s}{2L_dL_q}[\lambda_m L_q + \lambda_s(L_d - L_q)]\dot{\delta}, \qquad \text{at } t = 0, \tag{4.3.11}$$

which shows that the torque derivative is positive for a positive $\dot{\delta}$ if the term inside the brackets is positive. This results in the condition (Zhong *et al.* 1997)

$$\lambda_s < \frac{L_q}{L_q - L_d}\lambda_m = \lambda_{sL}, \tag{4.3.12}$$

which must be observed when deciding a flux linkage command to a DTC system. The DTC principles of eqn (4.1.16) are violated if the limit is not considered. The limit tends to infinity for surface-mounted PMS machines, since in these machines $L_d = L_q = L_s$.

The load angle, δ, in an IPM motor is also limited by a maximum value, which is obtained by differentiating the torque equation of eqn (4.1.8) with respect to δ and solving the resulting equation to get

$$\delta_L = \cos^{-1}\left\{\frac{\lambda_{sL}/\lambda_s - \sqrt{(\lambda_{sL}/\lambda_s)^2 + 8}}{4}\right\}. \tag{4.3.13}$$

Having found δ_L, the following condition must also be held in IPM motors:

$$\delta \leq \delta_L. \tag{4.3.14}$$

4.4 Flux control in DTC

A general discussion regarding flux weakening control in PMS motors is presented at the beginning of Section 3.6 before different types of flux weakening schemes under VC are presented. The discussion is equally valid for PMS motors under DTC. Here, flux weakening is used as a means for reaching high speed, as well as MTPA,

MTPV, unity power factor, and minimum loss operations. Flux control is mainly carried out in VC by adjusting a commanded signal of direct-axis component of stator current vector in different RFs. However, there is no such commanded current in DTC as presented in Section 4.2. Fortunately, the stator flux linkage is directly applied to the DTC system as a command signal. Therefore, it is possible to formulate the conditions for flux control schemes in terms of stator flux linkage. Flux linkage formulation in terms of motor torque and speed is carried out in this section for different operating schemes as follows.

4.4.1 Maximum torque per ampere control

MTPA control in DTC of PMS motors is achieved by adjusting the magnitude of stator flux linkage, λ_s, to minimize the stator current magnitude, i_s, for every torque reference. Therefore, a desired λ_s as a function of the developed torque, T_e, must be obtained. One way of obtaining such a function is to start from the MTPA condition in VC, which is recalled from eqn (3.6.5) as

$$i_d = \frac{\lambda_m}{2(L_q - L_d)} - \sqrt{\frac{\lambda_m^2}{4(L_q - L_d)^2} + i_q^2}. \tag{4.4.1}$$

The following flux linkage and torque equations are also held for a PMS motor in $d-q$ RF:

$$\lambda_d = L_d i_d + \lambda_m, \tag{4.4.2}$$

$$\lambda_q = L_q i_q, \tag{4.4.3}$$

$$T_e = \frac{3}{2} P \left[\lambda_m + (L_d - L_q) i_d\right] i_q. \tag{4.4.4}$$

When eqns (4.4.2)–(4.4.4) are substituted into eqn (4.4.1), the following two equations for flux linkage components are obtained (Hoang *et al.* 2012):

$$(L_q - L_d)^2 \lambda_q^4 + \frac{2}{3P} T_e \lambda_m L_q^3 \lambda_q - \left(\frac{2}{3P} T_e\right)^2 L_q^4 = 0, \tag{4.4.5}$$

$$\lambda_d = \frac{L_q}{(L_q - L_d)} \left(\lambda_m - \frac{2}{3} \frac{L_d T_e}{P \lambda_q}\right). \tag{4.4.6}$$

Also, it is known that

$$\lambda_s = \sqrt{\lambda_d^2 + \lambda_q^2}. \tag{4.4.7}$$

With eqns (4.4.5)–(4.4.7) it is possible to calculate λ_s as a function of T_e. Then, having the reference value of T_e, a reference value for λ_s can be calculated to meet MTPA operation. This procedure needs a solution for λ_q from eqn (4.4.5). However, since eqn (4.4.5) is a fourth-order equation of variable λ_q, a closed form solution for λ_q is cumbersome. Therefore, two alternatives solutions are presented. The first one is an approximate solution based on the Newton–Raphson approximation method, if the first term in eqn (4.4.5) is significantly smaller than the other terms. This is (Hoang *et al.* 2012)

$$\lambda_q \approx \frac{2L_q T_e}{3P\lambda_m} - \frac{2L_q T_e}{12P\lambda_m + \frac{27P^3\lambda_m^5}{4(L_q - L_d)^2 T_e^2}}. \tag{4.4.8}$$

When eqns (4.4.8), (4.4.6), and (4.4.7) are stored as part of the motor control system, the equations can be calculated online for every commanded value of T_e to produce a reference value of λ_s for use in DTC. The second solution is to calculate eqns (4.4.1)–(4.4.7) offline and store the results in a look-up table. MPTA control is then implemented according to the block diagram of Fig. 4.10, where the flux linkage command is determined as the output of the look-up table below the base speed in the so-called constant torque region.

MPTA for surface-mounted PMS motors with $L_d = L_q = L_s$ is simplified where $i_{d=0}$ and $i_q = i_s$, as mentioned in Section 3.6. This yields

$$\lambda_s = \sqrt{\lambda_d^2 + \lambda_q^2} = \sqrt{\lambda_m^2 + L_s^2 i_s^2}, \tag{4.4.9}$$

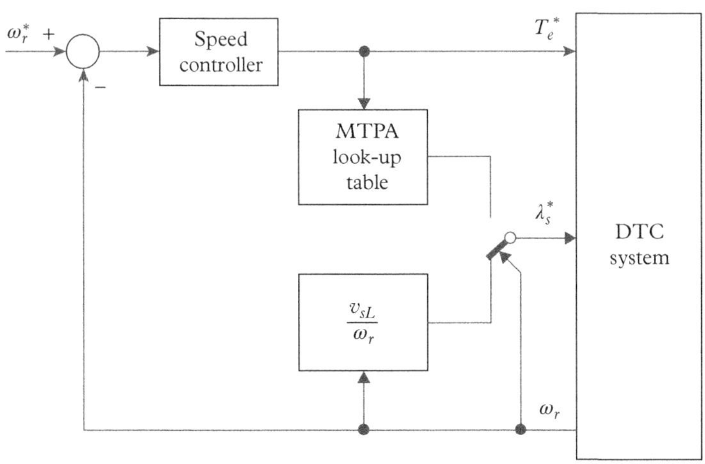

Figure 4.10 *Block diagram of MTPA control below the base speed and flux weakening control above the base speed.*

$$T_e = \frac{3}{2}P\lambda_m i_s \Rightarrow i_s = \frac{2}{3P}\cdot\frac{T_e}{\lambda_m}. \tag{4.4.10}$$

Substituting eqn (4.4.10) into eqn (4.4.9) results in

$$\lambda_s = \sqrt{\lambda_m^2 + \frac{4L_s^2}{9P^2}\cdot\frac{T_e^2}{\lambda_m^2}}. \tag{4.4.11}$$

The implementation of MTPA control for surface-mounted PMS motors is carried out with the help of the block diagram in Fig. 4.10, where eqn (4.4.11) replaces the MTPA look-up table.

4.4.2 Flux weakening at high speed

The voltage equation at steady-state operation, neglecting stator reference, is given by

$$v_s \approx \omega_r \lambda_s, \tag{4.4.12}$$

which states that the amplitude of the stator flux linkage varies inversely with the motor speed. This is used to produce the reference flux linkage above the base speed in the so-called flux weakening region. In this region, the voltage reaches the voltage limits, v_{sL}. Therefore, the reference flux linkage is determined by

$$\lambda_s = \frac{v_{sL}}{\omega_r}. \tag{4.4.13}$$

This is seen in Fig. 4.10.

4.4.3 Unity power factor control

Unity power factor (PF = 1) control in connection with DTC can be implemented by adjusting λ_s according to T_e. A function of λ_s in terms of T_e is thus necessary. This can be obtained by recalling the condition of PF = 1 in VC from Section 3.6 as

$$\frac{v_q}{v_d} = \frac{i_q}{i_d}. \tag{4.4.14}$$

The voltage and current components in eqn (4.4.14) can be found in terms of flux linkage components, λ_d and λ_q, from the following equations, neglecting the stator resistance:

$$v_d = -\omega_r \lambda_q, \tag{4.4.15}$$

$$v_q = \omega_r \lambda_d, \tag{4.4.16}$$

$$\lambda_d = \lambda_m + L_d i_d, \tag{4.4.17}$$

$$\lambda_q = L_q i_q. \tag{4.4.18}$$

Substituting eqns (4.4.15)–(4.4.18) into eqn (4.4.14) yields

$$\frac{\lambda_q^2}{L_q} + \frac{\lambda_d^2}{L_d} - \frac{\lambda_m \lambda_d}{L_d} = 0, \tag{4.4.19}$$

which together with the equation

$$\lambda_s^2 = \lambda_d^2 + \lambda_q^2, \tag{4.4.20}$$

can be solved to give λ_d and λ_q as functions of λ_s. Also,

$$i_s = \sqrt{i_d^2 + i_q^2}. \tag{4.4.21}$$

Substituting for i_d and i_q in eqn (4.4.21) from eqns (4.4.17) and (4.4.18) yields

$$i_s = \sqrt{\left(\frac{\lambda_d - \lambda_m}{L_d}\right)^2 + \left(\frac{\lambda_q}{L_q}\right)^2}. \tag{4.4.22}$$

Now, substituting already obtained λ_d and λ_q as functions of λ_s into eqn (4.4.22) we obtain i_s as a function of λ_s:

$$i_s = f(\lambda_s). \tag{4.4.23}$$

On the other hand, it is known that under unity power factor operation $\bar{i}_s$ is perpendicular to $\bar{\lambda}_s$ as shown in Section 3.7. Therefore,

$$T_e = \frac{3}{2} P \lambda_s i_s. \tag{4.4.24}$$

Now, substituting eqn (4.4.23) into eqn (4.4.24) gives the stator flux linkage in terms of the developed torque. If the reference torque, T_e^*, is known as is the case in DTC, then the reference stator flux linkage can be obtained as

$$\lambda_s^* = f\left(T_e^*\right). \tag{4.4.25}$$

It is possible to program eqn (4.4.24) in a software routine as part of control software. Then, for every T_e^*, an appropriate λ_s^* is calculated online by using eqn (4.4.24) as flux linkage command in DTC. However, a closed form equation like eqn (4.4.24) is rather difficult to obtain and implement. An alternative solution is to do the calculations offline and store the results in a look-up table as a range of λ_s^* versus torque reference. This can also be implemented as in Fig. 4.10.

4.5 DTC with replacement flux control

Basic DTC uses torque and stator flux linkage as command signal. However, it is possible to replace stator flux linkage with another relevant signal.

4.5.1 DTC with i_d control

An option is to use the direct-axis component of a stator current vector instead of the stator flux linkage magnitude. This replacement may be justified by the fact that i_d is the flux-producing component of a stator current vector. The corresponding control system block diagram is depicted in Fig. 4.11 (Vas 1998).

The system can still be called DTC as the torque is directly controlled and not through a current control loop. However, a particular feature of the basic DTC, i.e., not having a current transformation to the rotating RF, is violated in the control system of Fig. 4.11. This is because the motor phase current should be transformed into d–q current components to generate i_d for comparison with $i_d{}^*$. The transformation has already elaborated on in Section 2.5. Then, a hysteresis current controller receives the d-axis current error and generates a current flag, d, as an input to the switching table. The switching table received two flags, i.e., τ and d, in addition to the rotor position to determine the voltage vector (Vas 1998).

In this DTC system, the actual torque is estimated by the conventional torque equation in terms of i_d and i_q. The system does

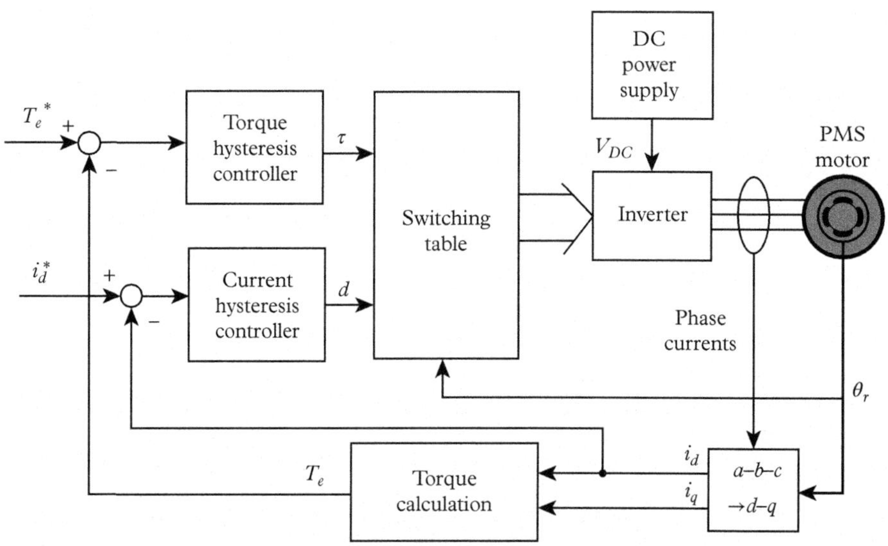

Figure 4.11 *Direct torque control system of PMS motors with i_d control.*

not need the rather problematic stator flux linkage calculation at the cost of having required the rotor position and the current transformation. As a result, the system cannot be recognized as sensorless in contrast to the basic DTC system.

4.5.2 DTC with reactive torque control

It is also possible to replace λ_s control with the reactive torque, T_Q, control, while keeping the electromagnetic torque, T_e, control intact, as the basic DTC. The reactive torque is related to the motor reactive power. The reactive power, in turn, is related to the flux linkage of the machine and can thus replace the stator flux linkage as a reference signal. The reactive torque is defined mathematically as

$$T_Q = \frac{3}{2} P \left[\lambda_D i_D + \lambda_Q i_Q\right]. \tag{4.5.1}$$

D–Q components of $\overline{\lambda}_s$ and $\overline{i}_s$ in eqn (4.5.1) are obtained by a–b–c to D–Q voltage and current transformations, where phase voltages and currents are given by voltage and current sensors, respectively, as shown in Fig. 4.12. The calculated reactive torque is then compared with its reference value to produce the corresponding error. The error is fed into a hysteresis controller, similar to that used for electromagnetic torque, as described in connection with the basic DTC system in Section 4.2. The hysteresis controller provides a flag, Q, to be used as an input to the switching table along with the other

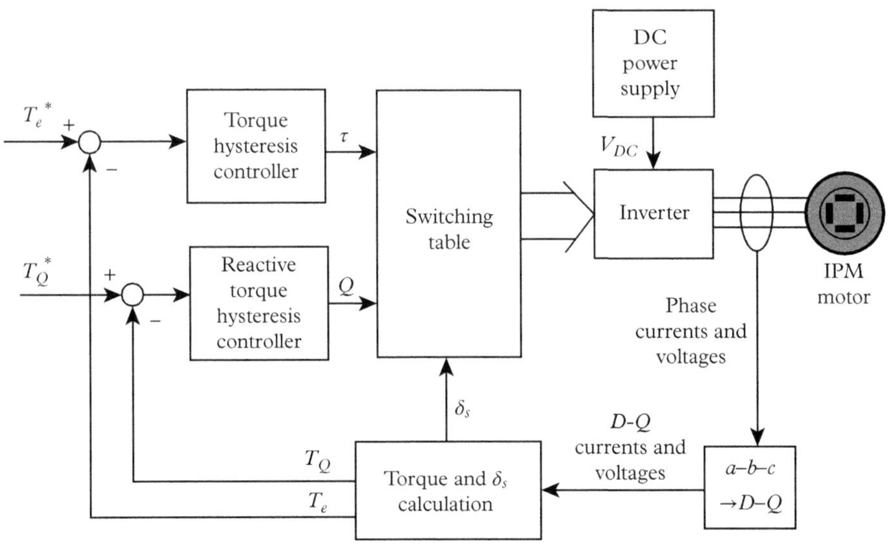

Figure 4.12 *Direct torque control system of PMS motors with reactive torque control.*

flag, τ, as seen in Fig. 4.12. The switching table is similar to that of the basic DTC presented in Section 4.2 (Vas 1998). The table also receives the angle of the stator flux linkage vector, which is calculated by the same procedure as that described in connection with the basic DTC system in Section 4.2.

It must be mentioned that the DTC system with reactive torque control does not need a rotor position signal, in contrast to the DTC system with i_d control. Therefore, the former control system is, in fact, a sensorless system.

4.6 SVM direct torque control

The conventional DTC schemes with hysteresis controllers and switching table, despite attractive features, such as providing fast machine dynamics, implementation in the stationary RF, and lack of RF transformation, sensorless nature, and ease of implementation, suffer from some drawbacks. The sampling period is very short due to the small hysteresis band of flux and torque controllers. The period, for instance, is about 10 times shorter than that for VC. This is due to the need for several comparisons of reference and actual values of the controlled signals, i.e., torque and flux linkage magnitude in the corresponding bands. This, in turn, requires very accurate estimation of the actual signals. The hysteresis controllers also trigger variable frequency inverter switching, usually with a high maximum frequency. This problem is severe if the bands are narrow. In the case that wider bands are selected to ease the problem of a shorter sampling period and high switching frequency, the motor torque, flux linkage, and current pulsations can become high, and restrict the system use in sensitive applications. Also, using switching tables contributes to current and torque pulsations caused by voltage vector switch over. Moreover, the motor experiences current and torque distortions upon the changes of flux linkage regions.

The above-mentioned difficulties can be avoided by replacing the hysteresis controllers with PI controllers, and the switching table with voltage vector calculation and space vector modulation. This can be done by different schemes as presented in this section.

4.6.1 SVM-DTC with closed-loop torque control in stationary RF

It is shown in eqn (4.1.9) that the torque deviation depends on δ and its deviation. This relationship makes it difficult to calculate the load angle deviation in terms of torque deviation. However, it is suggested

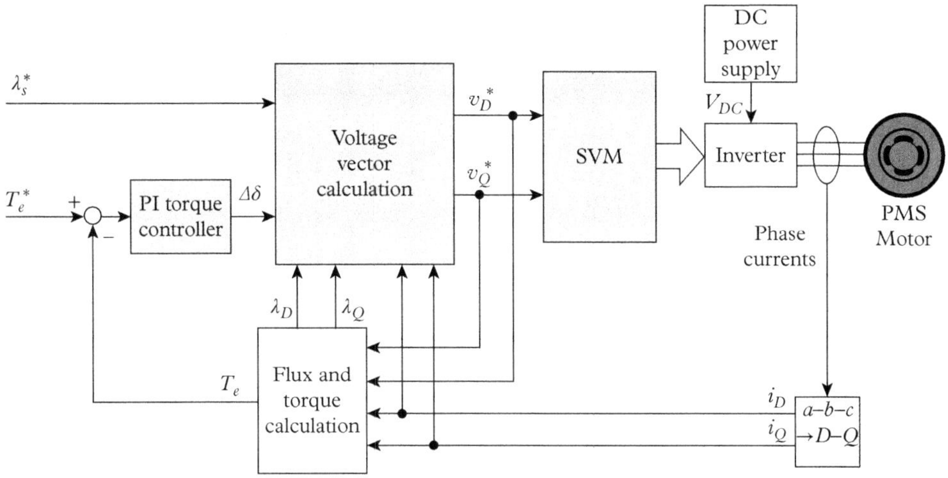

Figure 4.13 *SVM-DTC with closed-loop torque control.*

to get a linear load angle deviation by a linear torque controller, with torque error, ΔT_e, and the load angle deviation, $\Delta\delta$, as its input and output, respectively. In this way, the load angle deviation is obtained without using λ_s and δ as needed by eqn (4.1.9). This is used in an SVM-DTC as seen in Fig. 4.13, where λ_s^* and $\Delta\delta$ are applied to a voltage vector calculation block (Fu and Xu 1997, 1999). The block together with the SVM block substitutes for the switching table of the basic DTC, as seen in Fig. 4.13.

The voltage vector calculation block is shown in Fig. 4.14. In this block, $\Delta\delta$ is regarded as

$$\Delta\delta = \delta_s^* - \delta_s, \tag{4.6.1}$$

where δ_s^* and δ_s are the angles of the reference and the actual flux linkage vectors in the stationary RF. Equation (4.6.1) is written by assuming that the rotor position is unchanged during the load angle deviation, as assumed in Section 4.1. Therefore, having estimated the actual flux linkage angle, δ_s, by λ_Q and λ_D, as in Fig. 4.8, the reference angle, δ_s^*, is found from eqn (4.6.1). Now, λ_s^* and δ_s^* are used to calculate D- and Q-axis components of the reference flux linkage vector in stationary RF as

$$\lambda_D^* = \lambda_s^* \cos\delta_s^*, \tag{4.6.2}$$

$$\lambda_Q^* = \lambda_s^* \sin\delta_s^*. \tag{4.6.3}$$

These flux linkage components are compared with their calculated values to provide flux linkage errors along the D- and Q-axes, $\Delta\lambda_D$

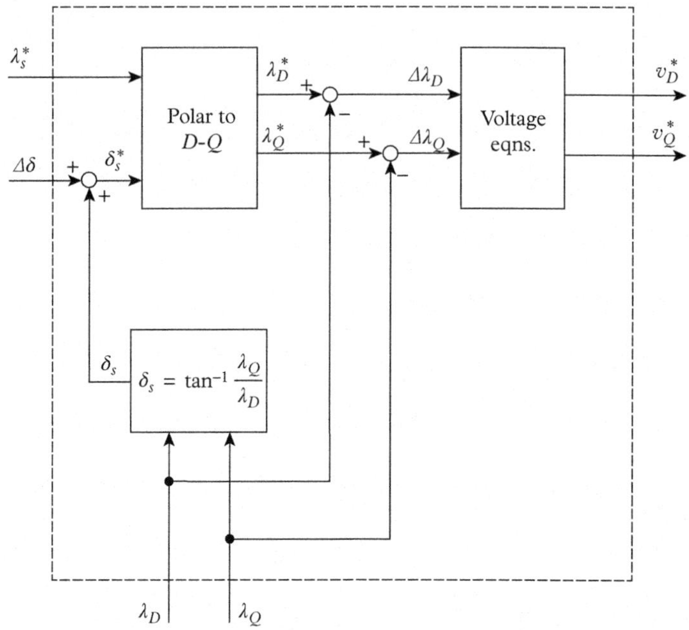

Figure 4.14 *Voltage vector calculation.*

and $\Delta\lambda_Q$, respectively. Finally, the voltage equations in two-axis stationary RF over a sampling time are employed to calculate the *D*- and *Q*-axes components of voltage vector as

$$v_D^* = R_s i_D + \frac{\Delta\lambda_D}{T_s}, \tag{4.6.4}$$

$$v_Q^* = R_s i_Q + \frac{\Delta\lambda_Q}{T_s}, \tag{4.6.5}$$

where T_s is the sampling time.

The flux linkage and torque calculations for use in the torque control loop and in the voltage vector calculations are obtained by eqns (4.2.4)–(4.2.8). These calculations need i_D and i_Q, which are obtained by a current reference transformation from *a–b–c* RF to *D–Q* RF as seen in Fig. 4.13. Also, the outputs of voltage vector calculation block are used as *D*- and *Q*-axis voltages required in the above-mentioned equations.

4.6.2 SVM-DTC with closed-loop torque and flux control in stator flux RF

A two-loop control system based on the DTC method and SVM technique is presented. The PMS machine model in stator flux linkage RF,

x–y, is used in justifying and implementing the system. The model is recalled here from Section 2.6 as

$$v_x = R_s i_x + p\lambda_x, \tag{4.6.6}$$

$$v_y = R_s i_y + \omega_s \lambda_x, \tag{4.6.7}$$

$$T_e = \frac{3}{2} P \lambda_s i_y, \tag{4.6.8}$$

$$\lambda_x = \lambda_s, \lambda_y = 0. \tag{4.6.9}$$

Substituting for i_y and λ_x in eqns (4.6.6) and (4.6.7) from eqns (4.6.8) and (4.6.9) yields

$$v_x = R_s i_x + p\lambda_s, \tag{4.6.10}$$

$$v_y = \frac{2R_s}{3P} \cdot \frac{T_e}{\lambda_s} + \omega_s \lambda_x. \tag{4.6.11}$$

The last two equations suggest that the motor flux linkage magnitude is regulated by v_x and the motor torque is regulated by v_y. As a result, the control system of Fig. 4.15 is presented (Xue *et al.* 1990; Foo and Rahman 2010). The torque and flux linkage errors are sent to the PI controllers to provide the voltage component signal commands along the x- and y-axes, v_x^* and v_y^*. A feed forward voltage signal, $\omega_s \lambda_s$, is added to the output of the torque controller in accordance with eqn (4.6.11). The signal uses the speed of the flux linkage vector, which is obtained by $\omega_s = p\delta_s$, where δ_s is calculated by the flux and torque calculation block. The voltage commands are transferred to v_D^* and v_Q^*.

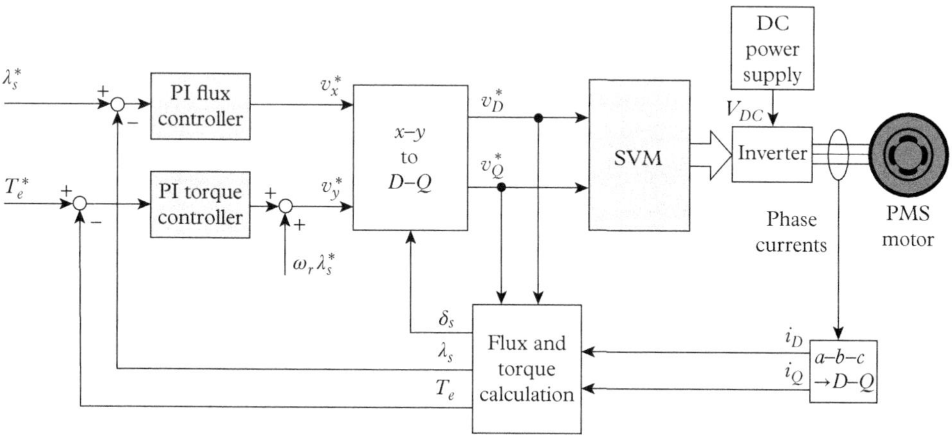

Figure 4.15 *SVM-DTC with closed-loop torque and flux linkage control.*

The voltage components provide switching signals to the inverter by the SVM technique. The flux linkage magnitude, angle, and torque are calculated by the same method as in Figs 4.13 and 4.14.

The control system, although not using hysteresis controllers and switching table, keeps the essence of DTC intact by directly controlling the torque and flux linkage, not via current control loops. The motor performance under the control system enjoys less current and torque ripples compared to the performance under the conventional DTC due to the PI controllers and SVM technique. The system benefits from $p\lambda_s = 0$ in eqn (4.6.10) at constant flux linkage magnitude at steady-state and also approximately at transient state, when the flux linkage command is fixed. This contributes to the smoother motor performance.

4.6.3 Optimal SVM-DTC

In this section, a DTC system that uses SVM as voltage modulation means is elaborated. The voltage command to SVM is calculated such that a torque response faster than that obtained under the conventional DTC is achieved.

4.6.3.1 Torque variations with flux linkage

The relationship between the magnitude and the angle of the stator flux linkage vector and the electromagnetic torque has already been discussed in sub-section 4.3.3. It has been shown that in order to achieve a fast torque response, the flux command needs to be correctly selected. This is discussed further here to provide a basis for an optimal voltage selection in DTC. Figure 4.16 shows the electromagnetic torque of a typical PMS machine with respect to the torque angle for different values of flux linkage magnitude (Ghassemi and Vaez-Zadeh 2005). It can be seen that small values of the flux linkage magnitude result in linear torque variations with respect to δ as presented before. However, the maximum achievable electromagnetic torque is smaller with smaller values of the flux linkage. This is a disadvantage, especially in the case where the machine is heavily loaded and needs bigger electromagnetic torque to start up rapidly. Therefore, it is important to vary the stator flux properly during torque transition to overcome the shortcoming mentioned. Thus, an effective strategy to find an optimum voltage vector for varying both the stator flux linkage magnitude and angle in order to obtain a maximum torque increase in each sampling period is desirable.

The dependency of the machine torque on the flux linkage magnitude and angle, presented in eqn (4.1.8), indicates that a bigger flux linkage command leads to a bigger torque. However, as the flux

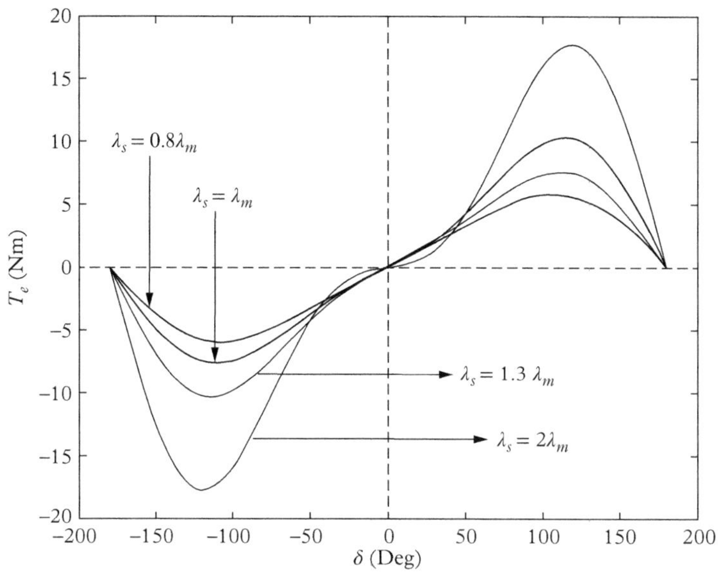

Figure 4.16 *PMS motor developed torque as a function of load angle for different values of stator flux linkage magnitude (Ghassemi and Vaez-Zadeh 2005).*

command increases, the torque response may become non-linear, which is undesirable. Therefore, any control method capable of providing both a high torque value and a linear response is desirable. This goal can be translated in DTC as finding a voltage vector capable of increasing the torque dynamics as fast as possible.

4.6.3.2 Optimal voltage selection

The optimal DTC can find voltage vectors to change both flux linkage angle and magnitude, which results in rapid torque response (Ghassemi and Vaez-Zadeh 2005). The closed form formula for optimal voltage vector will be derived to maximize torque change in each sampling time.

The torque of PMS machines is recalled as

$$T_e = \frac{3}{2} P\, i_q \left[\lambda_m + (L_d - L_q) i_d\right]. \tag{4.6.12}$$

It can differentiated with respect to i_d and i_q as

$$dT_e = \frac{3}{2} P \left\{(L_d - L_q)\, i_q\, d\, i_d + \left[\lambda_m + (L_d - L_q) i_d\right] di_q\right\}. \tag{4.6.13}$$

During each inverter switching period, T_s, the torque differential of eqn (4.6.13) can be approximated by

$$\Delta T_e = A\, \Delta i_d + B\, \Delta i_q, \tag{4.6.14}$$

where

$$A = \frac{3}{2} P(L_d - L_q) i_q, \tag{4.6.15}$$

$$B = \frac{3}{2} P \left[\lambda_m + (L_d - L_q) i_d \right]. \tag{4.6.16}$$

By neglecting the voltage drop across the stator resistance, the following equations are easily obtained from the machine voltage equations in rotor RF along the d- and q-axes:

$$\Delta i_d = \frac{v_d + e_d}{L_d} T_s, \quad e_d = \omega_e \lambda_q, \tag{4.6.17}$$

$$\Delta i_q = \frac{v_q - e_q}{L_q} T_s, \quad e_q = \omega_e \lambda_d. \tag{4.6.18}$$

Substituting eqns (4.6.17) and (4.6.18) into eqn (4.6.14) yields

$$\Delta T_e = f(v_d, v_q) = a v_d + b v_q + c, \tag{4.6.19}$$

where

$$a = \frac{AT_s}{L_d}, b = \frac{BT_s}{L_q}, \text{ and } c = \left(\frac{A e_d}{L_d} - \frac{B e_q}{L_q} \right) T_s. \tag{4.6.20}$$

A very fast torque response requires that ΔT_e has a maximum value at every switching period. This is possible by applying an optimal voltage vector to the machine in each period. This voltage is found in terms of its d and q components where

$$v_q = \sqrt{\left(\frac{2}{3} V_{DC} \right)^2 - v_d^2}. \tag{4.6.21}$$

Substituting eqn (4.6.21) into eqn (4.6.19), the torque change is given as a function of v_d only, i.e., $f(v_d)$. The optimal voltage is then obtained as a solution of the equation

$$\frac{\partial f(v_d)}{\partial v_d} = 0. \tag{4.6.22}$$

The last equation in connection with eqn (4.6.21) yields

$$v_d^* = \frac{2 a V_{DC}}{3 \sqrt{a^2 + b^2}}, \tag{4.6.23}$$

$$v_q^* = \frac{2 b V_{DC}}{3 \sqrt{a^2 + b^2}}, \tag{4.6.24}$$

which represent the d and q components of the optimal voltage vector $\overline{v}_s^*$, which is derived to maximize the torque change in each switching period T_s. Space vector modulation can be used to apply the calculated optimum voltage to PMS motors.

4.6.3.3 *Using SVM to apply optimum voltage to motor*

The optimal voltage vector was obtained in rotor RF previously. However, DTC is basically performed in a stationary RF. Therefore, a transformation from the rotor RF to the stationary RF is needed. This is done with a minimal requirement to the rotor angle. Then, the SVM is used to apply optimum voltage vector to the motor.

By using eqns (4.6.23) and (4.6.24), and having an initial rotor position, θ_r, the optimal voltage vector components in a two-axis stationary RF are found as presented in Chapter 2 as

$$\begin{bmatrix} v_D^* \\ v_Q^* \end{bmatrix} = \begin{bmatrix} \cos\theta_r & -\sin\theta_r \\ \sin\theta_r & \cos\theta_r \end{bmatrix} \begin{bmatrix} v_d^* \\ v_q^* \end{bmatrix}. \tag{4.6.25}$$

During a torque development duration, which typically takes less than 10 milliseconds, for small to medium size motors, rotor position, θ_r, does not change much and can be assumed constant. Therefore, by knowing the initial rotor position, continuous rotor position is not needed. The optimum voltage vector is realized by the SVM method, which produces a minimum harmonic distortion of a two-level inverter. In this method, a superposition is used as

$$\overline{v}_s^* \, T_s = \overline{v}_k \, t_a + \overline{v}_{k+1} \, t_b, \tag{4.6.26}$$

in which $\overline{v}_k$ and $\overline{v}_{k+1}$ are two voltage vectors adjacent to $\overline{v}_s^*$ and are 60° out of phase. The time duration of applying these voltage vectors, t_a and t_b, respectively, is obtained by solving eqn (4.6.26). The results are given by

$$t_a = \frac{3T_s}{2V_{DC}} \left[v_D^* - \frac{v_Q^*}{\sqrt{3}} \right], \tag{4.6.27}$$

$$t_b = \left[\frac{\sqrt{3}\, T_s}{V_{DC}} \right] v_Q^*. \tag{4.6.28}$$

Since the sum of the values calculated for t_a and t_b should be equal to the inverter switching period, T_s, they should be normalized as

$$t_{an} = \frac{t_a \, T_s}{t_a + t_b}, \tag{4.6.29}$$

$$t_{bn} = T_s - t_{an}. \tag{4.6.30}$$

4.6.3.4 *System block diagram*

Block diagram of optimal SVM-DTC is depicted in Fig. 4.17 (Ghassemi and Vaez-Zadeh 2005). Based on the equations derived for optimal voltage vector, eqns (4.6.23) and (4.6.24), instead of using constant flux linkage command as with conventional DTC,

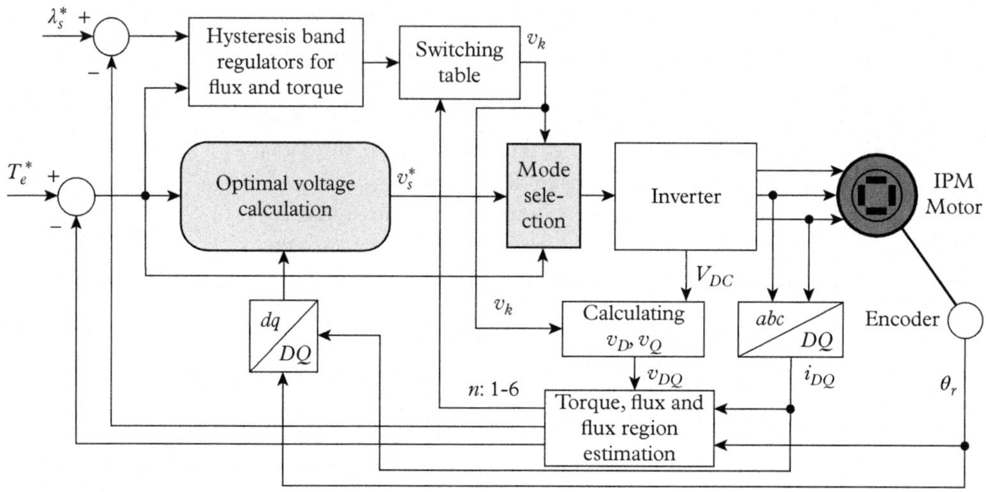

Figure 4.17 *Block diagram of optimal SVM-DTC system.*

a new block denoted as optimal voltage calculation and a software mode selection switch are used in the system. The optimal DTC has two modes of operation. During starting, in which there is a step change in the torque command, the switch is "1" and the optimal voltage in each switching period is calculated by using eqns (4.6.23) and (4.6.24) and is applied to the machine through the inverter. However, when the machine torque reaches its set point value, i.e., it is between the hysteresis bands, the switch turns to "0" and the hysteresis band controllers are used to maintain the flux linkage and torque control as in a conventional DTC system.

4.6.3.5 Performance evaluation

The motor performance under the optimal SVM-DTC strategy is compared with that obtained under the conventional DTC with constant flux linkage command. In both cases, the PMS motor with data from Table 3.1 is tested. The motor is supposed to start from standstill with the set point equal to 200% nominal torque (3.96 Nm). The minimum flux linkage command, which is used for conventional DTC for this motor, is $1.3\,\lambda_m$, as seen in Fig. 4.18 (Ghassemi and Vaez-Zadeh 2005). Choosing a flux linkage command less than $1.3\,\lambda_m$ causes system instability, because the maximum achievable machine torque is less than the torque command.

It can be seen in Fig. 4.18 that a faster torque response is achieved by the optimal SVM-DTC. The conventional DTC tends to increase both stator flux linkage and its angle during starting as seen

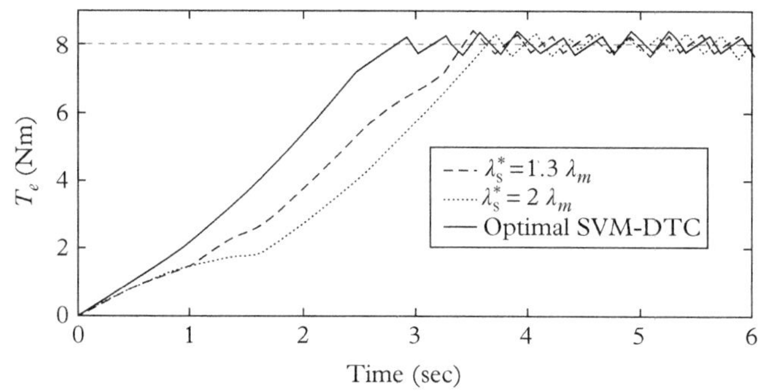

Figure 4.18 *Machine torque during motor start-up under the conventional DTC and the optimal SVM-DTC (Ghassemi and Vaez-Zadeh 2005).*

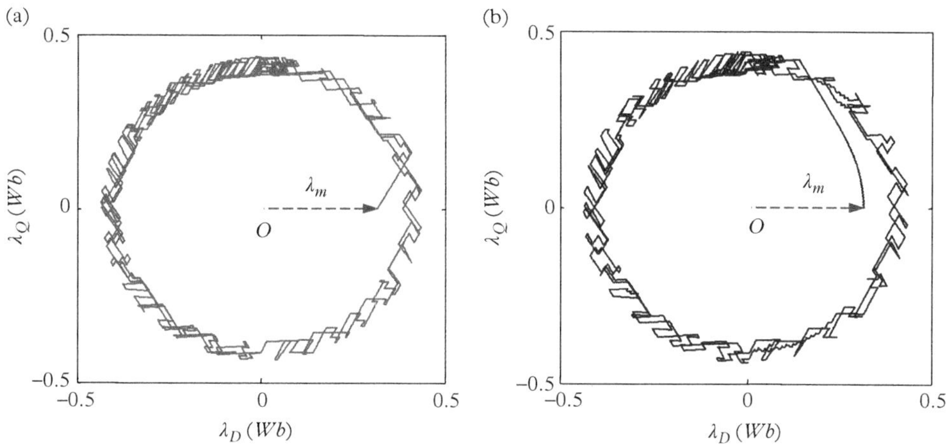

Figure 4.19 *Trajectory of flux linkage vector: (a) under the conventional DTC with* $\lambda_s^* = 1.3\,\lambda_m$, *and (b) under the optimal SVM-DTC (Ghassemi and Vaez-Zadeh 2005).*

in Fig. 4.19(a). However, the optimal DTC tends to mostly increase the flux linkage angle, rather than its magnitude, which can be seen in the flux linkage trajectory in Fig. 4.19(b). The motor draws positive i_d under the conventional DTC, as seen in Fig. 4.20. Since in IPM machines the direct-axis inductance, L_d, is smaller than the quadrature-axis inductance, L_q, a positive i_d produces a negative reluctance torque. In contrast, the motor absorbs a negative i_d, under the optimal SVM-DTC as seen in Fig. 4.20, which produces a positive reluctance torque resulting in a faster torque response.

Figure 4.21 shows the rotor position and speed. It is seen that the rotor position changes by less than 0.5° during the torque development duration, which is almost 3 milliseconds. This supports the assumption mentioned previously that the rotor position is almost constant during the torque development duration and thus the initial position is sufficient for RF transformation.

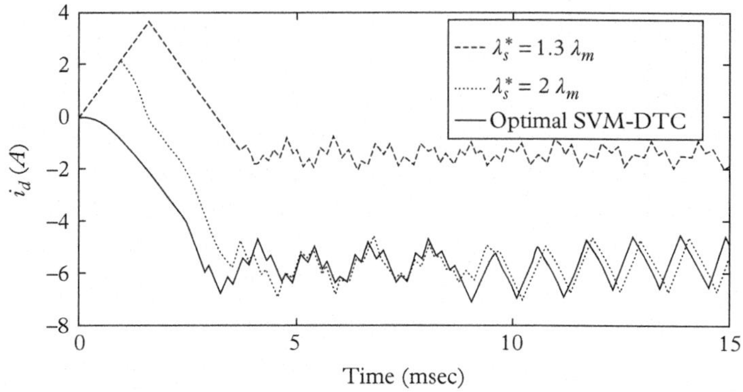

Figure 4.20 *Stator direct-axis current during motor start-up under the conventional DTC and the optimal SVM-DTC (Ghassemi and Vaez-Zadeh 2005).*

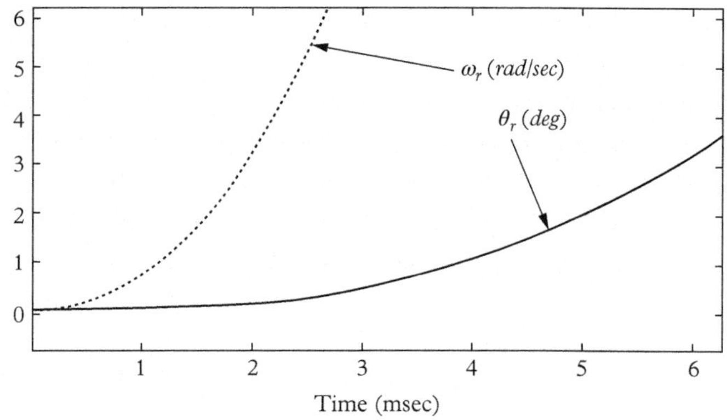

Figure 4.21 *Rotor angle and electrical speed under the optimal SVM-DTC (Ghassemi and Vaez-Zadeh 2005).*

4.7 Loss minimization DTC

An introduction about LMC is presented in Section 3.9, where it is mentioned that there exist two types of LMCs for providing loss minimization of electrical machines, i.e., offline and online. A third method of LMC as model searching LMC may be considered. In the offline LMC, based on the motor parameters and operating conditions, first a loss function is calculated offline. Then the minimization of the function results in an optimum control signal. The signal is applied to the machine as a command. This type of LMC is parameter-dependent. In online LMC, a search procedure is followed in which a control variable changes continuously or step-wise so that the minimum input power to the motor is found. This strategy is insensitive to motor parameters and it minimizes the total loss of the motor and the inverter. However, it is slow because the search procedure is carried out in the steady-state. Thus, in applications experiencing repetitive transient states, it may not result in considerable energy savings. In model searching LMC the motor loss is calculated online from the machine model for different values of a control variable. When the minimum loss is found, the corresponding value of the control variable is used as the command signal and applied to the motor control system. In this section, the offline LMC is presented first, followed by the model searching LMC in connection with DTC.

4.7.1 Offline loss minimization DTC

In offline loss minimization DTC, the optimal operating conditions are found in terms of the stator flux linkage magnitude (Siahbalaee and Vaez-Zadeh 2010). Considering a steady-state equivalent circuit of PMS motors with iron loss as presented in Section 2.8, the electrical loss can be expressed as

$$P_L = \frac{3}{2}\frac{(\omega_r\lambda_d)^2 + (\omega_r\lambda_q)^2}{R_c} + \frac{3}{2}R_s\left(i_d^2 + i_q^2\right), \tag{4.7.1}$$

where the following current equations can be derived from the equivalent circuit:

$$i_d = i_{dT} + i_{dc}, \quad i_q = i_{qT} + i_{qc}, \tag{4.7.2}$$

$$i_{dT} = -\frac{\omega_r\lambda_q}{R_c}, \quad i_{qT} = \frac{\omega_r\lambda_d}{R_c}, \tag{4.7.3}$$

$$i_{dT} = \frac{\lambda_d - \lambda_m}{L_d}, \quad i_{qT} = \frac{\lambda_q}{L_q}. \tag{4.7.4}$$

The substitution of eqns (4.7.3) and (4.7.4) into eqn (4.7.2) yields i_d and i_q; then from eqn (4.7.1) the loss function is obtained in terms of λ_d and λ_q as

$$P_L = a\lambda_d^2 + b\lambda_q^2 + c\lambda_d\lambda_q + d\lambda_d + e\lambda_q + f, \tag{4.7.5}$$

where the coefficients *a–f* are given in Table 4.2. Also, when eqn (4.7.4) is substituted into the torque equation (2.8.12), the *q*-axis component of flux linkage is calculated as

$$\lambda_q = \frac{2}{3P}\cdot\frac{L_dL_qT_e}{L_q\lambda_m + (L_d - L_q)\,\lambda_d}. \tag{4.7.6}$$

Substituting the above flux linkage component into the electrical loss of eqn (4.7.5) yields the electrical loss as a function of λ_d only. The minimum loss condition is then achieved as

$$\frac{\partial P_L}{\partial\lambda_d} = 0. \tag{4.7.7}$$

Table 4.2 *Parameters of loss function and loss minimization condition (Siahbalaee* et al. *2009a).*

$a = \frac{R_s}{L_d^2} + \frac{\omega_r^2}{R_c} + \frac{R_s\omega_r^2}{R_c^2}$	$d = -\frac{2R_s\lambda_m}{L_d^2}$	$k = 2T_eL_dL_q$	$K_0 = ckz^2 + dz^3 - uz - uekz$	$K_3 = 6au^2z + du^3$
$b = \frac{R_s}{L_q^2} + \frac{\omega_r^2}{R_c} + \frac{R_s\omega_r^2}{R_c^2}$	$e = \frac{2R_s\lambda_m\omega_r}{L_dR_c}$	$u = 3P(L_d - L_q)$	$K_1 = 2az^3 + 2ckuz + 3duz^2 - u^2 - u^2ek$	$K_4 = 2au^3$
$c = -\frac{2R_s\omega_r}{L_dR_c} + \frac{2R_s\omega_r}{L_qR_c}$	$f = \frac{R_s\lambda_m^2}{L_d^2}$	$z = 3PL_q\lambda_m$	$K_2 = 6az^2u + cku^2 + 3du^2$	

From the last two equations, one can find the minimum loss condition as

$$k_4\lambda^4 + k_3\lambda_d^3 + k_2\lambda_d^2 + k_1\lambda_d + k_0 = 0, \tag{4.7.8}$$

where the coefficients k_0–k_4 are given in Table 4.2 (Siahbalaee *et al.* 2009a).

The loss minimizing λ_d is found as a solution of eqn (4.7.8). Substituting this solution into eqn (4.7.6) yields the corresponding λ_q. Finally, the loss minimizing flux linkage is obtained by substituting the loss minimizing flux linkage components into the equation

$$\lambda_s = \sqrt{\lambda_d^2 + \lambda_q^2}. \tag{4.7.9}$$

A plot of the electrical loss, together with that of its copper and iron components, are depicted in Fig. 4.22 for a particular operating point for the PMS motor with the specifications given in Table 3.1. It is evident that the electrical loss has a convex shape, with a unique minimum corresponding to the loss minimizing flux linkage (Siahbalaee *et al.* 2009a).

Having the torque command and motor speed in each operating point, the loss minimizing flux linkage can be obtained for each operating point by following the above procedure. However, solving eqn (4.7.8) online is time consuming. Therefore, the calculations are carried out offline and the results stored as a look-up table with the torque command and the speed as the input and the flux linkage as the output. Such a look-up table is depicted in Fig. 4.23. The flux linkage is then used as the flux linkage command in any DTC system. The speed can be estimated instead of being measured. This needs little calculation since the motor speed at steady-state is the same as the rotation speed of stator flux linkage vector. The latter speed can be calculated from the flux linkage angle, which in turn is an essential part of the DTC system.

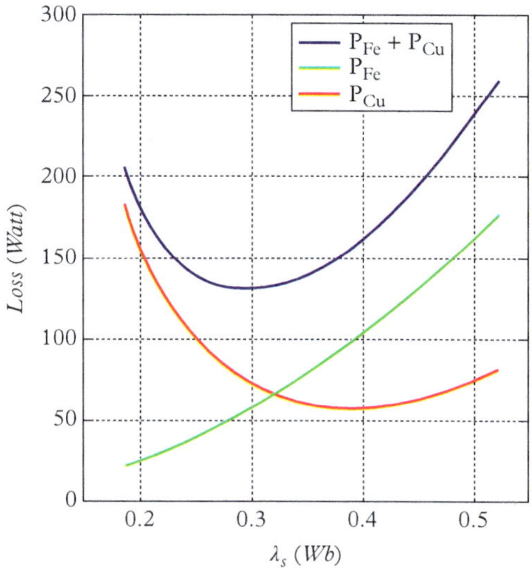

Figure 4.22 *Electrical, copper, and iron losses versus stator flux linkage (Siahbalaee* et al. *2009a).*

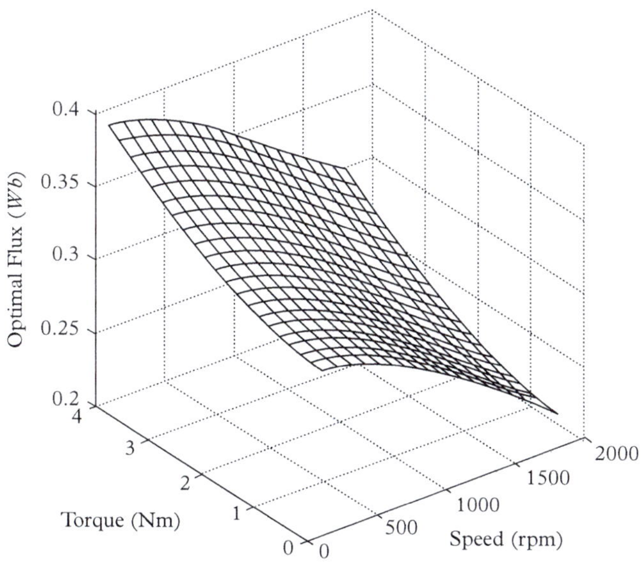

Figure 4.23 *Optimal loss minimizing flux linkage versus torque and speed (Siahbalaee* et al. *2009a).*

The offline LMC is applied to DTC PMS motors using SVM (Siahbalaee *et al.* 2009a). Despite parameter independency of DTC core, the machine parameters, as presented in Table 3.1, are needed to obtain the stator flux linkage for the look-up table. The dynamic and steady-state performances of the PMS motor are studied. A 2-kHz inverter switching frequency is employed.

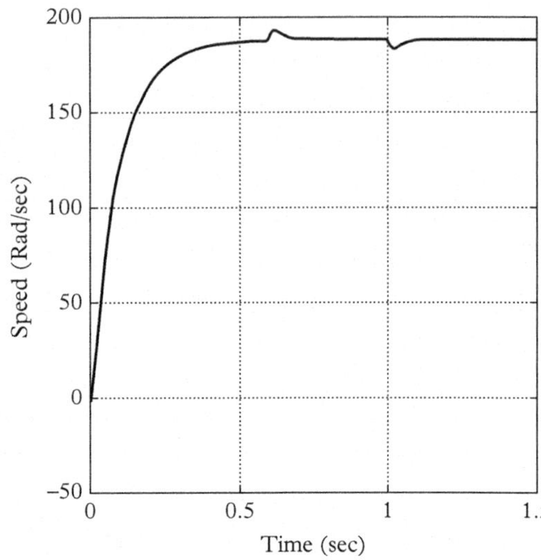

Figure 4.24 *Motor speed under the offline loss minimization DTC (Siahbalaee* et al. *2009a).*

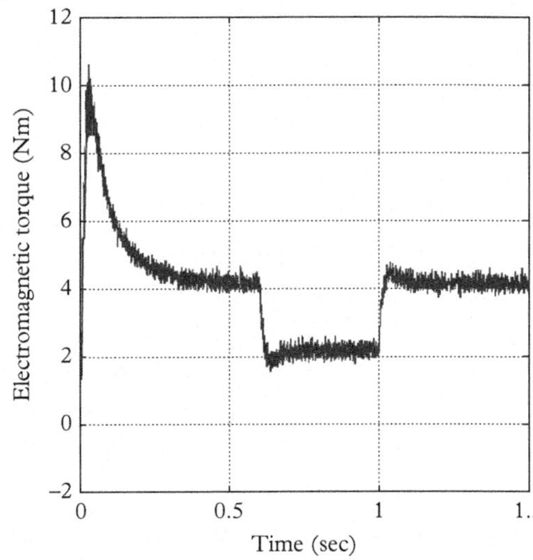

Figure 4.25 *Motor electromagnetic torque under the offline loss minimization DTC (Siahbalaee* et al. *2009a).*

The motor starts under a rated load of 3.96 Nm and reaches a steady-state speed of 1800 rpm in about 0.5 second as seen in Fig. 4.24 (Siahbalaee *et al.* 2009a). After 0.6 second from the start, the load reduces to 2 Nm and remains at this load until the rated load is re-applied after 1 second from the start. The developed torque is shown in Fig. 4.25. The torque in steady-state and under the rated load is 4.12 Nm. The stator flux linkage vector trajectory in the λ_D–λ_Q plane

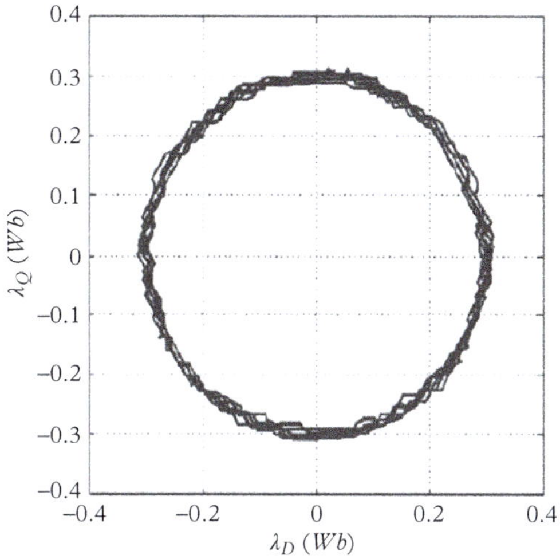

Figure 4.26 *Steady-state stator flux linkage vector trajectory under the offline loss minimization DTC (Siahbalaee* et al. *2009a).*

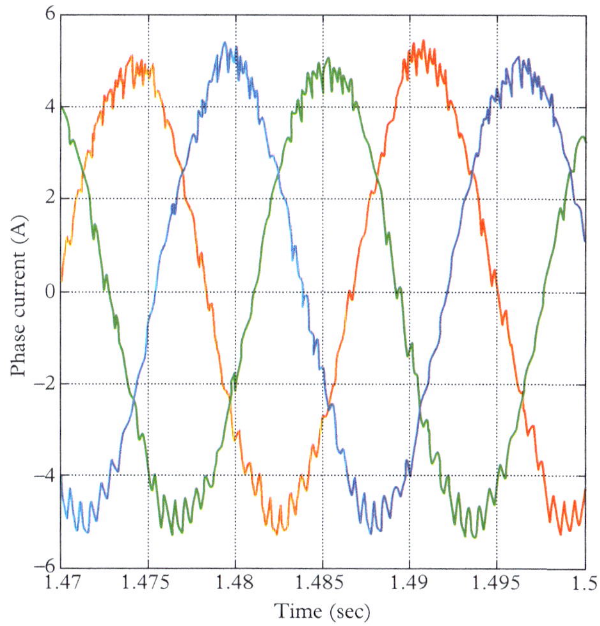

Figure 4.27 *Steady-state phase currents under the offline loss minimization DTC (Siahbalaee* et al. *2009a).*

and the phase currents at steady-state are shown in Figs 4.26 and 4.27, respectively (Siahbalaee *et al.* 2009a). The torque and flux linkage have moderate ripples due to the SVM scheme. Also, the phase currents are sinusoidal with some harmonics. The harmonics not only act against the smooth operation of the motor, but also contribute to the machine losses.

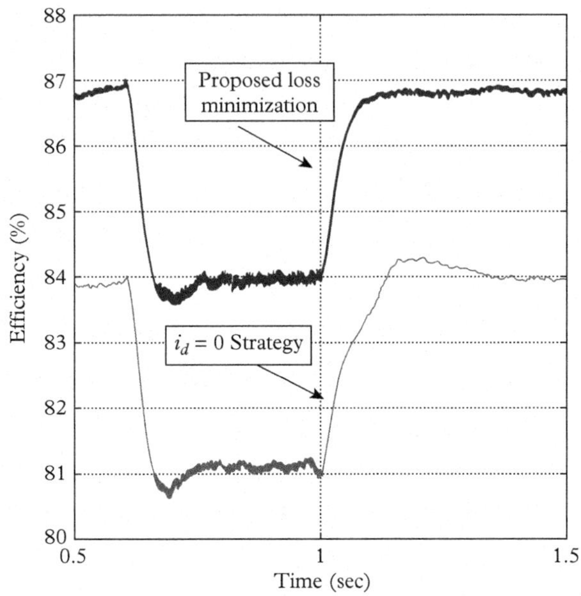

Figure 4.28 *Motor efficiency under the conventional DTC and the off-line loss minimization DTC (Siahbalaee* et al. *2009a).*

The machine efficiency under the LMC is seen in Fig. 4.28 in comparison with the motor efficiency under the condition of $\lambda_d = \lambda_m$ (corresponding to $i_d = 0$). The efficiency improvement under the LMC is evident in this figure.

4.7.2 Model searching loss minimization control under DTC

The model searching loss minimization control can be integrated into any motor control method including DTC. The loss minimization DTC method for PMS motors is presented here, in which the effect of change in the stator flux linkage on electrical loss is foreseen by using the model of machine loss. Subsequently, a selected voltage vector will force the stator flux linkage in a way that it can reduce motor electrical loss. The electrical loss is derived as a function of the stator flux magnitude, the load angle, and the operating point of the machine. The searching method for electrical loss minimization is presented next in connection with DTC (Siahbalaee *et al.* 2009b). The DTC is different from those presented earlier. Finally, the performance of a typical PMS motor under the loss minimization DTC is evaluated.

4.7.2.1 Loss calculation

The nature of the DTC method dictates that the stator flux linkage magnitude is the only control variable for handling electrical loss.

Hence, the electrical loss function is directly derived as a function of stator flux linkage. The electrical loss in eqn (4.7.1) is written as:

$$P_L = \frac{3}{2}\frac{(\omega_r\lambda_s)^2}{R_c} + \frac{3}{2}R_s\left(i_d^2 + i_q^2\right). \tag{4.7.10}$$

Using eqns (4.7.2)–(4.7.4) in connection with

$$\lambda_d = \lambda_s \cos\delta,$$
$$\lambda_q = \lambda_s \sin\delta \tag{4.7.11}$$

gives the loss function in terms of λ_s and δ, which must be estimated.

As the nature of DTC is discrete as mentioned previously, the load angle estimation can be better presented in discrete domain. In DTC, at each sampling time, stator flux linkage is estimated as follows:

$$\lambda_D(k) = \left\{v_D(k-1) - R_s i_D(k)\right\} T_s + \lambda_D(k-1), \tag{4.7.12}$$

$$\lambda_Q(k) = \left\{v_Q(k-1) - R_s i_Q(k)\right\} T_s + \lambda_Q(k-1), \tag{4.7.13}$$

$$\lambda_s(k) = \sqrt{\lambda_D^2(k) + \lambda_Q^2(k)}. \tag{4.7.14}$$

Also, the load angle can be estimated form a discrete version of eqn (4.1.9) as

$$\Delta\delta = \frac{2L_d L_q}{3P\lambda_s} \cdot \frac{T_e(k) - T_e^*(k-1)}{G(k-1)}, \tag{4.7.15}$$

where

$$G(k-1) = \lambda_m L_q \cos\delta(k-1) + (L_d - L_q)\lambda_s \cos 2\delta(k-1). \tag{4.7.16}$$

Figure 4.29 shows a block diagram of the load angle estimator. Thus, the block diagram of the electrical loss calculation will be similar to that in Fig. 4.30.

4.7.2.2 Loss minimization strategy

The main idea behind the loss minimization strategy is to predict the impact of a small change in the stator flux linkage amplitude at each sampling period to decrease electrical loss before any change on the flux linkage command is applied to the system. For this purpose, the effects of the change are predicted from the static model of the machine; then the flux linkage magnitude is commanded to be changed by a proper voltage vector in a way that the electrical loss decreases incrementally. Repetition of this procedure would minimize electrical loss. In other words, the gradient of loss with respect to the flux

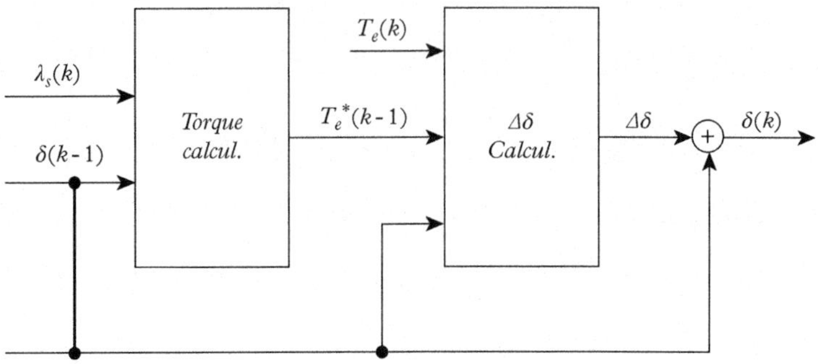

Figure 4.29 *Block diagram of load angle estimator in searching loss minimization DTC.*

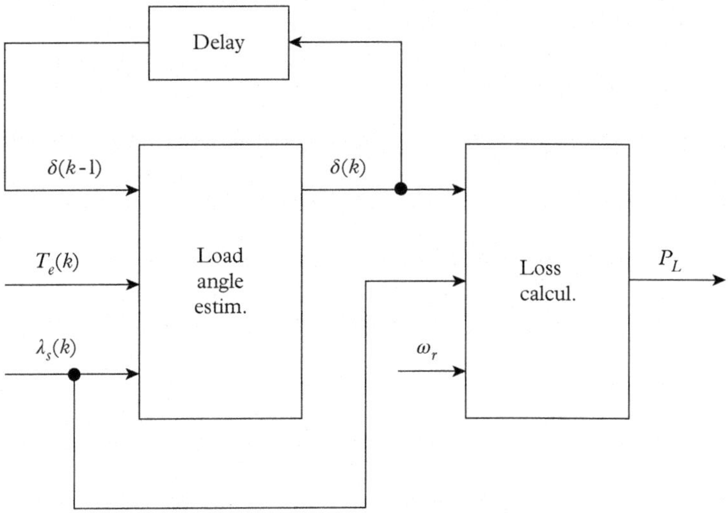

Figure 4.30 *Block diagram of IPM motor electrical loss calculation in searching loss minimization DTC.*

linkage magnitude is predicted and the voltage vector that changes the flux linkage magnitude to cause a reduction in the electrical loss is applied to the inverter. It is assumed that the operating point of the machine is constant during every sampling period. Therefore, electrical loss can be calculated from the steady-state machine model. Accordingly, at each sampling time the previous operating point is replaced by the new one.

This LMC scheme is different from the conventional online loss minimization control scheme presented in Section 3.9 in connection with VC where a step change in the stator flux linkage of a machine is applied in the steady-state repeatedly. Then, the resulting change in the input power of the machine caused by the variation in stator flux linkage is measured in a search for an optimal flux that corresponds to the minimum input power. However, in the model searching loss minimization control scheme, the effect of stator flux linkage variations on electrical loss is foreseen from the static model of the machine, before the flux linkage is actually adjusted. On the other hand, it is different from the offline or model-based loss minimization control scheme where the optimum value of the flux linkage is obtained from a minimum loss function as presented in sub-section 4.7.1.

Since, in DTC, flux linkage and torque controls are performed independently, both the commanded torque and the minimum loss may be achieved by torque control and flux linkage control, respectively. To follow the loss minimization procedure, assume that the stator flux linkage vector is located in the first region in a stationary plane, as shown in Fig. 4.5, where the electromagnetic torque is less than the reference torque. The DTC may apply one of the voltage vectors, $\bar{v}_2$ or $\bar{v}_3$, to the machine, where $\bar{v}_2$ increases and $\bar{v}_3$, decreases the stator flux linkage magnitude, respectively. The basic DTC, in this case, selects $\bar{v}_2$ to provide a larger rotation of the flux linkage vector and thus a larger toque increase. However, the loss minimization strategy does not necessarily select this voltage vector. The desired voltage vector among the two options is selected by considering which of the voltage vector adjusts the flux linkage such that it reduces the machine loss. This is shown in Fig. 4.31 with emphasis on the gradient of loss function with respect to the variation of flux linkage. If the gradient of the loss function with respect to the variation of flux linkage magnitude is positive, v_3 will be selected, and if it is negative, v_2 will be chosen.

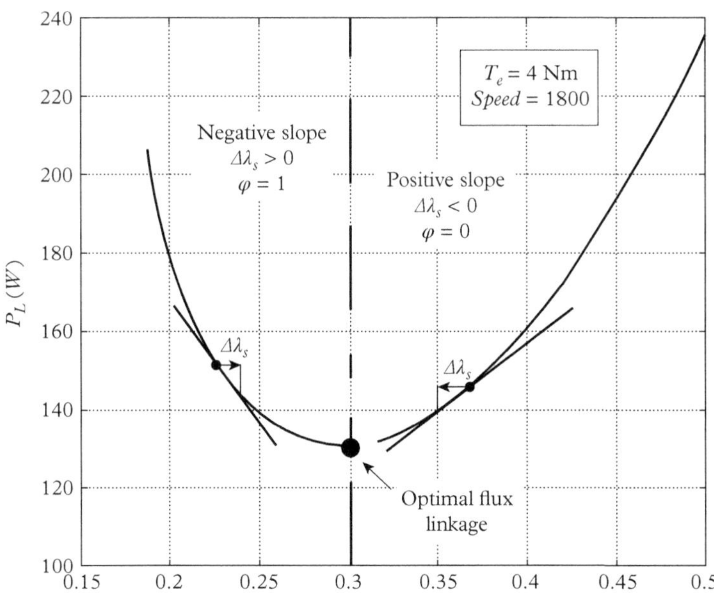

Figure 4.31 *Electrical loss versus flux linkage amplitude (Siahbalaee et al. 2009b).*

Therefore, in each sampling period, the loss minimization procedure is as following:

1. Determine two voltage vectors that can compensate for the torque error according to the conventional DTC principles.
2. Predict electrical loss for the two voltage vectors.
3. Select the voltage vector that causes a reduction in the loss with respect to the present loss, provided that the operating point is constant.
4. Apply the selected voltage to the machine via the inverter for a duration of T_s.

Repeating the above procedure in consecutive sampling periods adjusts the flux linkage toward its optimal value, corresponding to a minimum loss as shown in Fig. 4.31 (Siahbalaee *et al.* 2009b).

4.7.2.3 Control system

The block diagram of the loss minimization DTC is shown in Fig. 4.32. The loss minimization control is implemented in connection with a different DTC scheme. In comparison with the conventional DTC method, two comparators replace the flux linkage

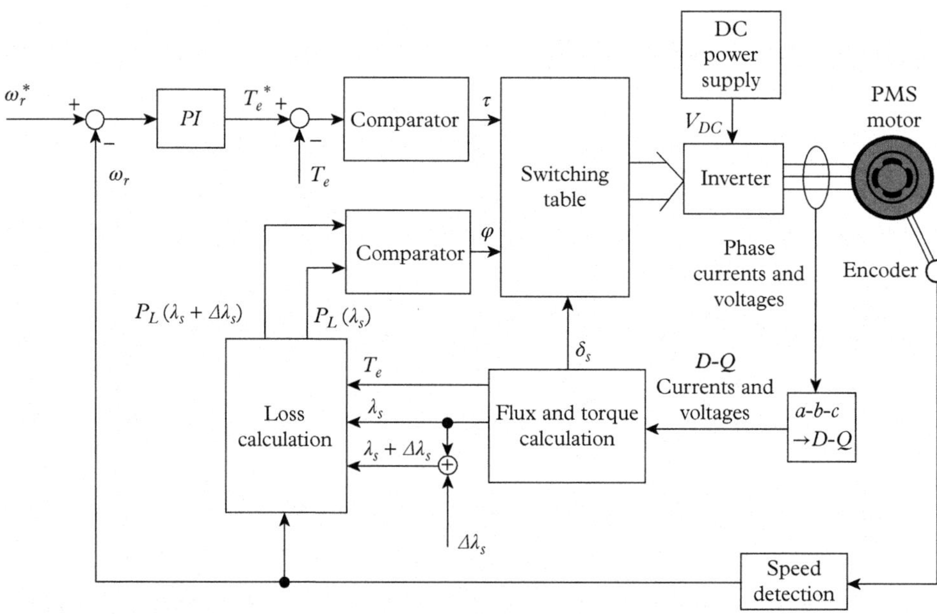

Figure 4.32 *Block diagram of model searching loss minimization DTC system.*

and torque hysteresis controllers, while the switching frequency is held constant (Tang *et al.* 2004).

In order to achieve speed control, the reference electromagnetic torque is applied to the motor through a speed controller. In PMS motors with an estimated amplitude of $\lambda_s(k)$ and its next value, $\lambda_s(k)+\Delta\lambda_s$, first, the machine electrical loss for each flux linkage is calculated. Then flags φ and τ are defined as in Fig. 4.32 by eqns (4.7.17) and (4.7.18):

$$\begin{aligned}&\text{if } P_L(\lambda_s(k)+\Delta\lambda_s)-P_L(\lambda_s(k))<0\rightarrow\ \varphi=1\ \ (\textit{i}\text{ncrement flux linkage}),\\&\text{if } P_L(\lambda_s(k)+\Delta\lambda_s)-P_L(\lambda_s(k))>0\rightarrow\ \varphi=0\ \ (\text{decrement flux linkage}),\end{aligned}\tag{4.7.17}$$

$$\begin{aligned}&\text{if } T_e<T_e^*\longrightarrow\tau=1\ \ (\text{increment torque}),\\&\text{if } T_e>T_e^*\longrightarrow\tau=0\ \ (\text{decrement torque}).\end{aligned}\tag{4.7.18}$$

In contrast to IPM motors, where the gradient of the loss function cannot be calculated directly, in non-salient pole PMS machines, where $L_d = L_q = L_s$, the calculation of the loss function gradient is straightforward and is given by (Siahbalaee *et al.* 2012)

$$\frac{\partial P_L}{\partial\lambda_s}=A\lambda_s+\frac{B}{\sqrt{1-\left(\frac{C}{\lambda_s}\right)^2}},\tag{4.7.19}$$

where the coefficients A, B, and C are dependent on machine parameters and operating conditions. Therefore, the flag, φ, is expressed as follows:

$$\begin{aligned}&\text{if }\frac{\partial P_L}{\partial\lambda_s}<0\longrightarrow\varphi=1\ \ (\text{increment flux linkage}),\\&\text{if }\frac{\partial P_L}{\partial\lambda_s}>0\longrightarrow\varphi=0\ \ (\text{decrement flux linkage}).\end{aligned}\tag{4.7.20}$$

The switching table is the same as that used in the conventional DTC scheme.

4.7.2.4 Sampling time

As mentioned previously, the comparators are chosen instead of flux linkage and torque controllers. Therefore, the inverter switching will be of constant frequency. As in the conventional DTC scheme, by reducing the sampling time, the flux and torque ripples reduce. However, data processing should be done faster. Moreover, the higher the switching frequency, the higher the losses of the inverter switches. On the other hand, any increase in the sampling time will cause the ripples of the flux linkage and torque to be increased. The permitted flux linkage and the torque bands are defined as $\Delta\lambda$ and ΔT, respectively. Also, T_{s1} and T_{s2} are the maximum switching times so that neither the flux linkage nor the torque violates their bands. They are given as

$$T_{s1} = \frac{\Delta\lambda_s^*}{v_s} = \frac{\Delta\lambda_s^*}{\frac{2}{3}V_{DC}}, \tag{4.7.21}$$

$$T_{s2} = \frac{\Delta T_e^* T_e}{T_0}, \tag{4.7.22}$$

where T_0 is the time required to accelerate the motor from standstill to T_e^*. Therefore, the minimum sampling time will be given as

$$T_s = \min\left(T_{s1},\ T_{s2}\right). \tag{4.7.23}$$

4.7.2.5 Motor performance

The performance of a PMS motor with the specifications and parameters presented in Table 3.1, under the present loss minimization DTC, is examined in comparison with a condition of $\lambda_d = \lambda_m$, corresponding to $i_d = 0$ control in VC. The variations of the machine parameters are ignored, and a sampling time of 50 μs is chosen.

It can be shown that under the nominal operating conditions with $T_e = 3.96$ Nm and $N = 1800$ rpm, the magnitude of the stator flux linkage is $\lambda_s = 0.468$ Wb. The electrical loss and efficiency of the machine under $i_d = 0$ and the loss minimization control strategies are depicted in Figs 4.33 and 4.34, respectively. It can be seen that, using the method described earlier, the electrical loss decreases and the efficiency of the machine increases in comparison with those of the $\lambda_d = \lambda_m$ condition. Figure 4.35 shows the flux linkage trajectories for both cases.

Assume that the machine starts with a nominal torque ($T_L = 3.96$ Nm). After $t = 1$ second, when the nominal speed ($N = 1800$ rpm) is reached, the load torque of the machine decreases to $T_L = 1$ Nm. Under nominal torque and speed ($T_L = 3.96$ Nm and $N = 1800$ rpm) and in the steady-state the electromagnetic torque will be equal to $T_e = 4.1108$ Nm. The motor speed, the electromagnetic torque, the minimized electrical losses, and the optimum efficiency of the machine are shown in Figs 4.36–4.39, respectively. It is seen that the loss minimization DTC method makes the loss minimization process fast and smooth.

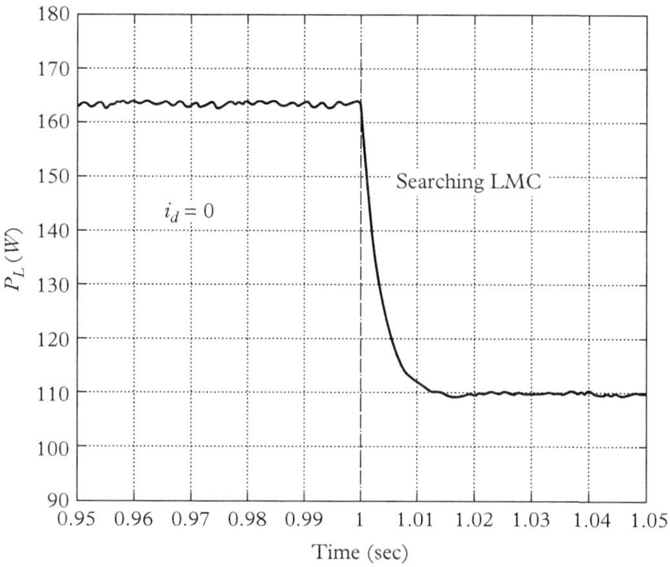

Figure 4.33 *The comparison of the electrical loss under a constant flux linkage corresponding to* $i_d = 0$ *and under the searching loss minimization control (Siahbalaee* et al. *2009b).*

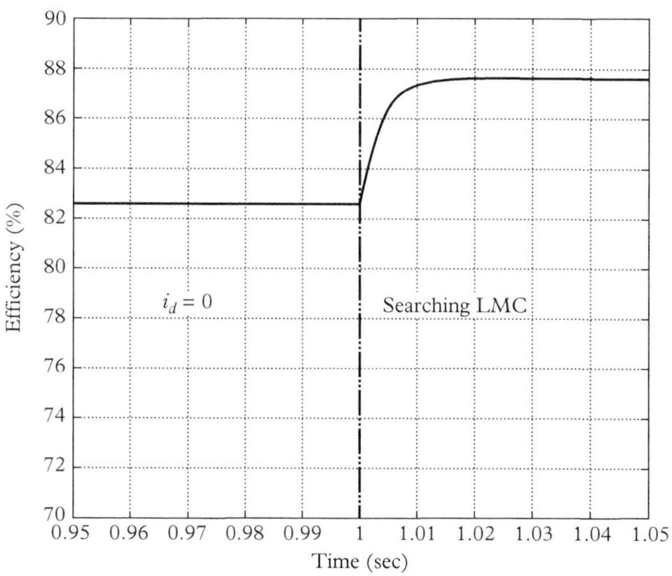

Figure 4.34 *The comparison of the machine efficiency under a constant flux linkage corresponding to* $i_d = 0$ *and under the searching loss minimization control (Siahbalaee* et al. *2009b).*

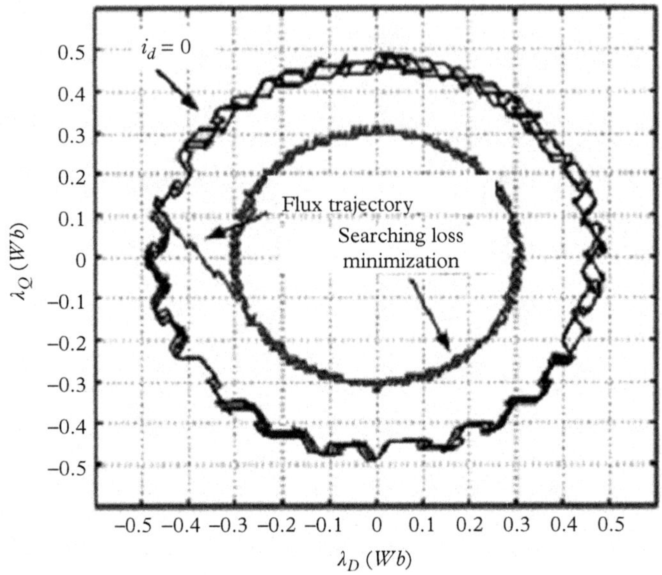

Figure 4.35 *The stator flux linkage trajectories under both control methods (Siahbalaee* et al. *2009b).*

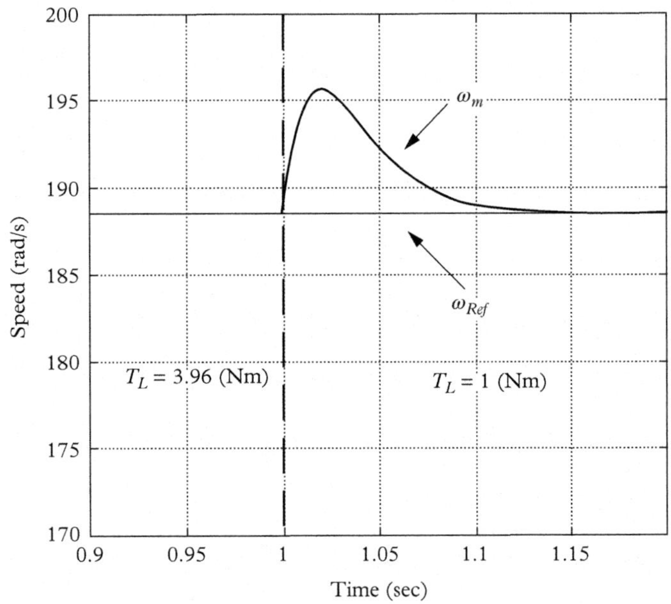

Figure 4.36 *The machine speed (Siahbalaee* et al. *2009b).*

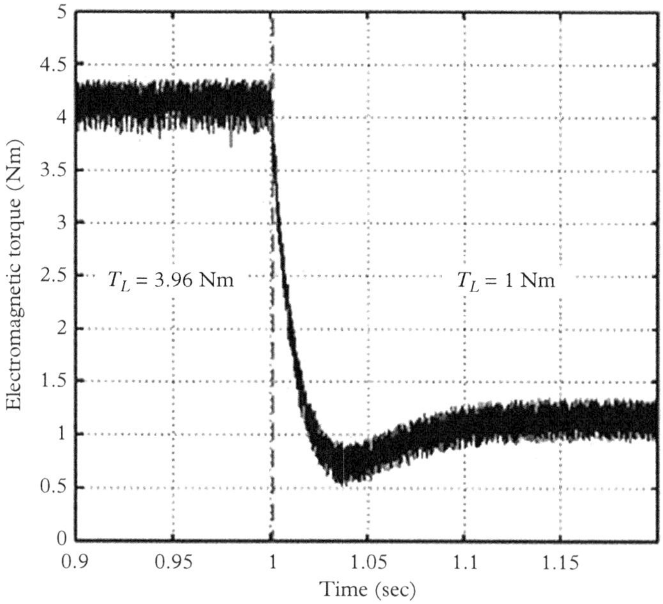

Figure 4.37 *The electromagnetic torque (unfiltered) (Siahbalaee* et al. *2009b).*

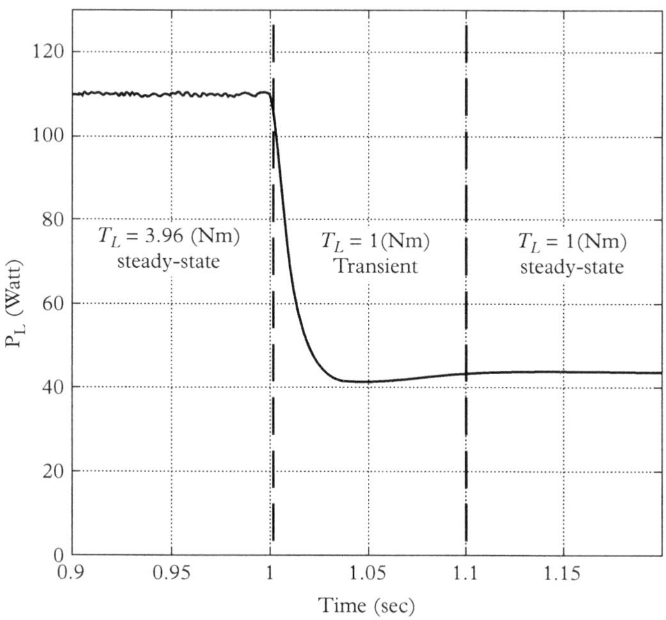

Figure 4.38 *The electrical loss (filtered) (Siahbalaee* et al. *2009b).*

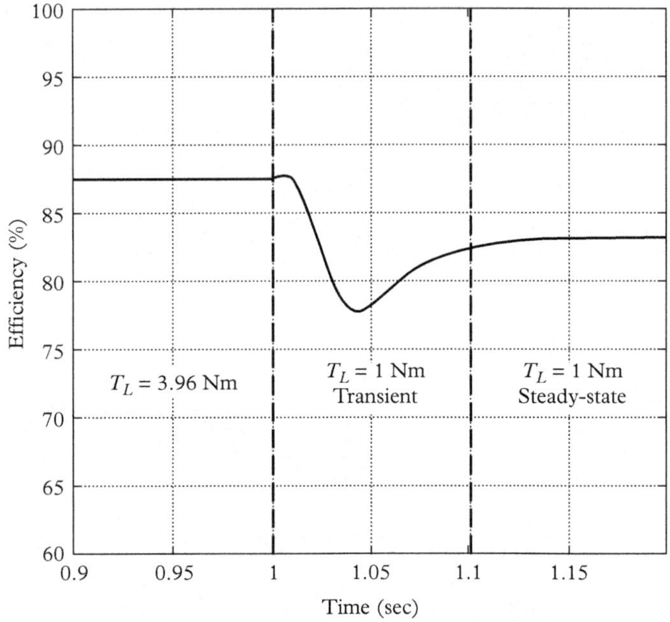

Figure 4.39 *The machine optimum efficiency (filtered) (Siahbalaee* et al. *2009b).*

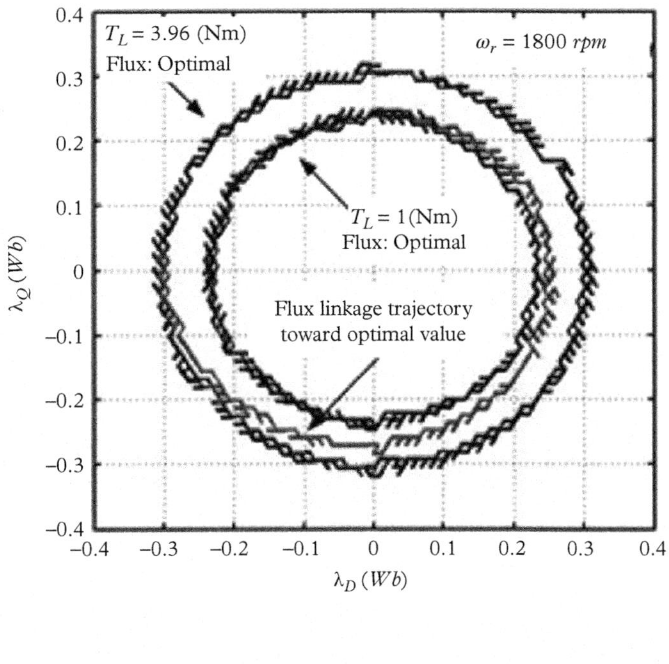

Figure 4.40 *The optimum flux trajectory upon the variation of load torque at t* = 1 *second (Siahbalaee* et al. *2009b).*

Figure 4.40, shows the trajectory of the stator flux linkage upon the variation of the load torque at $t = 1$ second, as it changes from an optimum steady-state value to a new optimum value in about 0.15 seconds.

4.8 Comparison of DTC and VC

It is worth having a deeper insight into basic DTC, particularly in comparison with VC. DTC takes the switching nature of motor supply into account in evolving the theory and practice of the control method. In fact, it considers the fundamental inverter behavior as an indispensable part of the control. Thus, it focuses on willingly controlling the machine at every inverter switching period by determining a specific voltage vector at the beginning of the period to be applied to the motor during the period, whereas neither the scalar control nor VC does so. The VC prefers a continuous supply of energy to the machine and copes with the discontinuity of the power supply anyway. DTC, on the other hand, enjoys discontinuous power supply and controls the machine accordingly. This can be seen by developing eqn (4.1.4) as the equation used for presenting the basis of control in which a torque deviation signal, instead of a torque signal, is investigated. This means that the focus of the control is on adjusting the torque variation for a short period, which is later defined as the inverter switching period. This signal is then related to the torque error to determine a flag. Torque and flux linkage flags are the main means of adapting the control system to a switching nature of the inverter as they take a value of 0 or 1, the same as an inverter switching state, which is either 0 or 1. A VC system basically does a similar choice indirectly, but is not so focused on each switching period.

All routines of DTC are performed in a stationary RF. Therefore, there is no need for a reference reframe transformation. The transformation needs an accurate rotor position as is provided in VC by high resolution optical encoders, and the corresponding encoder board and interface. The implementation of RF transformation also needs a special software routine and takes the processor memory and consumes processing time. This adds to the duration of the sampling period. Instead, a DTC system needs very approximate information regarding the stator flux linkage angle. It is enough to know the flux linkage vector is located in which region of 60° wide. In conclusion, DTC as a sensorless system and with no serous demands on flux linkage position can save implementation and processing time and money.

A DTC system uses no PI or other types of conventional controllers, but hysteresis controllers. A conventional controller usually goes through periods of transients that must be realized by software, thus taking time, while a hysteresis controller takes discrete values only with no dynamics. Therefore, it is fast and accurate.

The switching table in DTC can be regarded as a substitute for PWM block. However, the switching table needs no calculation to determine a voltage vector command for the inverter. The switching table provides predetermined voltage vectors according to the flags and flux linkage position, whereas a software PWM needs heavy online computation to determine switching signals to the inverter. A hardware PWM block may be included as an on-chip peripheral of a processors or as a separate chip, which in the latter case adds to the motor control system complexity. Nowadays, the burden of a software or a hardware PWM is not as serious as it has been previously. In fact, many processor chips include single or multiple on-chip hardware PWM blocks. Nevertheless, elimination of PWM is still a definite simplifying merited for DTC.

Torque and flux linkage controls are two objectives of most motor control systems. These controls are usually carried out via current control loops. Current signals are used in DTC to estimate flux linkage and torque. However, there is no current control loop in DTC. Thus, torque and flux linkage are controlled directly as the method's name states. This is an advantage, on one hand, while it may be regarded as a shortfall at the same time, as there is no direct control over the motor current.

It is also evident that the torque and flux linkage magnitudes must be estimated online in every sampling period. The sampling period in DTC is usually much shorter than that in VC. This is because the machine flux linkage error and torque error must be compared with the flux linkage and torque bands, respectively, several times, in a switching period. Therefore, the sampling period is much less than the switching period. With the variable switching period, there would be a strong constraint on the computation time available for the estimation of the machine flux linkage and torque. It can be said that DTC needs accurate and fast information of flux linkage magnitude (and torque), while VC needs fast and accurate information of stator flux linkage angle or rotor angle.

Estimation of flux linkage is carried out by integrating the induced voltages. It is seen that when the applied voltages to the machine are small at low speeds, the voltage drops on winding resistances are comparable to the applied voltages. Therefore, the integrated voltages are small. As a result, the integration error, relative to the applied voltage, is high. This may cause inaccuracy in control performance at low speeds.

Table 4.3 *Comparison of basic DTC and VC.*

Properties	*DTC*	*VC*
Reference frame	*Stationary, D–Q*	*Rotating*
Controlled variables	*Torque and flux linkage*	*Currents*
Sensed signals	*Current and voltage*	*Currents and rotor position*
Controller type	*Hysteresis*	*PI or hysteresis*
Parameter sensitivity	*To stator resistance*	*To all machine parameters*
Sampling frequency	*Very high*	*High*
Switching frequency	*Variable*	*Fixed*

DTC of PMS motors is less parameter dependent than the VC of the motors. In fact, it depends on one motor parameter, i.e., R_s. This is a major advantage, especially in the control of modern PMS motors, where high energy PM materials are used. A strong magnet flux increases the possibility of iron saturation and leakage fluxes. As a result, the motor inductances, L_d and L_q, are dependent on motor current and thus on motor load and speed. A DTC system with no need for information of L_d and L_q for control routines provides a more performance robustness against parameter variation.

Finally, the problem of increased ripples under DTC must be mentioned. The switching nature of hysteresis controllers in DTC is the main cause of ripples in many motor variables including torque. This is particularly problematic at low speeds when the ripple frequency is low due to a low inverter switching frequency. The ripples are too high to be damped out by the filtering nature of electric machines. Torque ripples cause audible noise, reduced motor efficiency, and loss of control accuracy. The latter, in particular, is an obstacle in high precision applications. In low power machines with small inertia constant, the ripples may be transferred to the machine speed, and disturb speed and position controls. Many methods of ripple reduction are presented in the literature, including SVM-DTC as described in this chapter. A concise comparison of basic DTC and VC is presented in Table 4.3.

4.9 Summary

The torque of an AC machine can be regarded as the outer product of rotor and stator flux linkages. In a PMS machine, the rotor flux linkage depends on the magnet poles and is rather fixed for a specific machine. Therefore, if the magnitude of the stator flux linkage is kept constant during the motor transient state, the torque response depends, largely, on the angle between the two flux linkages. Based on this reasoning, the fundamental principles of DTC as a stator flux linkage VC is presented in this chapter. The basic DTC system is then described by elaborating the hysteresis controllers for flux linkage and torque control. Also, the logic of selecting inverter voltage vectors and switching table to implement this logic is described. Finally, the flux linkage and torque estimations are given.

The operating limits of PMS machines under DTC are presented in terms of current limit, voltage limit, and flux linkage limit. Also, flux linkage control, including maximum torque per ampere, unity power factor, and flux weakening at high speed, is derived. Then, alternative DTC schemes, in which another machine variable instead of stator flux linkage is controlled, including d-axis current and reactive torque, are given. Also, different SVM-DTC schemes which provide smooth machine operation are presented. These include closed-loop torque control and closed-loop flux linkage and torque control schemes, in addition to the optimal SVM-DTC scheme. In line with increasing energy saving tendency in industrial applications, a major emphasis is placed on the loss minimization DTC. Two loss minimization methods, i.e., offline method and model searching method, are presented in detail, along with the machine performances under the control methods. Finally, a comprehensive comparison is made between the basic DTC and VC, emphasizing the pros and cons of DTC with respect to VC.

PROBLEMS

P.4.1. Derive the time derivative of the IPM motor torque equation by assuming that both the magnitude of stator flux linkage vector and load angle are time varying. Then, discuss the possibilities of torque control systems.

P.4.2 The principles of DTC state that the stator voltage vector is always in line with the flux deviation vector, $\Delta\overline{\lambda}_s$ during every switching period. Ignore the approximation that is involved in the principle due to neglecting the voltage drop on the winding resistance. Then consider its implication in selecting voltage vectors in DTC.

P.4.3 How optimal is a voltage vector, which is selected based on the golden rule of basic DTC in providing a fast torque transient during a switching interval?

P.4.4. DTC is based on rapid rotation of the stator flux linkage vector by applying a voltage vector to the machine as normal as possible to the flux linkage vector. Observe the fulfilment of this rule during the machine start-up, when the stator flux linkage magnitude is still much less than its reference value and is located off the hysteresis band.

P.4.5. Consider the trade-off between the burden of accurate calculations of flux linkage magnitude and load angle, which are required, respectively, in DTC and VC. Compare the severity of the requirements in DTC with respect to VC.

P.4.6. How useful is it to use invertor zero voltages in DTC together with $\bar{v}_1$–$\bar{v}_6$? Search the literature to find out how it can be done in DTC of PMS motors.

P.4.7. Assume a DTC with only a single switching in each flux linkage region. Draw the flux linkage vector trajectory for a whole rotation and the appropriate voltage vector in each region (like Fig. 4.5).

P. 4.8. How can the system of the above DTC be implemented?

P. 4.9. Dispose the flux linkage regions of the conventional DTC by 30°. Now, determine the corresponding switching table.

P.4.10. Consider a three-level hysteresis torque controller, instead of a two-level controller. Determine a switching table based on this controller. You may use inverter zero voltages, $\bar{v}_0$ *and* $\bar{v}_7$, as well as the non-zero voltage, $\bar{v}_1 - \bar{v}_6$. Discuss the pros and cons of the table.

P.4.11. The flux linkage calculation, as presented in Section 4.2, may not provide accurate results at low speeds, when the machine phase voltage magnitudes are small. Investigate this issue and its remedies by searching the literature. Write a technical report on this problem.

P.4.12. Demonstrate the current and voltage limits for an IPM motor in a $T_e - \lambda_s$ plane and also in a $\lambda_s - \delta$ plane.

P.4.13. Draw the MTPA trajectory on the planes of Problem 4.12. and discuss the performance constraints of the motor under the MPTA condition.

P.4.14. Calculate eqn (4.4.24) and depict it in the whole speed range.

P.4.15. Compare flux linkage trajectories under the optimal SVM-DTC and the basic DTC at transient state by plotting them on appropriate planes as depicted in sub-section 4.6.3.

BIBLIOGRAPHY

Choi, C.H., Seok, J.K., and Lorenz, R.D. (2013). Wide-speed direct torque and flux control for interior PM synchronous motors operating at voltage and current limits. *IEEE Trans. Ind. Appl.* 49(1), 109–117.

Depenbrock, M. (1986). Direct self-control (DSC) of inverter-fed induction machine. *IEEE Trans. Power Electron.* 3(4), 420–429.

Faiz, J. and Mohseni-Zonoozi, S.H. (2003). A novel technique for estimation and control of stator flux of a salient-pole PMSM in DTC method based on MTPF. *IEEE Trans. Ind. Electron.* 50(2), 262–271.

Fernandez-Bernal, F., Garcia-Cerrada, A., and Faure, R. (2001). Determination of parameters in interior permanent-magnet synchronous motors with iron losses without torque measurement. *IEEE Trans. Ind. Appl.* 37(5), 1265–1272.

Foo, G. and Rahman, M. (2010). Sensorless direct torque and flux-controlled IPM synchronous motor drive at very low speed without signal injection. *IEEE Trans. Ind. Electron.* 57(1), 395–403.

Foo, G., Sayeef, S., and Rahman, M. (2010). Low-speed and standstill operation of a sensorless direct torque and flux controlled IPM synchronous motor drive. *IEEE Trans. Energy Convers.* 25(1), 25–33.

Foo, G.H.B. and Zhang, X. (2016). Constant switching frequency based direct torque control of interior permanent magnet synchronous motors with reduced ripples and fast torque dynamics. *IEEE Trans. Power Electron.* 31(9), 6485–6493.

Fu, M. and Xu, L. (1997). A novel sensorless control technique for permanent magnet synchronous motor (PMSM) using digital signal processor. *IEEE National Aerospace and Electronics Conference, NAECON*, pp. 403–408. IEEE, Piscataway, NJ.

Fu, M. and Xu, L. (1999). A sensorless direct torque control technique for permanent magnet synchronous motors. *IEEE 34th IAS Annual Meeting*, pp. 159–164. IEEE, Piscataway, NJ.

Ghassemi, H. and Vaez-Zadeh, S. (2005). A very fast direct torque control for interior permanent magnet synchronous motors start up. *Energy Convers. Manag.* 46(5), 715–726.

Gulez, K., Adam, A.A., and Pastaci, H. (2007). A novel direct torque control algorithm for IPMSM with minimum harmonics and torque ripples. *IEEE/ASME Trans. Mechatron.* 12(2), 223–227.

Habibi, J. and Vaez-Zadeh, S. (2005). Efficiency-optimizing direct torque control of permanent magnet synchronous machines. *IEEE*

36th Power Electronics Specialists Conference, pp. 759–764. IEEE, Piscataway.

Haque, M.E., Zhong, L., and Rahman, M.F. (2003). A sensorless initial rotor position estimation scheme for a direct torque controlled interior permanent magnet synchronous motor drive. *IEEE Trans. Power Electron.* 18(6), 1376–1383.

Hoang, K.D., Zhu, Z.Q., and Foster, M. (2012). Online optimized stator flux reference approximation for maximum torque per ampere operation of interior permanent magnet machine drive under direct torque control. In: *Proceedings of the IET International Conference on Power Electronics, Machine Drives (PEMD)*, pp. 27–29, IET, UK.

Inoue, T., Inoue, Y., Morimoto, S., and Sanada, M. (2015). Mathematical model for MTPA control of permanent-magnet synchronous motor in stator flux linkage synchronous frame. *IEEE Trans. Ind. Appl.* 51(5), 3620–3628.

Inoue, T., Inoue, Y., Morimoto, S., and Sanada, M. (2016). Maximum torque per ampere control of a direct torque-controlled PMSM in a stator flux linkage synchronous frame. *IEEE Trans. Ind. Appl.* 52(3), 2360–2367.

Inoue, Y., Morimoto, S., and Sanada, M. (2008). Examination and linearization of torque control system for direct torque controlled IPMSM. In: *IEEE Industry Applications Society Annual Meeting*, pp. 1–7. IEEE, Piscataway, NJ.

Inoue, Y., Morimoto, S., and Sanada, M. (2012). Control method suitable for direct-torque-control-based motor drive system satisfying voltage and current limitations. *IEEE Trans. Ind. Appl.* 48(3), 970–976.

Jiaqun, X., Minggao, O., and Renyuan, T. (2006). Study on direct torque control of permanent magnet synchronous motor in electric vehicle drive. In: *9th IEEE International Workshop on Advanced Motion Control*, pp. 774–777. IEEE, Piscataway, NJ.

Koç, M., Wang, J., and Sun, T. (2017). An inverter nonlinearity-independent flux observer for direct torque-controlled high-performance interior permanent magnet brushless AC drives. *IEEE Trans. Power Electron.* 32(1), 490–502.

Llor, A., Retif, J., Lin-Shi, X., and Arnalte, S. (2003). Direct stator flux linkage control technique for a permanent magnet synchronous machine. In: *IEEE 34th Annual Power Electronics Specialists Conference*, pp. 246–250. IEEE, Piscataway, NJ.

Luukko, J. and Pyrhonen, J. (1998). Selection of the flux linkage reference in a direct torque controlled permanent magnet synchronous motor drive. In: *5th International Workshop on Advanced Motion Control*, pp. 198–203. IEEE, Piscataway, NJ.

Niu, F., Wang, B., Babel, A.S., Li, K., and Strangas, E.G. (2016). Comparative evaluation of direct torque control strategies for permanent magnet synchronous machines. *IEEE Trans. Power Electron.* 31(2), 1408–1424.

Paicu, M.C., Boldea, I., Andreescu, G.D., and Blaabjerg, F. (2009). Very low speed performance of active flux based sensorless control: interior permanent magnet synchronous motor vector control versus direct torque and flux control. *IET Electric Power Appl.* 3(6), 551–561.

Pellegrino, G., Armando, E., and Guglielmi, P. (2012). Direct-flux vector control of IPM motor drives in the maximum torque per voltage speed range. *IEEE Trans. Ind. Electron.* 59(10), 3780–3788.

Rahman, M.F. and Zhong, L. (1999). Voltage switching tables for DTC controlled interior permanent magnet motor. In: *25th Annual Conference of the IEEE Industrial Electronics Society*, pp. 1445–1451. IEEE, Piscataway, NJ.

Rahman, M.F., Zhong, L., and Lim, K.W. (1998). A direct torque-controlled interior permanent magnet synchronous motor drive incorporating field weakening. *IEEE Trans. Ind. Appl.* 34(6), 1246–1253.

Rahman, M.F., Zhong, L., and Lim, K.W. (1999). A direct torque control for permanent magnet synchronous motor drives. *IEEE Trans. Energy Convers.* 14(3), 637–642.

Rahman, M.F., Zhong, L., Haque, M.E., and Rahman, M. (2003). A direct torque-controlled interior permanent-magnet synchronous motor drive without a speed sensor. *IEEE Trans. Energy Convers.* 18(1), 17–22.

Rahman, M.F., Haque, M.E., Tang, L., and Zhong, L. (2004). Problems associated with the direct torque control of an interior permanent-magnet synchronous motor drive and their remedies. *IEEE Trans. Ind. Electron.* 51(4), 799–809.

Ren, Y., Zhu, Z.Q., and Liu, J. (2014). Direct torque control of permanent-magnet synchronous machine drives with a simple duty ratio regulator. *IEEE Trans. Ind. Electron.* 61(10), 5249–5258.

Sayeef, S., Foo, G., and Rahman, M.F. (2010). Rotor position and speed estimation of a variable structure direct-torque-controlled IPM synchronous motor drive at very low speeds including standstill. *IEEE Trans. Ind. Electron.* 57(11), 3715–3723.

Shinohara, A., Inoue, Y., Morimoto, S., and Sanada, M. (2014). Correction of reference flux for MTPA control in direct torque controlled interior permanent magnet synchronous motor drives. In: *International Power Electronics Conference*. Hiroshima, Japan, ECCE ASIA, pp. 324–329. IEEE.

Shinohara, A., Inoue, Y., Morimoto, S., and Sanada, M. (2017). Direct Calculation Method of Reference Flux Linkage for Maximum Torque per Ampere Control in DTC-Based IPMSM Drives. *IEEE Trans. Power Electron.* 32(3), 2114–2122.

Siahbalaee, J., Vaez-Zadeh, S., and Tahami, F. (2009a). A new loss minimization approach with flux and torque ripples reduction of direct torque controlled permanent magnet synchronous motors. In: *13th European Conference on Power Electronics and Applications*, pp. 1–8. IEEE, Piscataway, NJ.

Siahbalaee, J., Vaez-Zadeh, S., and Tahami, F. (2009b). A loss minimization control strategy for direct torque controlled interior permanent magnet synchronous motors. *J. Power Electron.* 9(6), 940–948.

Siahbalaee, J. and Vaez-Zadeh, S. (2010). Model-based loss minimization of direct torque controlled permanent magnet synchronous motors. In: *1st Power Electronic & Drive Systems & Technologies Conference*, pp. 273–278. IEEE, Piscataway, NJ.

Siahbalaee, J., Vaez-Zadeh, S., and Tahami, F. (2012). A predictive loss minimisation direct torque control of permanent magnet synchronous motors. *Aust. J. Electric. Electron. Eng.* 9(1), 89–98.

Sun, D. and He, Y-k. (2005). Space vector modulated based constant switching frequency direct torque control for permanent magnet synchronous motor. *Proc. Chin. Soc. Electric. Eng.* 25(12), 112.

Swierczynski, D. and Kazmierkowski, M.P. (2002). Direct torque control of permanent magnet synchronous motor (PMSM) using space vector modulation (DTC-SVM)-simulation and experimental results. In: *IEEE 28th Annual Conference of the Industrial Electronics Society*, pp. 751–755. IEEE, Piscataway, NJ.

Takahashi, I. and Noguchi, T. (1986). A new quick-response and high-efficiency control strategy of an induction motor. *IEEE Trans. Ind. Appl.* 22(5), 820–827.

Tang, L., Zhong, L., Rahman, M.F., and Hu, Y. (2003). A novel direct torque control for interior permanent-magnet synchronous machine drive with low ripple in torque and flux-a speed-sensorless approach. *IEEE Trans. Ind. Appl.* 39(6), 1748–1756.

Tang, L., Zhong, L., Rahman, M.F., and Hu, Y. (2004). A novel direct torque controlled interior permanent magnet synchronous machine drive with low ripple in flux and torque and fixed switching frequency. *IEEE Trans. Power Electron.* 19(2), 346–354.

Vaez, S., John, V., and Rahman, M. (1999). An on-line loss minimization controller for interior permanent magnet motor drives. *IEEE Trans. Energy Convers.* 14(4), 1435–1440.

Vaez-Zadeh, S. and Siahbalaee, J. (2010). A loss minimization strategy for PMS motors under direct torque control. In: *IEEE PES General Meeting*, pp. 1–4. IEEE, Piscataway, NJ.

Vas, P. (1998). *Sensorless Vector and Direct Torque Control.* Oxford University Press, Oxford.

Xu, Z., and Rahman, M.F. (2007). Direct torque and flux regulation of an IPM synchronous motor drive using variable structure control approach. *IEEE Trans. Power Electron.* 22(6), 2487–2498.

Xue, Y., Xu, X., Habetler, T. G., and Divan, D. M. (1990). A low cost stator flux oriented voltage source variable speed drive. In: *Conference Record of the 1990 IEEE Industry Applications Society Annual Meeting*, pp. 410–415. IEEE, Piscataway, NJ.

Zhang, Y. and Zhu, J. (2011). Direct torque control of permanent magnet synchronous motor with reduced torque ripple and commutation frequency. *IEEE Trans. Power Electron.* 26(1), 235–248.

Zhong, L., Rahman, M.F., Hu, W., Lim, K., and Rahman, M. (1999). A direct torque controller for permanent magnet synchronous motor drives. *IEEE Trans. Energy Convers.* 14(3), 637–642.

Zhong, L., Rahman, M.F., Hu, W.Y., and Lim, K. (1997). Analysis of direct torque control in permanent magnet synchronous motor drives. *IEEE Trans. Power Electron.* 12(3), 528–536.

Zolghadri, M.R., Guiraud, J., Davoine, J., and Roye, D. (1998). A DSP based direct torque controller for permanent magnet synchronous motor drives. In: *PESC 98 Record. 29th Annual IEEE Power Electronics Specialists Conference, 1998.* Vol. 2, pp. 2055–2061. IEEE, Piscataway, NJ.

Predictive, Deadbeat, and Combined Controls

5

VC as the first developed high performance control method for AC motors has evolved through the decades and retained its solid domination in the field of motor control. DTC as the second widely developed motor control method for AC motors has been researched extensively. It has made its way into the industry through many applications. The field of AC motor control, however, has evolved since then with an exciting diversity. The practice of selecting an inverter voltage vector in each switching interval as established by DTC turns out to be the common practice of many newer control methods. Among them, predictive control (PC), though not very recent in its origination, has received attention recently. The control method tests all possible inverter voltage vectors on the motor model, then compares the motor performances under the vectors, and finally selects the best one according to a desired criterion to be applied to the motor. The method has found many schemes for induction and PMS motors. In contrast, the deadbeat control (DBC), as another AC motor control method, solves a reverse model of the machine to determine the voltage vector that meets the torque command in just one switching interval. It is interesting in the sense that it does not need to test different voltage options on the motor model. Combined control (CC) uses current control together with the switching table to determine the desired inverter voltage vector in each instance. It combines selective elements of VC and DTC to achieve salient performance features of the two control methods. These methods, along with some other methods, are still evolving to serve the field of motor control with ease of implementation and performance superiorities. The three control methods mentioned are presented in this chapter in connection with PMS motors.

5.1 Predictive control

Predictive control has been known since the 1960s as a part of optimal control theory. It first determines the possible performances of

Control of Permanent Magnet Synchronous Motors. Sadegh Vaez-Zadeh. © Sadegh Vaez-Zadeh 2018.
Published in 2018 by Oxford University Press. DOI 10.1093/oso/9780198742968.001.0001

the system in advance under different actuating inputs. Second, it decides on the best actuating input to the system to realize the optimal performance. The method chooses the best input in order to keep a control variable, like current, within a band or make it to follow a desired trajectory. In a more general sense, it chooses the input to optimize an objective function and comes up with a compromise among different goals. In addition, it may impose restrictions on variables other than the controlled ones.

PC encompasses a family of control schemes. Among them, the model predictive control (MPC) uses a dynamic model of the system under control for the prediction of the system's future performance. Thus, the accuracy of the model directly influences the system performance. Also, parameter variations may drift the performance from the desired objectives. In this case parameter adaptation or online parameter estimation may be necessary. MPC, in turn, can be realized in different approaches. A fact that contributes to the application of the method in some areas is the possibility of using the method in connection with other control methods. When it comes to motor control, PC can be used with VC, DTC, etc.

Despite the mentioned advantages, PC faces some challenges including calculation intensity. It stems from the essence of the method, which is to find the system performance under many possible actuating signals, though none of them are actually used except one. Since the method is implemented by microprocessors, it is needed to do all the calculations numerically during a short sampling time. This puts computation burden on the processor. To reduce the burden, some means are proposed in the literature, including the partial transfer of the calculations to offline computation. Under this solution, the results of calculations are saved in a look-up table as a function of the system state. They are used according to the current state of the system (Mariethoz *et al.* 2009). An alternative way is to solve the optimization problem analytically. This can be carried out when the system is linear or the model can be linearized with some sort of approximation (Kennel *et al.* 2001; Hassaine *et al.* 2007).

PC in motor drives reduces to the prediction of system performances under the available voltage vectors as the actuating inputs. Considering the finite set of inverter voltages, fast processors can handle online calculations of the predictive control for motor drive applications. As the switching states increase in multilevel converters, however, the processor may be overloaded, considering the short sampling time.

5.1.1 Principles of model-based predictive control

The general presentation of the MPC uses a discrete-time state-space model of the system to calculate the system's future performances as (Rodriguez and Cortes 2012)

$$x(k+1) = Ax(k) + Bu(k), \tag{5.1.1}$$

$$y(k) = Cx(k) + Du(k), \tag{5.1.2}$$

where $x(k)$ and $x(k+1)$ are the system state vector at the current and the next instants, respectively. Also, $u(k)$ and $y(k)$ are the input and the output vectors, respectively, at the current instant. A, B, C, and D are the system, input, output, and disturbance matrices, respectively. The model, which is presented by eqns (5.1.1) and (5.1.2), is used to predict the system performance.

An objective function, J, as a function of system states and inputs is defined to formulate the favorite performance of the system as

$$J = f(x(k), u(k), \ldots, u(k+N)), \tag{5.1.3}$$

where N is a positive number known as the prediction horizon and is the number of future instances over which the control can predict the system performance. A long horizon strengthens the predictability of the system behavior at a cost of higher calculations. The vector $u(k+N)$ is the system input at instance $k+N$. The sequence of the inputs, previous to $u(k+N)$, is also included in J as shown in Fig. 5.1 (Rodriguez and Cortes 2012). The figure shows the reference and actual states of the system, in addition to the discrete inputs in consecutive instances from the past up to instance $k+N$. It is seen that the system error reduces as the time goes on and the actual state gets closer to the reference.

The favorite performance of the system, which is included in J, depends on the system under control and the applicant desire. The common part of J, however, is some measure of system output error, i.e., the error between a reference and a predicted value of the system output as shown in Fig. 5.1. A square error is frequently used for this purpose. By including this part to the objective function, the regulatory function of the control system is fulfilled. Therefore, there is no need for controllers like PI. The objective function may include other parts to fulfill non-regulatory objectives like high efficiency, high power factor, and limited switching frequency. It is common to define the objective function as a weighing sum of different objectives. Each weight is determined according to the importance of a corresponding

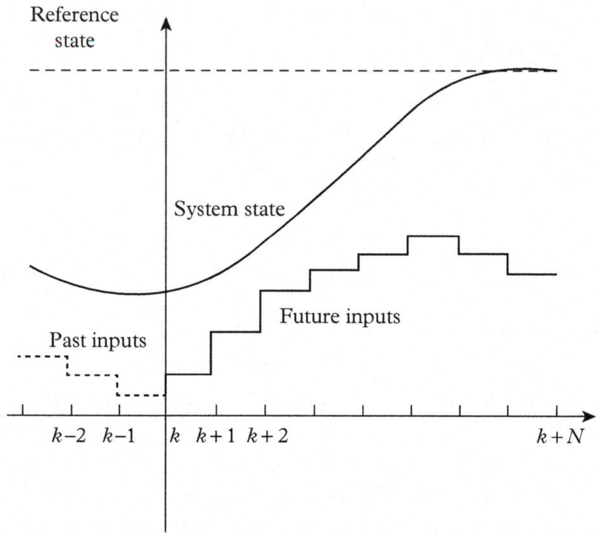

Figure 5.1 *Working principles of model-based predictive control.*

objective. More complicated objective functions may also be defined as in many optimization methods.

The optimization problem is to find the inputs such that J is minimized, subject to restrictions of the system. The system model presented by eqns (5.1.1) and (5.1.2) is used in finding the solution. A general solution is complicated. A simple solution, however, is to examine the value of J for all possible inputs in each sampling. Then, the least value of J determines the best input in each sampling. The solution consists of a sequence of N optimal input signals. However, the controller will apply only the first element of the sequence. Once the selected input is applied to the system, the system model and the objective function are updated by using the most recent data obtained by measurement. Therefore, the predictive horizon is regarded as a moving window of N instances as shown in Fig. 5.2.

According to the descriptions presented previously, the model-based predictive control performs two main tasks as shown in Fig. 5.3. The tasks are defined as

- *Prediction:* model of the system, known as the predictive model, is used to predict the behavior of the controlled variables, usually

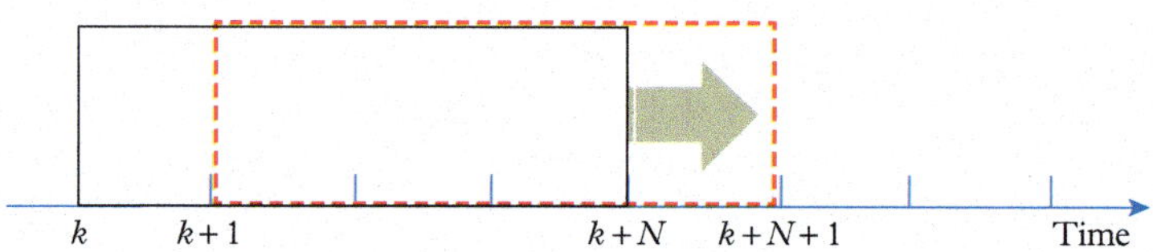

Figure 5.2 *The prediction horizon as a moving window of N sampling times.*

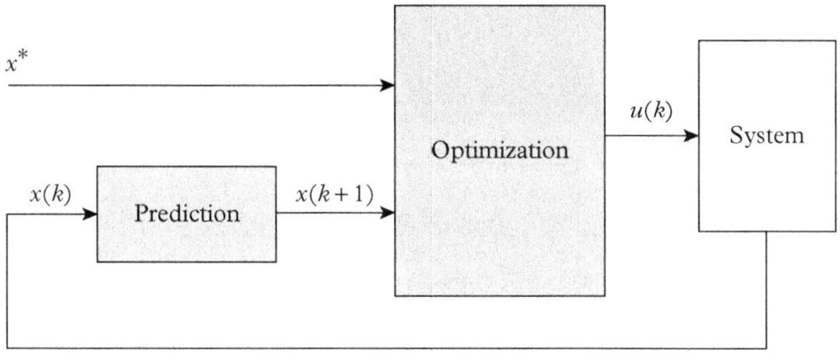

Figure 5.3 *Functional block diagram of general model-based predictive control.*

the system outputs, for all possible actuating inputs in the prediction horizon. The present and past values of the system inputs and actual outputs are needed for performing the task.

- *Optimization:* the objective function is formed and minimized to find the best actuating input. A representative of the errors between the predicted and the reference variables is a part of the objective function as mentioned earlier, in addition to possible other terms to optimize the system performance as desired. The best actuating input is then applied to the system.

5.1.2 Predictive current control of PMS motors

A model-based predictive current control for a PMS motor drive is presented in Fig. 5.4, where the discrete-time state variable vector in rotor reference frame is defined as

$$\boldsymbol{x}(k) = \left[i_d(k) \;\; i_q(k) \right]^T. \tag{5.1.4}$$

In this system, the reference or the output to be tracked is the same as the state vector. Therefore, a current VC is to be implemented by utilizing MPC, instead of separate PI controllers for i_q and i_d. The predictive model here is a discrete-time state-space model of the PMS machine presented by the equations

$$i_d^p(k+1) = \left(1 - \frac{R_s T_s}{L_s}\right) i_d(k) + T_s \omega_r i_q(k) + \frac{T_s}{L_s} v_d, \tag{5.1.5}$$

$$i_q^p(k+1) = \left(1 - \frac{R_s T_s}{L_s}\right) i_q(k) - T_s \omega_r i_d(k) - \lambda_m T_s \omega_r + \frac{T_s}{L_s} v_q. \tag{5.1.6}$$

The machine equations, as the predictive model, predict the stator current components for each of the possible actuating inputs. If

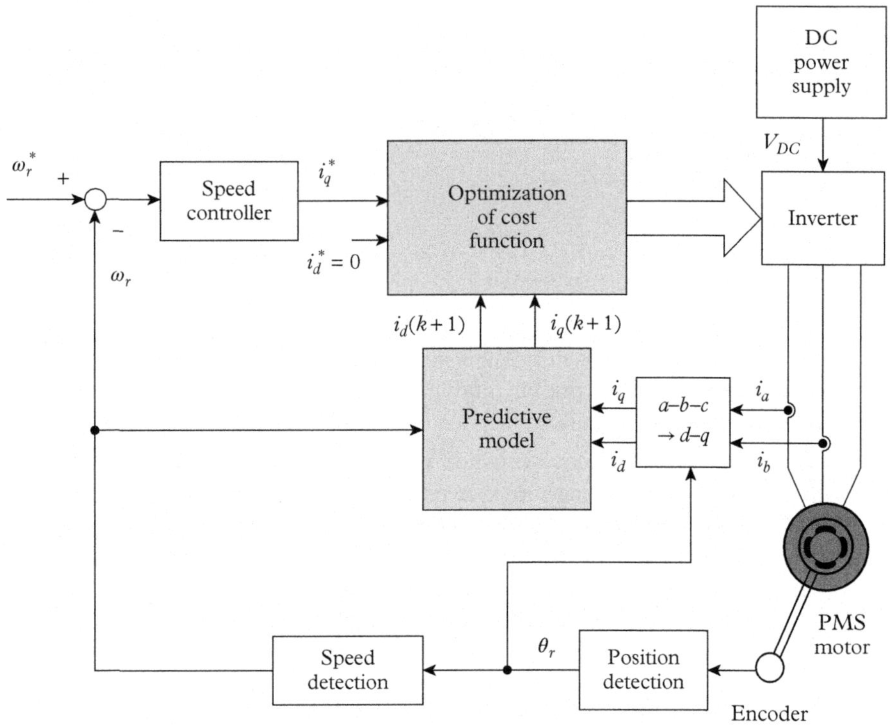

Figure 5.4 *Block diagram of model-based predictive current control of PMS motors.*

a two-level inverter is considered, eight switching states generating seven different voltage vectors are possible. The inverter voltages are transformed from the stationary reference frame to the rotor reference frame.

The speed controller generates the q-axis current command, $i_q{}^*$, as in VC. This current is compared with the predicted q-axis current, i_q^p, to produce the q-axis current error. Also, the d-axis current error is equal to the predicted d-axis current, i_d^p, because $i_d{}^* = 0$. Having the errors, the optimizer calculates the value of an objective function for all seven possible inverter voltage vectors. The objective function includes functions of the current errors. Alternatively, it may contain other terms relating to the desired steady-state and/or dynamic performances of the machine and inverter. A choice for the objective function may be presented as (Rodriguez and Cortes 2012)

$$J = (i_q^* - i_q^p(k+1))^2 + (i_d^p(k+1))^2 + \hat{f}(i_d^p(k+1), i_q^p(k+1)), \tag{5.1.7}$$

where the first term is the square of the q-axis current error. Minimization of this term guarantees a desirable tracking of the torque-producing component of the machine stator current, resulting in good dynamic and steady-state torque responses. The second term reduces the d-axis component of stator current to 0, resulting in maximum torque per ampere operation. The last term is a non-linear function for limiting the magnitude of the stator current vector by its maximum allowed value as explained in Section 3.5. It is defined as

$$\hat{f}(i_d^p(k+1), i_q^p(k+1)) = \begin{cases} \infty & if \; |i_q^p| > i_{sL} \;\text{ or }\; |i_d^p| > i_{sL} \\ 0 & if \; |i_q^p| \le i_{sL} \;\text{ and }\; |i_d^p| \le i_{sL} \end{cases}, \tag{5.1.8}$$

where i_{sL} is the value of the maximum allowed stator current magnitude. By this term, the voltage vectors causing overcurrent are not selected as the best input, since J tends to infinity under overcurrent conditions. On the other hand, when there is no overcurrent, only the first two terms of J take effect and the voltage vector that minimizes the current error will be selected. The selected voltage vector is then applied to the inverter during the next sampling interval. The selected voltage vector may be an active voltage vector or a zero one.

It is also possible to apply successively two voltage vectors in a single sampling interval, one active voltage and one zero voltage (Morel *et al.* 2009). In this case, the duty ratio of applying the active voltage is determined to be similar to the space vector modulation presented in Section 4.6.

5.1.3 Predictive flux and torque control of PMS motors

In predictive flux and torque control (PFTC), the same principles as those of the predictive current control are used. But in PFTC schemes, the future values of the stator flux linkage and torque, instead of current components, are predicted. Hence, the desired conditions, which are formulated by the objective function, concern the behavior of flux linkage and torque. Predictions are made for every possible actuating voltages of the inverter. The optimization block then selects the voltage vector that minimizes the objective function.

A block diagram of PFTC is shown in Fig. 5.5. The current values of estimated stator flux linkage, $\hat{\lambda}_s(k)$, and torque, $\hat{T}_e(k)$, are obtained by using the motor current and voltages as in DTC, presented in sub-section 4.2.3. Then, the predictive model containing the machine equations computes the future values of the controlled variables at the next instant, i.e., $\lambda_s^p(k+1)$ and $T_e^p(k+1)$, by using the current and

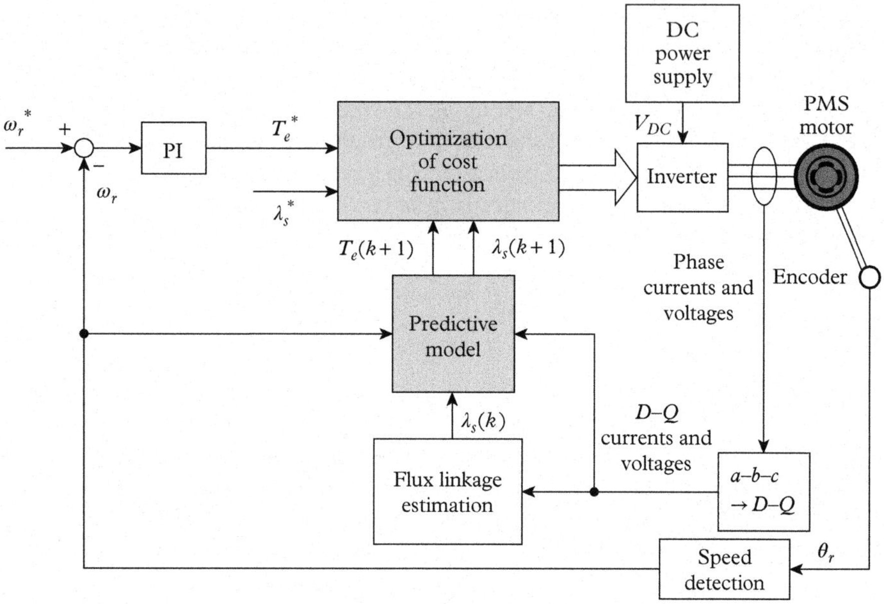

Figure 5.5 *Block diagram of model-based predictive flux and torque control of PMS motors.*

voltage components in the two-axis stationary reference frame. The flux linkage components are calculated as

$$\lambda_D^p(k+1) = \hat{\lambda}_D(k) + T_s v_D(k) - R_s T_s i_D(k), \tag{5.1.9}$$

$$\lambda_Q^p(k+1) = \hat{\lambda}_Q(k) + T_s v_Q(k) - R_s T_s i_Q(k). \tag{5.1.10}$$

A prediction of the machine torque at instance $(k+1)$ needs the predicted current components at this instance. Therefore, recall from eqn (2.10.10) that

$$\bar{\lambda}_s = \bar{\lambda}_m + L_s \bar{i}_s. \tag{5.1.11}$$

Assuming that the rotor flux linkage vector does not change during a sampling time, a discrete form of eqn (5.1.11) in two-axis stationary reference frame yields

$$i_D^p(k+1) = i_D(k) + \frac{\lambda_D^p(k+1) - \hat{\lambda}_D(k)}{L_s}, \tag{5.1.12}$$

$$i_Q^p(k+1) = i_Q(k) + \frac{\lambda_Q^p(k+1) - \hat{\lambda}_Q(k)}{L_s}. \tag{5.1.13}$$

The predicted torque is obtained as

$$T_e^p(k+1) = \frac{3}{2}P(\lambda_D^p(k+1)\,i_Q^p(k+1) - \lambda_Q^p(k+1)\,i_D^p(k+1)). \tag{5.1.14}$$

Now, the torque is predicted by substituting eqns (5.1.9)–(5.1.10) and eqns (5.1.12)–(5.1.13) into eqn (5.1.14). The flux linkage, currents, and torque predictions are made for every possible voltage vector generated by the inverter, i.e., seven different voltage vectors for a two-level inverter.

Having found these values the optimization block chooses the optimum switching state of the inverter, corresponding to the best voltage vector, which minimizes the objective function. This function contains the control law in order to achieve an appropriate torque and flux regulation. A simple choice for the objective function is the square sum of torque error and flux linkage error,

$$J = \left(T_e^* - T_e^p\right)^2 + w_\lambda\left(\lambda_s^* - \lambda_s^p\right)^2, \tag{5.1.15}$$

where w_λ is a weighting factor that puts emphasis on either torque or flux linkage. In the case of equally emphasized torque and flux linkage errors, the weighting factor reduces to the ratio of rated torque to rated flux linkage to provide unit consistency for J. A refined objective function can be formed by normalizing the torque and flux linkage errors by their rated values, respectively, as (Rodriguez and Cortes 2012)

$$J = \frac{\left(T_e^* - T_e^p\right)^2}{T_{er}^2} + w_\lambda\frac{\left(\lambda_s^* - \lambda_s^p\right)^2}{\lambda_{sr}^2}, \tag{5.1.16}$$

where T_{er} and λ_{sr} are the rated torque and flux linkage, respectively. The weighting factor in eqn (5.1.16) does not have any unit, because the normalized terms have no unit. Therefore, it is equal to unity, when equal emphasis is put on the torque and flux linkage. More sophisticated objective functions can be defined by taking into account extra performance objectives.

It is now useful to have a look at MPC in comparison with DTC. In DTC, the desired voltage is selected according to a table that is the result of a heuristic approach, while MPC selects the voltage vector based on mathematical optimization. DTC makes use of hysteresis controllers, which leads to a variable inverter switching frequency, while MPC has fixed switching frequency. Finally, in DTC, long estimation times may result in the torque and flux linkage exceeding the hysteresis bands and causing unwanted errors, while in MPC, the controller errors are predicted in advance and unwanted errors do not occur (Morel *et al.* 2009). Nevertheless, MPC needs much more computations.

5.2 Deadbeat control

DBC is a kind of predictive control that provides a very fast dynamic response for the system under control. The main difference between this control and the conventional predictive control schemes is that the reference signal in the former method is met by the system in one sampling time. It implies that the reference signal is observed by one beat. The control outputs, which are the inputs to the system under control, are obtained by solving an inverse model of the system. As a result, an accurate model of the system is required.

DBC can be applied to motor drives as a current VC or as a direct torque and flux control. In both schemes, the deadbeat controller calculates the desired voltage components to be applied to the machine in order to reach the reference signals in the next sampling interval. In the former scheme, however, the reference signals are the machine stator current components, while in the latter scheme the reference signals are the reference torque and flux linkage. In both schemes, a modulator like a sinusoidal PWM or SVM is needed to generate the switching signals for the inverter.

Direct torque and flux control by DBC is much more common than the current control by DBC in motor drives (Kenny and Lorenz 2001). Therefore, the former is presented in this section in full detail. The concept of DBC in electric machines, however, can be understood more easily in connection with the current control. Thus, a simple current controlling deadbeat controller is presented briefly, before embarking to the torque and flux controlling deadbeat controller.

5.2.1 Basic principles of DBC

A current-controlled PMS motor drive with a deadbeat controller is shown in Fig. 5.6 (Rodriguez and Cortes 2012). Considering a non-salient PMS machine, the state variable model of the machine in a two-axis stationary reference frame is presented in matrix notation as

$$\begin{bmatrix} \dot{i}_D \\ \dot{i}_Q \end{bmatrix} = \begin{bmatrix} -\frac{R_s}{L_s} & 0 \\ 0 & -\frac{R_s}{L_s} \end{bmatrix} \begin{bmatrix} i_D \\ i_Q \end{bmatrix} + \begin{bmatrix} \frac{1}{L_s} & 0 \\ 0 & \frac{1}{L_s} \end{bmatrix} \begin{bmatrix} v_D \\ v_Q \end{bmatrix}. \quad (5.2.1)$$

The model has been derived in detail in Section 2.4, where the same notation is used. A discrete-time version of eqn (5.2.1) in matrix form can be presented in eqn (5.2.2) by assuming constant variables during a sampling interval, T_s, as

$$\boldsymbol{i}_s(k+1) = \boldsymbol{\Phi}\boldsymbol{i}_s(k) + \boldsymbol{\Gamma}\boldsymbol{v}_s(k), \quad (5.2.2)$$

where $\boldsymbol{v}_s$ and $\boldsymbol{i}_s$ are the input and the output vectors, respectively, and

$$\boldsymbol{\Phi} = e^{-\left(R/L\right)T_s}, \tag{5.2.3}$$

$$\boldsymbol{\Gamma} = \int_0^{T_s} e^{-\left(R/L\right)\tau} d\tau . \frac{1}{L}. \tag{5.2.4}$$

An inverse of (5.2.2) is solved to find the voltage vector in order to reach a zero current error at one sampling interval. The resulting voltage vector as the input to the machine is obtained as

$$\boldsymbol{v}_s^* (k) = \frac{1}{\boldsymbol{\Gamma}} \left(\boldsymbol{i}_s^* (k+1) - \boldsymbol{\Phi} \boldsymbol{i}_s (k) \right). \tag{5.2.5}$$

It is seen in Fig. 5.6 that by applying the obtained voltage as in eqn (5.2.5) to the modulator, the exact reference current will flow into the machine and thus a zero current error is achieved by the feedback loop.

5.2.2 DBC of torque and flux linkage

In direct torque and flux linkage control of PMS machines by DBC, an inverse model of the machine is solved for the inputs voltages to the machine (outputs of the deadbeat controller), where the input to the deadbeat controller are the torque and the flux linkage errors as seen in Fig. 5.7 (Lee *et al.* 2009; Lee *et al.* 2011a). The method will be described next by referring to the machine model in discrete time domain.

Recall that the torque in rotor reference frame can be presented as a function of flux linkage and current components as

$$T_e = \frac{3}{2} P \left[\lambda_d i_q - \lambda_q i_d \right]. \tag{5.2.6}$$

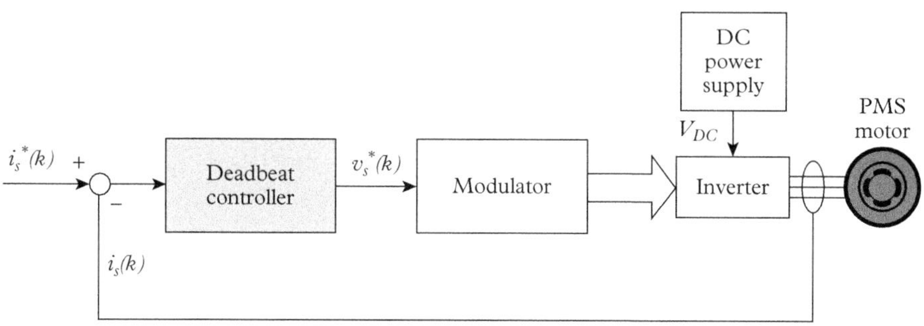

Figure 5.6 *Basic principles of DBC for a current-controlled PMS motor.*

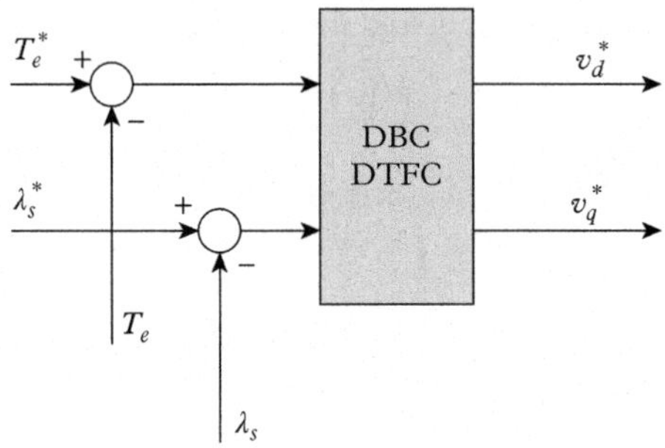

Figure 5.7 *DBC with the estimated flux linkage and torque as the inputs and the voltage components in rotor reference frame as the outputs.*

The torque differential is obtained from eqn (2.5.18) as

$$\dot{T}_e = \frac{3}{2} P \left[\dot{\lambda}_d i_q + \lambda_d \dot{i}_q - \dot{\lambda}_q i_d - \lambda_q \dot{i}_d \right]. \tag{5.2.7}$$

Also the voltage equations in this reference frame are recalled from Section 2.5 as

$$v_d = R_s i_d + p\lambda_d - \omega_r \lambda_q, \tag{5.2.8}$$

$$v_q = R_s i_q + p\lambda_q + \omega_r \lambda_d. \tag{5.2.9}$$

Finding current components from the last two equations, substituting them into eqn (5.2.7), and converting the result into the discrete form yields

$$\begin{aligned} &\frac{T_e(k+1) - T_e(k)}{T_s} \\ &= \frac{3}{2} P \Bigg[v_d(k) \lambda_q(k) \frac{L_d - L_q}{L_d L_q} + v_q(k) \frac{(L_d - L_q)\lambda_d(k) + \lambda_m L_q}{L_d L_q} \\ &+ \frac{\omega_r(k)}{L_d L_q} \left((L_q - L_d) \left(\lambda_d(k)^2 - \lambda_q(k)^2 \right) - \lambda_d(k) \lambda_m L_q \right) \\ &+ \frac{R_s \lambda_q(k)}{(L_d L_q)^2} \left(\left(L_q^2 - L_d^2 \right) \lambda_d(k) - L_q^2 \lambda_m \right) \Bigg], \end{aligned} \tag{5.2.10}$$

where

$$\Delta T_e(k) = T_e(k+1) - T_e(k). \tag{5.2.11}$$

The discrete equation (5.2.10) in connection with eqn (5.2.11) can be rearranged as a voltage equation containing torque deviation as

$$v_q(k) T_s = M v_d(k) T_s + B, \tag{5.2.12}$$

where

$$M = \frac{\left(L_q - L_d\right)\lambda_q\left(k\right)}{\left(L_d - L_q\right)\lambda_d\left(k\right) + \lambda_m L_q},$$

$$B = \frac{L_q L_d}{(L_q - L_d)\lambda_q\left(k\right) + \lambda_m L_q}$$

$$\times \left[\frac{2\Delta T_e}{3P} - \frac{\omega_r T_s}{L_d L_q}\left((L_q - L_d)(\lambda_d(k)^2 - \lambda_q(k)^2) - \lambda_q(k)\lambda_m L_q\right)\right.$$

$$\left. - \frac{R_s \lambda_q(k) T_s}{(L_d L_q)^2}\left((L_q^2 - L_d^2)\lambda_d(k) - L_q^2\lambda_m\right)\right]. \quad (5.2.13)$$

Equation (5.2.12) shows a linear locus of the torque in terms of d–q volt-sec components as depicted in Fig. 5.8. To solve for the components of the voltage vector in eqn (5.2.12), additional equations are needed since M and B contain flux linkage components. Therefore, a discrete form of eqns (5.2.8) and (5.2.9), where the voltage drops due to the stator resistance are ignored and the cross-coupling terms of the stator flux linkages are decoupled, is formed as

$$\lambda_d\left(k+1\right) = \lambda_d\left(k\right) + v_d\left(k\right)T_s, \quad (5.2.14)$$

$$\lambda_q\left(k+1\right) = \lambda_q\left(k\right) + v_q\left(k\right)T_s. \quad (5.2.15)$$

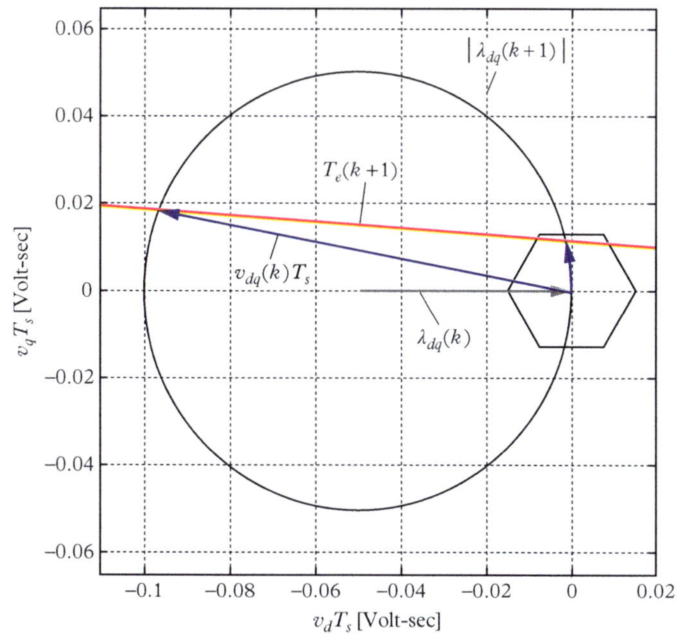

Figure 5.8 *Graphical representation of deadbeat solution for direct torque and flux control (Lee et al. 2011a).*

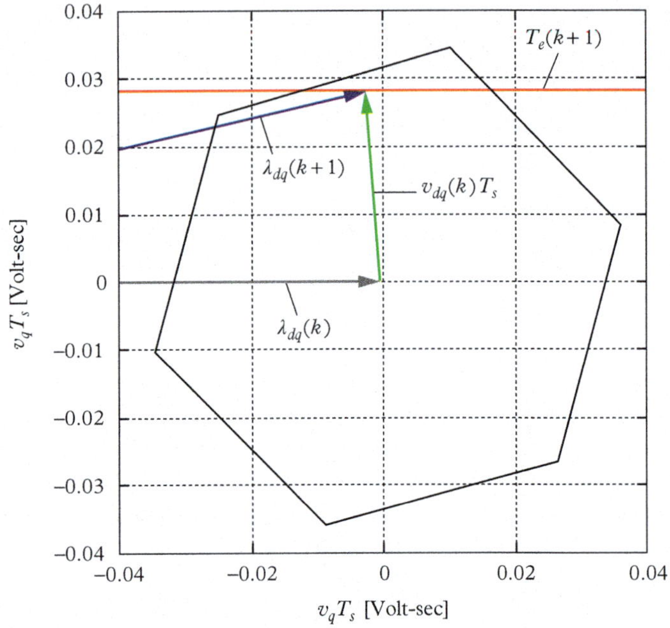

Figure 5.9 *Selected voltage of DBC represented by its volt-sec (Lee et al. 2011a).*

The flux linkage components must satisfy the commanded value of the total flux linkage magnitude. This condition can be depicted by a circle of constant radius equal to the commanded flux linkage magnitude as in Fig. 5.8. This is also presented mathematically by the equation

$$\lambda_s^* (k)^2 = \lambda_d (k+1)^2 - \lambda_q (k+1)^2 = (\lambda_d (k) + v_d (k)\, T_s)^2 + \left(\lambda_q (k) + v_q (k)\, T_s\right)^2 . \qquad (5.2.16)$$

Equations (5.2.12)–(5.2.16) are solved to give the values of the voltage components for every pair of torque and flux linkage errors, as shown in Fig. 5.8. The figure depicts graphically the solution as the cross-section of the torque line (eqn (5.2.12)) and the flux linkage circle (eqn (5.2.16)).

It is seen that among the possible inverter voltage vectors, two voltage vectors shown by their volt-sec end on the constant stator flux linkage locus. The voltage limit of the inverter is also shown by a hexagon in Fig. 5.8. The voltage vector within the hexagon is chosen as the output of the deadbeat controller. A close-up representation of the solution for a practical case is shown in Fig. 5.9, where the selected voltage vector is clearly shown by its volt-sec representation.

5.3 Combined control

Different schemes of VC and DTC as presented in Chapters 3 and 4 in connection with PMS motors provide excellent performances such as rapid dynamics and accurate responses. VC and DTC, however, differ substantially in the way they are implemented. VC deals primarily with a machine current VC, and controls the machine torque and flux linkages through the control loops, while DTC controls the machine flux linkage amplitude and torque directly without any current control. VC generates the corresponding voltage commands through a pulse width modulation (PWM) technique, like sinusoidal PWM or space vector modulation, while DTC determines desirable voltage vector commands by a predefined inverter switching table. The information of the machine rotor position is essential for normal operation of motors under VC, while DTC does not need such information. Instead, DTC calculates the stator flux linkage position by using the current and voltage measurement. In fact DTC can be regarded as a sensorless motor control method.

More importantly, the mathematical foundations and the principles of VC and DTC are far apart. While the motors under VC are formulated in a rotating reference frame, oriented along either rotor or stator flux linkage, DTC requires a motor formulation in a stationary reference frame. VC is a current VC as inferred by its name, while DTC can be regarded a flux linkage VC.

The major differences between VC and DTC have been the focus of many studies so far. The studies usually try to translate the differences into the methods' detailed performance discrepancies, including torque and flux linkage pulsations, inverter switching frequency, etc. In fact, the theoretical and practical disagreements of VC and DTC have overshadowed the basic performance similarities of the two methods, i.e., swiftness and accuracy. The machines' basic performances under these two methods are so close that it is hard to come to a general conclusion in favor of either method unless a specific application under a particular operating condition is sought. Even, in many applications, both methods can provide superior machine performance with minor modifications to the original VC and DTC.

The performance similarity of both VC and DTC, despite their theoretical and practical differences, raises a serious question regarding the fundamentals of VC and DTC: is there a common basic for VC and DTC which provides the performance excellence of the two methods despite all their differences? The answer is yes. The two methods share a fundamental basis theoretically and practically. The basis cannot be easily obtained by using convention machine models, however. By using a deviation model of the machine torque and flux linkage, it is possible to show that there is an analogy between VC and DTC and reach a common basis of the control methods (Vaez-Zadeh and Jalali 2007; Shafaie and Vaez-Zadeh 2011). The basis shows how it is possible to replace current control with flux linkage control and vice versa. Therefore, consistent parts of VC and DTC are selected to build a CC system that maintains the common performance features of the VC and DTC, while overcoming some structural and performance deficiencies of the methods.

5.3.1 Common basis of CC

The main function of a control method is to improve the dynamic performance of the system under control. The dynamics of the system is determined by the mutual deviations of system variables with respect to time. Therefore, it is convenient to work with deviation of machine variables rather than with the variables themselves when considering a control method. The switching nature of power electronic converters provides the opportunity to consider variable deviations in switching intervals. Considering the high frequency of the switching, it is convenient to assume linear deviation of machine variables during a switching interval. Therefore, a linearized deviation model of the machine is used when analysing the machine to reach the basis of the control methods. This simplified model provides clear insight into the fundamental machine dynamics and makes it possible to find the common basis.

Recall the machine developed torque from Chapter 2 as

$$T_e = \frac{3}{2}P\left[\lambda_d i_q - \lambda_d i_d\right], \tag{5.3.1}$$

where the d- and q-axis flux linkage components are given as

$$\lambda_d = L_d i_d + \lambda_m, \quad \lambda_q = L_q i_q, \tag{5.3.2}$$

where L_d and L_q stand for the direct- and the quadrature-axis inductances, respectively.

Substituting for d- and q-axis flux linkages from eqn (5.3.2) into eqn (5.3.1) yields

$$T_e = \frac{3}{2}P\left[\lambda_m + (L_d - L_q)i_d\right] i_q\,. \tag{5.3.3}$$

A deviation model of eqn (5.5.3), assuming linear variations over a short time interval, is obtained as

$$\Delta T_e = K_1 \Delta i_d + K_2 \Delta i_q, \tag{5.3.4}$$

where

$$K_1 = \frac{3}{2}P\,(L_d - L_q)\,i_q,$$

$$K_2 = \frac{3}{2}P\,[\lambda_m + (L_d - L_q)i_d]\,. \tag{5.3.5}$$

This means that the small torque deviation around an operating point, specified by the current components i_d and i_q, is a weighted function of the current component deviations, Δi_d and Δi_q.

Therefore, it is possible to control T_e through controlling these two current deviations. In other words, the machine torque can be controlled by controlling the current deviation vector, $\Delta\bar{i}_s$, where

$$\Delta\bar{i}_s = \Delta i_d + j\Delta i_q. \tag{5.3.6}$$

On the other hand, it is possible to find the d- and q-axis currents from (5.3.2) and substitute them into (5.3.1) to obtain

$$T_e = \frac{3}{2}P\left[\frac{\lambda_m}{L_d} + \left(\frac{1}{L_q} - \frac{1}{L_d}\right)\lambda_d\right]\lambda_q. \tag{5.3.7}$$

A deviation model of (5.3.7), assuming linear variations in a short time interval, yields

$$\Delta T_e = K_3\Delta\lambda_d + K_4\Delta\lambda_q, \tag{5.3.8}$$

where

$$K_3 = \frac{3}{2}P\left(\frac{1}{L_q} - \frac{1}{L_d}\right)\lambda_q,$$

$$K_4 = \frac{3}{2}P\left[\frac{\lambda_m}{L_d} + \left(\frac{1}{L_q} - \frac{1}{L_d}\right)\lambda_d\right]. \tag{5.3.9}$$

This means that the small torque deviation around an operating point, specified by the flux linkage components, λ_d and λ_q, is a weighted sum of the component deviations, $\Delta\lambda_d$ and $\Delta\lambda_q$.

Therefore, it is possible to control T_e through controlling these two flux linkage deviations. In other words, the machine torque can be controlled by controlling the flux linkage deviation vector, $\Delta\bar{\lambda}_s$, where

$$\Delta\bar{\lambda}_s = \Delta\lambda_d + j\Delta\lambda_q. \tag{5.3.10}$$

Comparing eqns (5.3.4) and (5.3.5) with eqns (5.3.8) and (5.3.9), respectively, the following equalities are evident:

$$\frac{\Delta\lambda_d}{\Delta i_d} = \frac{K_1}{K_3} = L_d, \qquad \frac{\Delta\lambda_q}{\Delta i_q} = \frac{K_2}{K_4} = L_q. \tag{5.3.11}$$

It means that there are linear relationships between K_1–K_3 and K_2–K_4 as there are between $\Delta\lambda_d$–Δi_d and $\Delta\lambda_q$–Δi_q, respectively. This implies that the apparent contrast in expressing torque dynamics in terms of current deviations or flux linkage deviations is superficial and they are basically the same as is shown by eqn (5.3.11). The common basis is that the machine torque deviation is equal to the weighted sum of two deviation variables, which can be presented as either current components or equivalent flux linkage components. This common basis provides the same fundamental motor performances under VC and DTC, despite the differences in mathematical presentation and practical implantation of the two methods.

The common basis of VC and DTC can be proved in stator flux linkage reference frame, too. The basis in this RF is explained now in connection with surface-mounted PMS motors. The torque equation in these motors is recalled as

$$T_e = \frac{3}{2} P \lambda_m i_q = \frac{3}{2}\frac{P}{L_s} \lambda_m \lambda_s \sin\delta. \tag{5.3.12}$$

The linearized torque deviation is then obtained as

$$\Delta T_e = \frac{3}{2}\frac{P}{L_s} \lambda_m (\sin\delta\ \Delta\lambda_s + \lambda_s \cos\delta\ \Delta\delta). \tag{5.3.13}$$

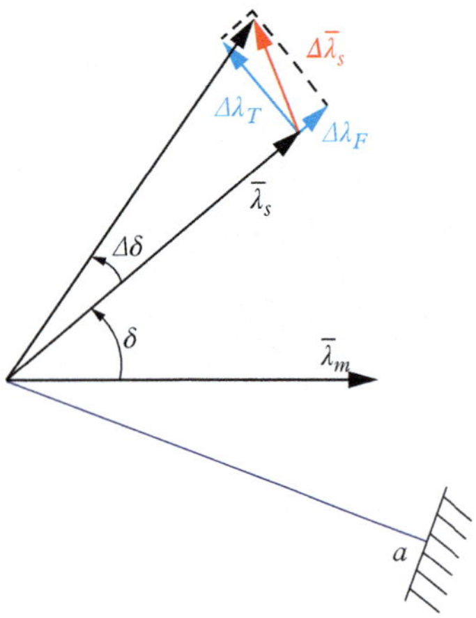

Figure 5.10 *Deviation of flux linkage vector as a function of its components.*

Referring to Fig. 5.10, it is possible to define the vector of flux linkage deviation, $\Delta\bar{\lambda}_s$, with a real component, $\Delta\lambda_F$, and an imaginary component, $\Delta\lambda_T$, along the flux linkage vector and perpendicular to the flux linkage vector, respectively, as

$$\Delta\bar{\lambda}_s = \Delta\lambda_F + j\Delta\lambda_T, \tag{5.3.14}$$

where the components are given with good approximation as

$$\Delta\lambda_F = \Delta\lambda_s, \qquad \Delta\lambda_T = \lambda_s\,\Delta\delta\,. \tag{5.3.15}$$

Substituting (5.3.15) into (5.3.13) yields

$$\Delta T_e = K_5\,\Delta\lambda_F + K_6\,\Delta\lambda_T, \tag{5.3.16}$$

where the coefficients K_5 and K_6 are constants at the operating point around which the linearization takes place, and they are presented by

$$K_5 = \frac{3}{2}\frac{P}{L_s}\lambda_m \sin\delta, \quad K_6 = \frac{3}{2}\frac{P}{L_s}\lambda_m \cos\delta\,. \tag{5.3.17}$$

This means that the small torque deviation around an operating point, specified by the flux linkage components, λ_F and λ_T, is a weighted sum of the component deviations. Therefore, it is possible to control T_e by controlling $\Delta\bar{\lambda}_s$ through its components.

Also, it is possible to obtain the following linearized relationship from eqn (5.1.11) for surface-mounted PMS motors, where it is assumed at the start that the rotor flux linkage vector does not change noticeably during the short interval of inverter switching period:

$$\Delta\bar{\lambda}_s = L_s\,\Delta\bar{i}_s\,. \tag{5.3.18}$$

Using eqn (5.3.18), the following flux linkage deviations are obtained:

$$\Delta\lambda_F = L_s\Delta i_F, \qquad \Delta\lambda_T = L_s\Delta i_T. \tag{5.3.19}$$

Substituting the flux linkage deviations from eqn (5.3.19) into eqn (5.3.16) yields

$$\Delta T_e = K_7\Delta\, i_F + K_8\Delta\, i_T, \tag{5.3.20}$$

where

$$K_7 = \frac{3}{2}P\lambda_m \sin\delta, \qquad K_8 = \frac{3}{2}P\lambda_m \cos\delta\,. \tag{5.3.21}$$

This means that the small torque deviation around an operating point is a linear function of the current component deviations, Δi_F and Δi_T. Therefore, it is possible to control T_e through controlling these two current deviations. In other words, the machine torque can be controlled by controlling the current vector deviation, $\Delta\bar{i}_s$, where

$$\Delta\bar{i}_s = \Delta i_F + j\Delta i_T. \tag{5.3.22}$$

When eqns (5.3.16) and (5.3.17) are compared with eqns (5.3.20) and (5.3.21), respectively, the following equalities are evident:

$$\frac{K_7}{K_5} = \frac{K_8}{K_6} = L_s. \tag{5.3.23}$$

They mean that there are linear relationships between K_5–K_7 and K_6–K_8. This implies again that expressing torque dynamics in terms of current deviations or flux linkage deviations is basically the same. It provides a common basis for current VC as in VC and flux linkage VC as in DTC, despite the differences in mathematical presentation and practical implantation of the two methods.

The common basis of VC and DTC proved by eqns (5.3.11) and (5.3.23) offers the possibility of replacing flux linkage control with current control and vice versa in implementing the motor torque control system. This is explained next as CC.

5.3.2 Combined control of PMS motors

Comparing eqn (5.3.16) with eqn (5.3.20) in connection with eqns (5.3.17) and (5.3.21), respectively, proves the linear dependency of flux linkage deviation components on current deviation components, i.e.,

$$\Delta\lambda_F \propto \Delta i_F, \qquad \Delta\lambda_T \propto \Delta i_T. \tag{5.3.24}$$

The simple dependency of eqn (5.3.24) is revealed upon the use of the machine torque, currents, and flux linkage deviations instead of the variables themselves. The relationship of eqn (5.3.24) is valid regardless of any control method. Therefore, it implies the analogy of current VC and flux linkage VC in general and VC and DTC in particular by the help of deviation signals. This common basis for the interpretation of VC and DTC allows for the combined VC and DTC by selecting appropriate parts of the two methods.

It is well understood that in the DTC method, torque changes by selecting proper voltage vectors according to the amplitude and angle of the stator flux vector, which provides a faster torque response due to a faster selection of the stator voltage vectors. This is due to use of a predetermined switching table instead of a much more time-consuming PWM. Also, using a fast hysteresis controller providing the inputs of the switching table results in a faster dynamic response for DTC. Therefore, the hysteresis controllers and the switching table can be selected for the CC instead of PI current controllers and PWM to retain a fast dynamics.

On the other hand, VC uses current controllers, instead of flux linkage and torque controllers, to generate voltage commands. Therefore, there is no need for torque and flux linkage feedback. Moreover, flux linkage and torque estimation are not required. Thus, CC utilizes the current controllers, instead of flux linkage and torque controllers.

By combining the selected parts of VC and DTC in a single implementation, a CC system is formed as in Fig. 5.11 (Shafaie and Vaez-Zadeh 2012). It is seen that the torque and flux linkage commands are used to generate current component commands according to any desired flux weakening scheme. The commanded current components are compared with the feedback currents to provide current errors as in VC. The errors are applied to hysteresis controllers to determine the switching flags τ and φ. These flags, together with the angle of the stator flux linkage vector, δ_s, are used as the inputs to a switching table to select the inverter voltage vector. The switching table is the same as that in the conventional DTC, except for the inputs of current flags instead of flux linkage and torque flags. However, the current flags are produced by the current errors, which are

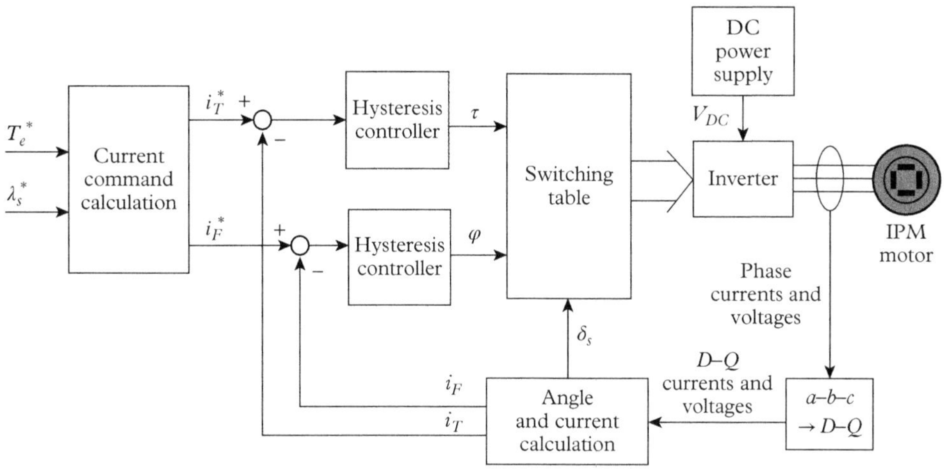

Figure 5.11 *Block diagram of CC applied to a PMS motor.*

proportional to the flux linkage and torque errors as presented by eqn (5.3.24). The proportionalities given in eqn (5.3.24) have a profound role in CC since they allow the same DTC switching table used in connection with current controllers. The stator flux linkage angle is calculated by the scheme presented in Section 4.2. Also, the current commands can be found by using eqn (2.6.17) in connection with eqns (2.6.22) and (2.6.23), where the torque and flux linkage commands are used in the equations.

It is also possible to use the angle of stator current space vector, α_s, as an input to the switching table, instead of δ_s. The switching table then should be modified for current vector.

The performance of a PMS motor under CC can be seen in Figs 5.12–5.14. The motor parameters are given in Table 3.1. The

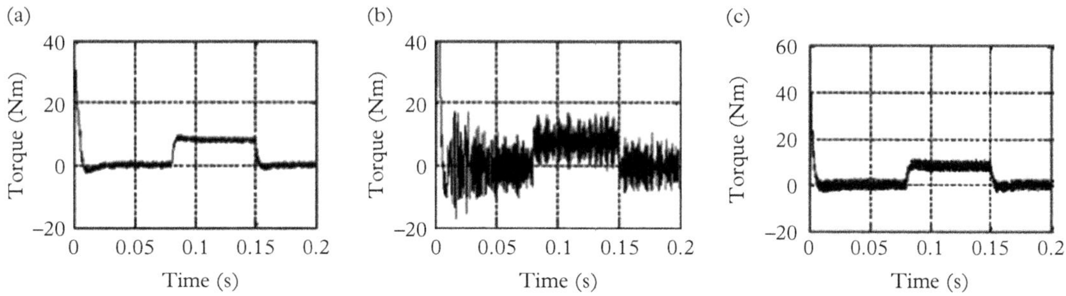

Figure 5.12 *The motor torque dynamics under: (a) VC; (b) DTC; (c) CC (Shafaie and Vaez-Zadeh 2012).*

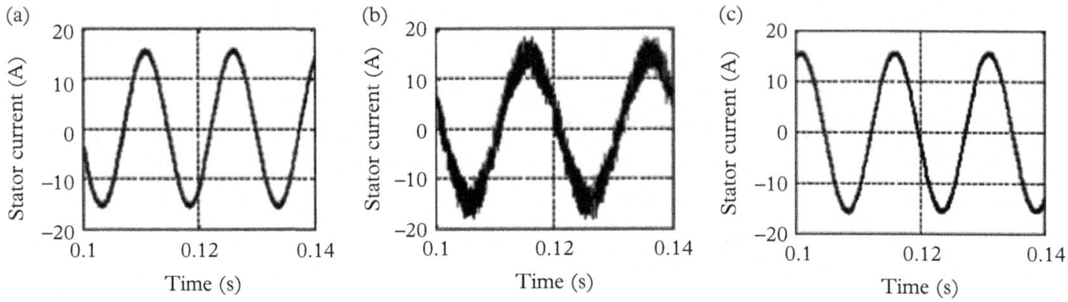

Figure 5.13 *The stator phase current under: (a) VC; (b) DTC; (c) CC (Shafaie and Vaez-Zadeh 2012).*

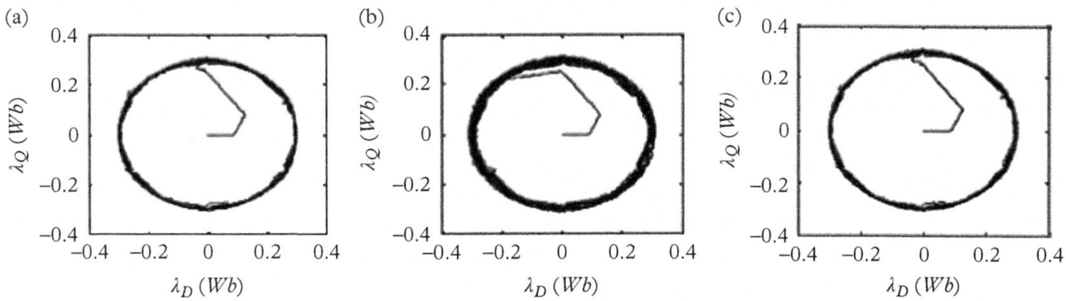

Figure 5.14 *The stator flux linkage trajectory under: (a) VC; (b) DTC; (c) CC (Shafaie and Vaez-Zadeh 2012).*

same motor performance under VC and DTC is also presented for the sake of comparison. Identical operating conditions are sought (Shafaie and Vaez-Zadeh 2012). Figure 5.12 shows the motor developed torque under the three control methods. The torque development under the CC has less pulsation with respect to DTC and more pulsations with respect to VC. These features are due to the indirect torque control and use of the current hysteresis controller, respectively. Figure 5.13 shows the stator phase current under these three control systems. It is seen that the current ripples under the proposed control method are less than those under DTC. This results in less torque pulsations as reported in Fig. 5.12. Figure 5.14 shows the stator flux linkage trajectory under the three control methods. It is evident that the trajectory under CC tends to rotate the flux linkage vector faster than that under VC and slower than that under DTC. Also, the ripples are less than those under DTC and higher than those under VC. This is due to the direct control of the stator flux linkage in DTC and indirect control of the stator flux linkage in the CC and VC.

5.4 Summary

In this chapter, three control methods for PMS motors are presented. The methods include MPC, DBC, and CC. Predictive control as a general control method is adapted to PMS motors. The fundamental principles of the method are explained in discrete time domain first. The method predicts the plant output in a time window for all possible inputs. Then, it calculates an objective function, including a term corresponding to the output error for all the predicted outputs, and selects the inputs, which minimize the objective function in the window. Finally, it applies the input for the next instance to the plant. The method is applied to a surface-mounted PMS motor by presenting the discrete-time model of PMS machines. The objective function is defined as a weighted sum of q-axis current error square and square of maximum torque per ampere conditions.

DBC uses an inverse model of the machine to calculate the desired input in each sampling instance to reach the reference signal in just one sampling interval. It is explained for an inverse current model of a PMS motor to calculate the desired voltage vector components. Then, it is adapted to a direct flux and torque control. The desired voltage components are calculated such that a pair of flux linkage and torque errors is met.

The presentation of CC begins by proving that the machine torque deviation can be formulated by a weighted sum of current component deviations or flux linkage component deviations. It is further proved that the current and flux linkage deviations are proportional. It is then concluded that the torque controls by current VC and flux linkage control are basically the same. Thus, it is possible to select particular parts of VC and DTC for a single control system. Following this reasoning, a CC system including hysteresis current controllers and a switching table is applied to a PMS machine. The system lacks the calculation intensive PWM and torque estimation parts. The motor performance under the control system is presented. It is shown that the motor dynamics is faster than that under VC and smoother than that under DTC.

PROBLEMS

P.5.1. Consider taking the current and voltage limits into account in CC.

P.5.2. Design an objective function for predictive current control of PMS motors by taking unity power factor condition into account.

P.5.3. Design an objective function for predictive flux linkage and torque control by taking MTPA condition into account.

P.5.4. Provide a rough comparison of computation burden in DTC and model-based predictive control.

P.5.5. Derive a formulation of DBC to obtain the appropriate voltage vector for PMS motors in stator flux reference frame.

P.5.6. It is possible to design a CC system by a switching table based on current vector regions instead of flux linkage vector regions. Discuss the system advantages in comparison with the flux linkage vector one.

P.5.7. Consider proofing the common basis of VC and DTC in stator flux linkage reference frame for IPM motors.

P.5.8. Determine the current commands in CC in terms of torque and stator flux linkage command.

P.5.9. Compare CC with DTC system with *d*-axis current command as presented in Chapter 4.

P.5.10. Compare CC with a VC system in stator flux linkage reference frame, which uses hysteresis current controller.

P.5.11. Compare CC with space vector modulation direct torque control (SVM-DTC) having closed-loop torque and flux linkage control as presented in Chapter 4.

BIBLIOGRAPHY

Abbaszadeh, A., Khaburi, D.A., Kennel, R., and Rodríguez, J. (2017). Hybrid exploration state for the simplified finite control set-model predictive control with a deadbeat solution for reducing the current ripple in permanent magnet synchronous motor. *IET Electric Power Appl.* 11(5), 823–835.

Alexandrou, A.D., Adamopoulos, N.K., and Kladas, A.G. (2016). Development of a constant switching frequency deadbeat predictive control technique for field-oriented synchronous permanent-magnet motor drive. *IEEE Trans. Ind. Electron.* 63(8), 5167–5175.

Boulghasoul, Z., Elbacha, A., and Elwarraki, E. (2012). Intelligent control for torque ripple minimization in combined vector and direct controls for high performance of IM drive. *J. Electric. Eng. Technol.* 7(4), 546–557.

Boulghasoul, Z., Elbacha, A., Elwarraki, E., and Yousfi, D. (2011). Combined vector control and direct torque control an experimental review and evaluation. In: *International Conference on Multimedia Computing and Systems*, pp. 1–6. IEEE, Piscataway, NJ.

Chen, X-a., Shan, W-t., He, Y., Lu, Y-y., and Liu, J-f. (2013). The research on hybrid drive and control technique of high-speed motorized spindle based on adaptive fuzzy neural network control. *Proc. Inst. Mech. Eng., Part I, J. Syst. Control Eng.* 228(6), 3–11.

Chen, Y., Sun, D., Lin, B., Ching, T., and Li, W. (2015). Dead-beat direct torque and flux control based on sliding-mode stator flux observer for PMSM in electric vehicles. In: *41st Annual Conference of the IEEE Industrial Electronics Society*, pp. 2270–2275. IEEE, Piscataway, NJ.

Chiang, G.T., Itoh, J-i., Lee, J.S., and Lorenz, R.D. (2012). Performance evaluation of interior permanent magnet synchronous machines using deadbeat-direct torque flux control in an indirect matrix converter with a reactor free boost converter. In: *IEEE Energy Conversion Congress and Exposition*, pp. 4008–4014. IEEE, Piscataway, NJ.

Errouissi, R., Al-Durra, A., Muyeen, S.M., and Leng, S. (2017). Continuous-time model predictive control of a permanent magnet synchronous motor drive with disturbance decoupling. *IET Electric Power Applications*, 22 pp.

Farasat, M. and Karaman, E. (2011). Efficiency-optimized hybrid field oriented and direct torque control of induction motor drive. In: *International Conference on Electrical Machines and Systems*, pp. 1–4. IEEE, Piscataway, NJ.

Farasat, M., Karaman, E., Trzynadlowski, A.M., and Fadali, M.S. (2012). Hybrid field orientation and direct torque control for electric vehicle motor drive with an extended Kalman filter. In: *Energytech, IEEE*, pp. 1–6. IEEE, Piscataway, NJ.

Farasat, M., Trzynadlowski, A.M., and Fadali, M.S. (2014). Efficiency improved sensorless control scheme for electric vehicle induction motors. *IET Electric. Syst. Transport.* 4(4), 122–131.

Fuentes, E., Kalise, D., Rodríguez, J., and Kennel, R. M. (2014). Cascade-free predictive speed control for electrical drives. *IEEE Trans. Ind. Electron.* 61(5), 2176–2184.

Fuentes, E., Rodriguez, J., Silva, C., Diaz, S., and Quevedo, D. (2009). Speed control of a permanent magnet synchronous motor using predictive current control. In: *Proceedings of the International Power Electronics and Motion Control Conference*, pp. 390–395. IEEE, Piscataway, NJ.

Fuentes, E.J., Silva, C., Quevedo, D.E., and Silva, E.I. (2009). Predictive speed control of a synchronous permanent magnet motor. In: *IEEE International Conference on Industrial Technology*, pp. 1–6. IEEE, Piscataway, NJ.

Gehlot, N. and Alsina, P. (1991). Deadbeat controlled field oriented induction motor with reduced order rotor flux observer. In:

International Conference on Industrial Electronics, Control and Instrumentation, pp. 573–578. IEEE, Piscataway, NJ.

Hassaine, S., Moreau, S., Ogab, C., and Mazari, B. (2007). Robust speed control of PMSM using generalized predictive and direct torque control techniques. In: *IEEE International Symposium on Industrial Electronics*. IEEE, Piscataway, NJ, pp. 1213–1218.

Karimi, H., Vaez-Zadeh, S., and Salmasi, F.R. (2016). Combined vector and direct thrust control of linear induction motors with end effect compensation. *IEEE Trans. Energy Convers.* 31(1), 196–205.

Kennel, R., Linder, A., and Linke, M. (2001). Generalized predictive control (GPC)-ready for use in drive applications? In: *2001 IEEE 32nd Annual Power Electronics Specialists Conference*, pp. 1839–1844. IEEE, Piscataway, NJ.

Kenny, B. H. (2001). *Deadbeat Direct Torque Control of Induction Machines Using Self-sensing at Low and Zero Speed.* University of Wisconsin, Madison, WI.

Kenny, B.H. and Lorenz, R.D. (2001). Stator and rotor flux based deadbeat direct torque control of induction machines. In: *Conference Record of the 2001 IEEE Industry Applications Conference, 36th IAS Annual Meeting*, pp. 133–139. IEEE, Piscataway, NJ.

Lee, J-H., Kim, C-G., and Youn, M-J. (2002). A dead-beat type digital controller for the direct torque control of an induction motor. *IEEE Trans. Power Electron.* 17(5), 739–746.

Lee, J.S., Choi, C-H., Seok, J-K., and Lorenz, R.D. (2009). Deadbeat direct torque and flux control of interior permanent magnet machines with discrete time stator current and stator flux linkage observer. In: *IEEE Energy Conversion Congress and Exposition*, pp. 2504–2511. IEEE, Piscataway, NJ.

Lee, J.S., Choi, C-H., Seok, J-K., and Lorenz, R.D. (2011a). Deadbeat-direct torque and flux control of interior permanent magnet synchronous machines with discrete time stator current and stator flux linkage observer. *IEEE Trans. Ind. Appl.* 47(4), 1749–1758.

Lee, J.S., Xu, W., Hurst, Z., Bradley, B., Quattrone, F., and Lorenz, R.D. (2011b). Time optimal torque control and loss minimization in AC machines using deadbeat direct torque and flux control. In: *International Conference on Electrical Machines and Systems*, pp. 1–6. IEEE, Piscataway, NJ.

Lee, J.S., and Lorenz, R.D. (2014). Deadbeat direct torque and flux control of IPMSM drives using a minimum time ramp trajectory method at voltage and current limits. *IEEE Trans. Ind. Appl.* 50(6), 3795–3804.

Lee, J.S., Lorenz, R.D., and Valenzuela, M.A. (2014). Time-optimal and loss-minimizing deadbeat-direct torque and flux control for interior permanent-magnet synchronous machines. *IEEE Trans. Ind. Appl.* 50(3), 1880–1890.

Lee, J.S. and Lorenz, R.D. (2016). Robustness analysis of deadbeat-direct torque and flux control for IPMSM drives. *IEEE Trans. Ind. Electron.* 63(5), 2775–2784.

Lee, K-B., Blaabjerg, F., and Lee, K-W. (2005). A simple DTC-SVM method for matrix converter drives using a deadbeat scheme. In: *European Conference on Power Electronics and Applications*, pp. 1–10. IEEE, Piscataway, NJ.

Lee, K-B., Bae, C-H., and Blaabjerg, F. (2006). An improved DTC-SVM method for matrix converter drives using a deadbeat scheme. *Int. J. Electron.* 93(11), 737–753.

Le-Huy, H., Slimani, K., and Viarouge, P. (1994). Analysis and implementation of a real-time predictive current controller for permanent-magnet synchronous servo drives. *IEEE Trans. Ind. Electron.* 41(1), 110–117.

Linder, A., Kanchan, R., Kennel, R., and Stolze, P. (2010). *Model-based Predictive Control of Electric Drives.* Cuvillier, Göttingen.

Lorenz, R. (2008). The emerging role of dead-beat, direct torque and flux control in the future of induction machine drives. In: *11th International Conference on Optimization of Electrical and Electronic Equipment*, pp. xix–xxvii. IEEE, Piscataway, NJ.

Mariethoz, S., Domahidi, A., and Morari, M. (2009). Sensorless explicit model predictive control of permanent magnet synchronous motors. In: *IEEE International Electric Machines and Drives Conference*, pp. 1250–1257. IEEE, Piscataway, NJ.

Mohammadi, J., Vaez-Zadeh, S., Afsharnia, S., and Daryabeigi, E. (2014). A combined vector and direct power control for DFIG-based wind turbines. *IEEE Trans. Sustain. Energy* 5(3), 767–775.

Moon, H-T., Kim, H-S., and Youn, M-J. (2003). A discrete-time predictive current control for PMSM. *IEEE Trans. Power Electron.* 18(1), 464–472.

Morel, F., Lin-Shi, X., Retif, J-M., and Allard, B. (2008). A predictive current control applied to a permanent magnet synchronous machine, comparison with a classical direct torque control. *Electric Power Syst. Res.* 78(8), 1437–1447.

Morel, F., Lin-Shi, X., Rétif, J-M., Allard, B., and Buttay, C. (2009). A comparative study of predictive current control schemes for a permanent-magnet synchronous machine drive. *IEEE Trans. Ind. Electron.* 56(7), 2715–2728.

Neves, F., Menezes, B., and Silva, S. (2004). A stator flux oriented induction motor drive with deadbeat direct torque and flux control. *Electric Power Components Syst.* 32(12), 1319–1330.

Obermann, T.R., Hurst, Z.D., Bradley, B.F., and Lorenz, R.D. (2010). Deadbeat-direct torque & flux control motor drive using a single control law to minimize motor losses. In: *International Conference on Electrical Machines and Systems*, pp. 742–747. IEEE, Piscataway, NJ.

Obermann, T.R., Hurst, Z.D., and Lorenz, R.D. (2010). Deadbeat-direct torque & flux control motor drive over a wide speed, torque and flux operating space using a single control law. In: *IEEE Energy Conversion Congress and Exposition*, pp. 215–222. IEEE, Piscataway, NJ.

Olavarría, G.D., Fernández, F.M., and Alaküla, M. (2014). FPGA implementation of a deadbeat direct torque and flux control scheme for induction machines. In: *International Conference on Electrical Machines*, pp. 776–782. IEEE, Piscataway, NJ.

Pacas, M. and Weber, J. (2005). Predictive direct torque control for the PM synchronous machine. *IEEE Trans. Ind. Electron.* 52(5), 1350–1356.

Richter, J. and Doppelbauer, M. (2016). Predictive trajectory control of permanent-magnet synchronous machines with nonlinear magnetics. *IEEE Trans. Ind. Electron.* 63(6), 3915–3924.

Rodriguez, J. and Cortes, P. (2012). *Predictive Control of Power Converters and Electrical Drives*, John Wiley & Sons, Chichester.

Saur, M., Piepenbreier, B., Xu, W., and Lorenz, R.D. (2014). Implementation and evaluation of inverter loss modeling as part of DB-DTFC for loss minimization each switching period. In: *16th European Conference on Power Electronics and Applications*, pp. 1–10. IEEE, Piscataway, NJ.

Shafaie, R. and Vaez-Zadeh, S. (2011). Toward a common framework for analysis of high performance controls of PMS motor drives. In: *2nd Power Electronics, Drive Systems and Technologies Conference*, pp. 241–245. IEEE, Piscataway, NJ.

Shafaie, R. and Vaez-Zadeh, S. (2012). A novel control method based on common framework of VC and DTC for IPMS motor drives. *J. Basic Appl. Sci. Res.* 2(4), 4251–4257.

Siami, M., Khaburi, D.A., Abbaszadeh, A., and Rodríguez, J. (2016). Robustness improvement of predictive current control using prediction error correction for permanent-magnet synchronous machines. *IEEE Trans. Ind. Electron.* 63(6), 3458–3466.

Suryawanshi, H., Patil, U., Renge, M., and Kulat, K. (2013). Modified combined DTC and FOC based control for medium voltage induction motor drive in SVM controlled DCMLI. *EPE J.* 23(4), 23–32.

Vaez-Zadeh, S. and Jalali, E. (2007). A combined vector control and direct torque control method for high performance induction motor drives. *Energy Convers. Manag.* 48(12), 3095–3101.

Vaez-Zadeh, S. and Daryabeigi, E. (2011). Combined vector and direct torque control methods for IPM motor drives using emotional controller (BELBIC). In: *2nd Power Electronics, Drive Systems and Technologies Conference*, pp. 145–150. IEEE, Piscataway, NJ.

Wang, B., Guo, W., Wang, Y., and Wang, Z. (2008). A deadbeat direct torque control of surface permanent magnet synchronous machines using space vector modulation. In:–*International Conference on Electrical Machines and Systems*, pp. 1086–1088.IEEE, Piscataway, NJ.

Wang, Y., Flieh, H., Lee, S-C., and Lorenz, R.D. (2015). Implementation issues and performance evaluation of deadbeat-direct torque and flux control drives. In: *IEEE International Electric Machines & Drives Conference*, pp. 953–959. IEEE, Piscataway, NJ.

Wipasuramonton, P., Zhu, Z., and Howe, D. (2006). Predictive current control with current-error correction for PM brushless AC drives. *IEEE Trans. Ind. Appl.* 42(4), 1071–1079.

Xu, W. and Lorenz, R.D. (2014). Dynamic loss minimization using improved deadbeat-direct torque and flux control for interior permanent-magnet synchronous machines. *IEEE Trans. Ind. Appl.* 50(2), 1053–1065.

Xu, W. and Lorenz, R.D. (2014). High-frequency injection-based stator flux linkage and torque estimation for DB-DTFC implementation on IPMSMs considering cross-saturation effects. *IEEE Trans. Ind. Appl.* 50(6), 3805–3815.

Xu, W. and Lorenz, R.D. (2014). Low switching frequency stator flux linkage observer for interior permanent magnet synchronous machines. In: *IEEE Energy Conversion Congress and Exposition*, pp. 5184–5191. IEEE, Piscataway, NJ.

Xu, W. and Lorenz, R.D. (2014). Reduced parameter sensitivity stator flux linkage observer in deadbeat-direct torque and flux control for IPMSMs. *IEEE Trans. Ind. Appl.* 50(4), 2626–2636.

Yang, M., Lang, X., Long, J., and Xu, D. G. (2017). A flux immunity robust predictive current control with incremental model and extended state observer for PMSM drive. *IEEE Trans. Power Electron.* 32(12), 9267–9279.

Zarei, M. E. and Asaei, B. (2013). Combined vector control and direct power control methods for DFIG under normal and unbalanced and distorted grid voltage conditions. In: *4th Electronics, Drive Systems and Technologies Conference*. IEEE, Piscataway, NJ, pp. 107–112.

6 Rotor Position and Speed Estimation

Sensorless control currently forms a major part of the literature on PMS machine control. It is also an essential feature of modern commercial products in the field of motor drives. As a result, the field enjoys overwhelming research in the academic arena and emerging innovations in the market.

Motor position and speed control are the most needed control loops in motor control applications. The control loops of these types need actual rotor position and/or speed signals. There are several means for sensing these variables for motor control systems. Through the use of different laws and rules, they can all be classified under mechanical sensors. Tachometers, resolvers, and encoders are among the most used devices of this kind. A wide variety are readily available in the market for sensing rotor position or speed with a varying degree of accuracy, depending on the application requirement. They often reduce the overall rigidness of the motor drive system, however, and add to its cost. In addition, they need connecting wires from the motor to the control system. The industry usually regards this wiring as cumbersome, particularly if the motor and the control system are far apart. In addition, the connections may be subject to electromagnetic noises, which usually exist in the industrial environment. The solution is to shield the wires, which in turn imposes extra cost to the system.

Sensorless motor control, therefore, is a cost-saving practical alternative to the motor control with mechanical sensors. Most commercial motor drives now have a sensorless option embedded in the control system. Using the option adds to the motor system reliability and rigidity and reduces primary and maintenance costs of the motor drive system. There are many sensorless control methods used in commercial products or presented in the literature. The figure of merit in selecting a sensorless method among the existing ones lies in its ability to function under standstill, steady-state, and transient conditions including acceleration and braking. The low speed transient operation is especially important. The position and speed signal quality, including the required accuracy and noiselessness, are also of

Control of Permanent Magnet Synchronous Motors. Sadegh Vaez-Zadeh. © Sadegh Vaez-Zadeh 2018. Published in 2018 by Oxford University Press. DOI 10.1093/oso/9780198742968.001.0001

major concern. The ability to work under changing reference speed and load, including step reversal and ramp references, is another criterion. Finally, robustness against parameter variations must be mentioned as a figure of merit.

6.1 Methods of rotor position estimation

Sensorless motor control is attractive in practice due to the reasons mentioned previously. It is also challenging due to its demanding criteria like accuracy, robustness, swiftness, and capability of working over the entire range of motor operation. Consequently, position and speed estimation has appeared as an active research area over the past two decades, providing a dense literature. This is also due to various opportunities to tackle the challenges being available. As a result, various position and speed estimation methods have been presented in the literature, many being applicable to PMS motors, or specifically developed for these motors. The requirement of an initial rotor position signal, even for open-loop operation of PMS motors, which is not demanded by induction motors, has brought a new dimension to the rotor position estimation of PMS machines and broadened the corresponding research agenda, compared with that of induction machines. Therefore, a comprehensive investigation of position and speed estimation methods for PMS machines is cumbersome. Also, the classification of the methods by a single arrangement may not be useful or even possible. Nevertheless, the chapter aims at studying selected major developments in the field based on a cross classification of many schemes.

It is common to categorize the position estimation of PMS machines into two main groups: back EMF-based methods and saliency-based methods. This is justified partly because of the distinct cause of the two groups. Back EMF depends on the motion of rotor (motor operation) and saliency is a motor structure property. Based on the research expanding the methods, however, it is difficult to confine the methods to these categories, one reason being the implementation of the back EMF methods and saliency methods by different techniques, which appear to be more important than either of the two main categories. Observer-based estimation, for instance, is capable of using both back EMF and saliency methods. Observer-based estimation, however, has emerged as a major category of position and speed estimation. The other reason is the development of new methods that cannot be categorized under either of the two old categories. Hypothetical position estimation is an example. In fact, the wide diversity of different position and speed estimation methods, on one hand, and their overlap through various aspects, on the other, make categorization of the field a difficult task. Nevertheless, a big picture of the field, even if it is not a matter of consensus, is useful for understanding the methods' fundamental principles, before embarking on the details. Also, it helps the comparison and selection of the methods to be consistent with applications. Therefore, a categorization of the position and speed estimation method is attempted here despite the mentioned difficulties.

The rotor position and speed estimation methods can be divided into five main categories, as seen in the chart of Fig. 6.1. The back EMF-based method, as the most traditional, is still attractive for both salient and non-salient PMS motors as it considers directly the interaction of PM poles with the stator windings during the motor operation. The EMF is position dependent; thus, it can be manipulated by different schemes to deliver the rotor position. The back EMF is induced if there is relative motion between the PM poles and the windings. Therefore, it is not applicable under zero and low speed conditions. Two schemes using the machine equations in a two-axis stationary reference frame and the equations in terms of original machine variables are presented in the second section of this chapter to elaborate the back EMF rotor position estimation.

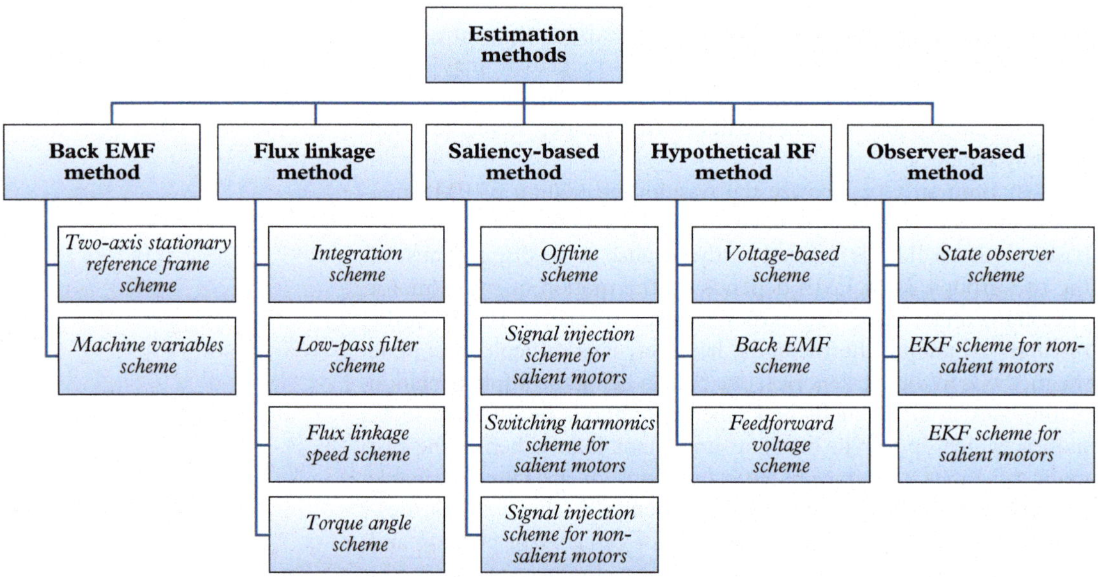

Figure 6.1 *A classification of position and speed estimation methods and schemes.*

The second method is based on the estimation of position and speed of the flux linkage vector. In this method, the magnet-produced EMF is not considered alone, but the total flux linkage provided by the magnet pole and the winding currents is considered. This method is applicable to both salient and non-salient PMS motors. For a reason similar to that mentioned with the previous method, however, it cannot be used for position estimation at zero speed. It is also not accurate under low speed conditions. The method is presented in four different schemes, starting with the traditional integration scheme, followed by its improved version, i.e., low-pass filter scheme. In addition, flux linkage speed estimation, avoiding the derivative of the flux position, is presented. If the torque angle is added to the flux linkage angle, the rotor position is obtained. Therefore, a torque angle estimation scheme is also presented.

The fact that the winding inductance is a function of rotor position is the basis of saliency-based methods. This method is capable of estimating rotor position over the entire speed range, including zero speed. The method is considered by many different schemes in the literature. Traditionally, it is treated by the offline scheme for salient motors only. Online schemes, however, employing high-frequency signals are more attractive. The rotor position can be extracted from high-frequency signals, superimposed on the normal machine variables, e.g., a current component. The high-frequency signals can be produced by injecting an extra high-frequency reference voltage into the inverter. This scheme is applicable to both salient and non-salient PMS motors since the non-salient motors also experience saliency at high frequencies due to the magnetic saturation. The saliency method can be implemented by using high-frequency harmonics caused by the usual inverter switching, instead of the injecting extra signal.

The fourth main category of rotor position estimation is referred to as the hypothetical rotor position or RF method. The main idea behind the method is to present the machine model in a reference frame oriented along a hypothetical rotor position. The error between the hypothetical rotor position and the unknown actual positions is estimated by the machine model using the measured motor signals. As the calculations go on, the error reduces by a PI controller; thus, the hypothetical position approaches the actual rotor position. Different rotor estimation schemes based on this method are presented in this chapter, including voltage-based and back EMF schemes. In addition, a similar feedforward voltage scheme is discussed for non-salient motors.

The last method of rotor position estimation is based on the closed-loop observers. Many different schemes are available under this category, often having overlaps with basic principles of other categories reviewed previously. A simple observer-based rotor position scheme is the state observer scheme. A state model of PMS motors is employed to design an estimator by determining a feedback gain. The estimator is subject to parameter variation error and noise. Therefore, EKF is used as the basis of the most common observer-based schemes for rotor and speed estimation. It can be used for an entire speed range, including zero speed. Despite the theoretical complexity and computation intensity of the method, the robustness against inaccurate and noisy measurements has made the method attractive for commercial products. Two EKF estimation schemes for non-salient and salient motors are presented in the chapter. Other methods like fuzzy logic and neural network estimation methods are also investigated in the literature, but are not elaborated on further in this chapter (Vas 1998, 1999).

6.2 Back EMF-based method

In PMS machines, the movement of rotor magnets induces an alternating back electromotive force in the armature windings according to Faraday's law. The instantaneous magnitude of back EMF depends on the magnets' positions with respect to the windings positions. Therefore, the rotor position can be determined from the back EMF, when the rotor is not stalled. This method of position estimation needs to extract the induced back EMF in a stator phase winding from the phase voltage when the machine is operating. In this respect, the method is best fitted to brushless DC motors, where the inverter does not always supply a phase winding. Thus, the back EMF induced in that phase is easily detected by monitoring the terminal voltage of that phase. In PMS machines, however, it is required to distinguish the back EMF from the current dependent voltages in the windings, i.e., the ohmic voltage drops and the induced voltages from winding currents. Many different schemes have been proposed in the literature for this purpose; from these, two are presented in this section. Also, a third scheme is presented in sub-section 6.4.2 under the hypothetical rotor position method. The rotor position estimation by the back EMF detection method is not capable of the position estimation at zero speed since there is no back EMF at standstill. Also, the position estimation at low speed deteriorates due to low back EMF and noise domination.

6.2.1 Two-axis stationary reference frame scheme

The machine stator voltage equations in two-axis stationary reference frame are presented by eqns (2.4.4) and (2.4.5). The derivative of flux linkage components in these equations contain the D- and Q-axis components of magnet-produced back EMF, i.e., e_{mD} and e_{mQ}, as

$$e_{mD} = -\omega_r \lambda_m \sin\theta_r, \tag{6.2.1}$$

$$e_{mQ} = \omega_r \lambda_m \cos\theta_r. \tag{6.2.2}$$

It is seen that they contain rotor position information in terms of $\sin\theta_r$ and $\cos\theta_r$. The back EMF terms can be calculated from eqns (6.2.1) and (6.2.2) online, if D- and Q-axis current and voltage components are available from measured phase currents and voltages. It is possible to replace the reference voltage with the inverter with the actual voltages, thus saving voltage sensors if the switching interval of the inverter is short compared to the electrical time constant of the motor. Of course, it eliminates the required manipulations of the sensed voltage like filtering, delay compensation, and DC offset compensation. Also, it is possible to determine the D- and Q-axis voltage components in eqns (6.2.1) and (6.2.2) by using the DC link voltage together with switching functions of the inverter. In this case, a single voltage sensor on the DC link replaces the phase voltage sensors. Having calculated the back EMF components, the rotor position is obtained by

$$\theta_r = \tan^{-1}\frac{-e_{mD}}{e_{mQ}}. \tag{6.2.3}$$

This method of rotor position estimation can be applied to both surface-mounted PM machines and IPM machines. In surface-mounted PM machines, where $L_d = L_q$, the terms having the difference of L_d and L_q in eqns (6.2.1) and (6.2.2) vanish. As a result, the motor speed does not interfere with the rotor position estimation. In IPM machines, however, estimation is recursive; i.e., the motor speed must be calculated from the derivative of rotor position and is fed back to the estimation process to update the rotor position.

The back EMF method of position estimation is theoretically straightforward and practically fast and simple, without needing extra measurement other than the ones usually carried out for basic motor control systems. The accuracy of the method, however, depends on the measuring sensors. It is also affected by the noise of the measured machine signals. The calculations of eqns (6.2.1)–(6.2.3) show that the estimation is largely influenced by the motor parameter variations. The sensorless control system by this method needs some means for providing initial rotor position at the motor start.

6.2.2 Machine variables scheme

The rotor position can be estimated by calculating back EMF components in terms of actual measured currents and voltages, without transforming the actual variable to a two-axis reference frame. The final relationship is obtained as (Hoque and Rahman 1994)

$$\theta_r = \tan^{-1}\left(\frac{v_b - v_c - R_s\left(i_b - i_c\right) - L_d\frac{d(i_b - i_c)}{dt} - \sqrt{3}\omega_r\left(L_d - L_q\right)i_a}{\sqrt{3}\left(v_a - R_s i_a - L_d\frac{di_a}{dt}\right) + \omega_r\left(L_d - L_q\right)\left(i_b - i_c\right)}\right), \tag{6.2.4}$$

which can be derived offline by using eqn (6.2.3), where the back EMF components are transferred from the two-axis stationary reference frame to the actual machine variables. A merit of this scheme in practice is the reduced online computations, since the measured currents and voltages are directly used. Nevertheless, some critical online signal manipulations are still required for finding current derivatives to be used in eqn (6.2.4).

6.3 Flux linkage method

The angular position of the stator flux linkage is needed in many high-performance motor control systems including the VC system in stator flux reference frame and the DTC system, as presented in Chapters 3 and 4, respectively. By using this angle, the rotor position is no longer needed in the control system. Therefore, the system is regarded as sensorless. There are many schemes of obtaining this position; a selection of them is presented in this section.

6.3.1 Integration scheme

A straightforward scheme is to calculate the flux linkage angle by using the voltage equations of a machine in a stationary reference frame (Wu and Slemon 1991). This requires sensing the phase voltages and currents of the motor by voltage and current sensors, respectively, and transforming them to two-axis stationary reference frame, the *D*–*Q* reference frame, as elaborated in Chapter 2. They are employed then in calculating the *D*- and *Q*-axis flux linkage components as

$$\lambda_D = \int \left(v_D - R_s\, i_D\right) dt, \tag{6.3.1}$$

$$\lambda_Q = \int \left(v_Q - R_s\, i_Q\right) dt. \tag{6.3.2}$$

Finally, the angle of the stator flux linkage space vector is calculated as

$$\delta_s = tag^{-1}\frac{\lambda_Q}{\lambda_D}. \tag{6.3.3}$$

Having obtained the stator flux linkage vector components by eqns (6.3.1) and (6.3.2), it is also possible to calculate the magnitude of the vector as it is needed in flux control loops, for instance, in DTC systems. Figure 4.8 shows a synthesis of the angle and magnitude of the stator flux linkage, plus the torque.

The sensorless performance of a PMS motor under this method of flux position calculation depends greatly on the accuracy of the calculated stator flux linkage components and these, in turn, depend on the accuracy of the monitored voltages and currents. Therefore, care must be taken to compensate for possible DC offset, phase shift, quantization error, and non-linearity of the sensing devices. Also, the method depends on the stator resistance. Therefore, an accurate value must be used for the nominal value of this parameter and it must be adapted to temperature changes especially under heavy loading.

In addition, the integrations of eqns (6.3.1) and (6.3.2) are prone to errors at low frequencies, where the applied voltages to the motor and thus the *D*- and *Q*-axis voltage components become comparable with the ohmic voltage drops. The voltage drops of the inverter switches must also be taken into account at low frequencies. An incorrect flux angle causes unwanted pulsations in the electromagnetic torque of the motor. The angle estimation error will also cause oscillation in calculated speed values. Nevertheless, with professional care, the open-loop stator flux linkage calculation is satisfactory down to 1–2 Hz. Other methods including closed-loop position estimation methods must be used for frequencies below this value. The reference voltages or the DC link voltage together with the switching functions of the inverter instead of actual voltages may be used to avoid voltage sensors, as mentioned in the previous section.

The method presented in this sub-section can be used in different systems including some VC systems. Two VC systems in the stator flux reference frame are presented in Section 3.7. The systems use current transformations from *a–b–c* RF to *x–y* RF and vice versa. The transformations need the stator flux linkage angle, which is obtained by the help of a rotor position sensor. The systems can become sensorless by implementing the stator flux linkage angle estimation method just presented.

6.3.2 Low-pass filter scheme

Low-pass filters, instead of the pure integrators of eqns (6.3.1) and (6.3.2), can estimate the stator flux linkage components. In this scheme, single-stage or cascaded-stage filters may be used. Each stage may include a filter of the form $T/(1 + pT)$ as presented in Fig. 6.2 (Vas 1998). T is the filter time constant and must be adjusted for best

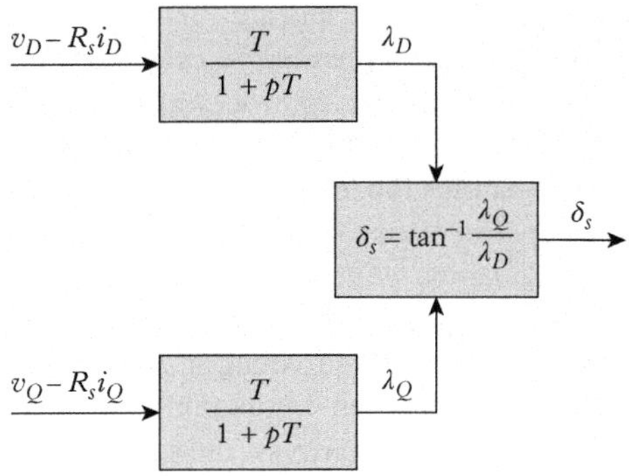

Figure 6.2 *Stator flux linkage estimation by using low-pass filters.*

performance. A large T contributes to accurate flux estimates at low stator frequencies by decreasing the phase shift of the estimated flux linkages with respect to the actual ones. A large T adversely affects the swiftness of the estimation by decreasing the damping characteristic of the filter. Therefore, T must be designed such that an overall best performance is achieved. It is done by compromising between accuracy and swiftness as desired by the applicant. This degree of freedom is an advantage of using filters instead of pure integrators. Also, the effect of initial conditions on the integration reduces by using filters instead of pure integrators. Moreover, filters with an adaptive time constant can be designed and implemented to achieve good accuracy and swiftness over a range of frequencies.

6.3.3 Flux linkage speed estimation scheme

In PMS machines, the flux linkage angle can be represented by

$$\delta_s = \theta_r + \delta, \tag{6.3.4}$$

where θ_r is the rotor magnet pole position and δ is the stator flux linkage position with respect to the pole position or the load angle. The first time derivative of the flux linkage position gives the rotating speed of the flux linkage vector, i.e.,

$$\omega_s = p\delta_s. \tag{6.3.5}$$

At steady-state operation, where the flux linkage position is fixed, eqn (6.3.5) gives the rotor speed too. At transient state, however, where there is a change in the reference electromagnetic torque, the stator flux linkage space vector moves relative to the rotor (to produce a new torque level as stated in Section 4.2), and this influences the rotor speed. This effect can be neglected if the rate of change of the electromagnetic torque is limited (Vas 1998). Otherwise, the rotor position must be estimated as will be presented in the next sub-section. The speed calculation by eqn (6.3.5) can be problematic due to the discontinuity in the position of the stator flux linkage vector. Therefore, it is desirable to avoid derivation of this signal. This is possible by substituting eqn (6.3.3) into eqn (6.3.5) to yield

$$\omega_s = \frac{\lambda_D p\lambda_Q - \lambda_Q p\lambda_D}{\lambda_D^2 + \lambda_Q^2}, \qquad (6.3.6)$$

where the derivatives of the flux linkage components in eqn (6.3.6) can be replaced by the corresponding voltage components minus the resistive voltage drops. The replacements result in

$$\omega_s = \frac{\lambda_D(v_Q - R_s i_Q) - \lambda_Q(v_D - R_s i_D)}{\lambda_D^2 + \lambda_Q^2}. \qquad (6.3.7)$$

The flux linkage components in eqn (6.3.7) are given by eqns (6.3.1) and (6.3.2). The considerations presented with the integration scheme presented in sub-section 6.3.1 must be re-emphasized here. It is especially recalled that the speed estimation by this scheme deteriorates at low speeds due to the inaccuracy of the flux linkage integrations of eqns (6.3.1) and (6.3.2).

6.3.4 Torque angle estimation scheme

Both rotor position and speed are required in many motor control systems. The rotor position and speed can be estimated by using the estimated stator flux linkage vector position and speed, as presented by different schemes in the previous section. The relation between the rotor position and the stator flux linkage position shown in eqn (6.3.4) can be re-presented as

$$\theta_r = \delta_s - \delta. \qquad (6.3.8)$$

Therefore, having estimated δ_s, the rotor position can be calculated by estimating the torque angle, δ. The torque angle in PMS machines, in general, depends on the current vector components and is given by the following equation as investigated thoroughly in Chapter 3:

$$\delta = tag^{-1}\frac{\lambda_q}{\lambda_d} = \frac{L_q i_q}{\lambda_m + L_d i_d}. \qquad (6.3.9)$$

Substituting eqns (6.3.3) and (6.3.9) into eqn (6.3.8) results in the rotor position estimation. The rotor position thus needs the measurement of phase currents and transforms them into the rotor RF.

The transformation itself requires an online estimation of the rotor position, however. Therefore, the rotor position estimation by this method, in contrast to the estimation of δ_s, is recursive. It is possible to design a recursive estimation of the rotor position by using reference currents instead of actual ones. This is possible in VC of PMS motors, where the reference stator current components, i_d^* and i_q^*, are generated by flux linkage and torque (speed) controllers, respectively. This position estimation provides satisfactory results if the current controllers work accurately and rapidly. Otherwise, compensating means is necessary (Genduso *et al.* 2010). Having estimated the rotor position, the rotor speed can then be obtained by differentiating the position signal. The limitations considered with the flux linkage angle estimation in sub-section 6.3.1 also exist in this scheme.

6.4 Hypothetical reference frame method

The main idea behind this method is to present the surface-mounted PMS motor model in a hypothetical reference frame oriented along an estimated rotor position. The error between the estimated rotor position and the unknown actual position is formulated as part of a motor electrical signal, i.e., a voltage component. The signal is estimated by the machine model using the measured motor variables. The error is then applied to a PI controller. The controller reduces the error; thus, the hypothetical position approaches the actual rotor position. In addition, a back EMF scheme and a feedforward voltage scheme are discussed using the hypothetical RF.

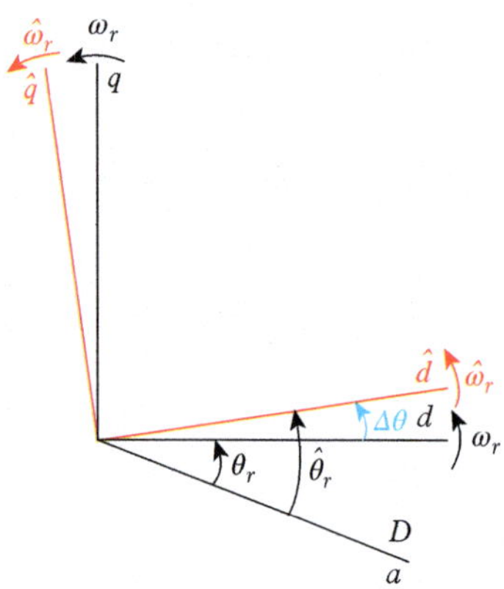

Figure 6.3 *Hypothetical rotor reference frame in counterclockwise direction of rotation.*

6.4.1 Voltage-based scheme

Transformation of a PMS motor model from *a-b-c* stationary reference frame to *d-q* rotor RF is convenient for the purpose of the motor analysis and control. This is done by using the information of rotor position. In the case of sensorless control, however, where the actual rotor position, θ_r, is not available, the transformation may be carried out from *a–b–c* RF to a hypothetical two-axis RF rotating with an estimated rotor position of $\hat{\theta}_r$ with respect to the *a-b-c* RF. This RF can be referred to as $\hat{d}$–$\hat{q}$ RF. The angle between the *d*–*q* and the $\hat{d}$–$\hat{q}$ reference frames would be $\Delta\theta = \hat{\theta}_r - \theta_r$ as seen in Fig. 6.3.

The voltage equations in $\hat{d}$–$\hat{q}$ RF is given by

$$\begin{bmatrix} \hat{v}_d \\ \hat{v}_q \end{bmatrix} = \begin{bmatrix} R_s + L_s p & -L_s p\hat{\theta}_r \\ L_s p\hat{\theta}_r & R_s + L_s p \end{bmatrix} \begin{bmatrix} \hat{i}_d \\ \hat{i}_q \end{bmatrix} + \lambda_m p\theta_r \begin{bmatrix} \sin\Delta\theta \\ \cos\Delta\theta \end{bmatrix}. \quad (6.4.1)$$

Also, the voltage equations in the actual rotor RF, if the actual position were known, could be presented by

$$\begin{bmatrix} v_d \\ v_q \end{bmatrix} = \begin{bmatrix} R_s + L_s p & -L_s p\theta_r \\ L_s p\theta_r & R_s + L_s p \end{bmatrix} \begin{bmatrix} i_d \\ i_q \end{bmatrix} + \lambda_m p\theta_r \begin{bmatrix} 0 \\ 1 \end{bmatrix}. \qquad (6.4.2)$$

The current components in the actual rotor RF, i_d and i_q, can be related to the hypothetical RF by

$$\begin{bmatrix} i_d \\ i_q \end{bmatrix} = \begin{bmatrix} \cos\Delta\theta & -\sin\Delta\theta \\ \sin\Delta\theta & \cos\Delta\theta \end{bmatrix} \begin{bmatrix} \hat{\imath}_d \\ \hat{\imath}_q \end{bmatrix}. \qquad (6.4.3)$$

Substituting eqn (6.4.3) into eqn (6.4.2) yields

$$\begin{bmatrix} v_d \\ v_q \end{bmatrix} = \begin{bmatrix} R_s + L_s p & -L_s p\hat{\theta}_r \\ L_s p\hat{\theta}_r & R_s + L_s p \end{bmatrix} \begin{bmatrix} \hat{\imath}_d \\ \hat{\imath}_q \end{bmatrix} + \lambda_m p\theta_r \begin{bmatrix} 0 \\ 1 \end{bmatrix}. \qquad (6.4.4)$$

When comparing the longitudinal voltage components in eqns (6.4.1) and (6.4.4), the d-axis voltage difference can be given by (Matsui and Shigyo 1992)

$$\Delta v_d = \hat{v}_d - v_d = \lambda_m p\theta_r \sin\Delta\theta. \qquad (6.4.5)$$

Assuming that $p\theta_r \neq 0$ and $\Delta\theta$ is small, the following approximation is valid:

$$\Delta v_d = \hat{v}_d - v_d \approx \lambda_m p\theta_r \Delta\theta \propto \Delta\theta, \qquad (6.4.6)$$

which indicates that the rotor position error can be calculated by having the d-axis voltage difference, Δv_d, through detecting the actual voltage and calculating the estimated voltage using the detected current. It is possible to calculate the actual voltage, using the DC link voltage and the inverter switching function. The motor speed also needs to be estimated in order to implement the position estimation. This can be obtained from the $\hat{q}$-axis voltage equation (6.4.1), assuming that $\Delta\theta \approx 0$ to yield

$$p\theta_r \approx p\hat{\theta}_r = \frac{\hat{v}_q - (R_s + L_s p)\,\hat{\imath}_q}{\lambda_m + L_s \hat{\imath}_d}. \qquad (6.4.7)$$

Since the rotor position can be obtained by integrating the rotor speed, a PI controller acting on the speed error signal of $p\hat{\theta}_r - p\theta_r$ reduces the error signal and gives the estimated position. The position estimation scheme can be implemented by the control system of Fig. 6.4.

It is evident from eqns (6.4.6) and (6.4.7) that the position and speed estimations depend on the motor parameters. Therefore, the estimation accuracy depends on the accuracy on the parameters.

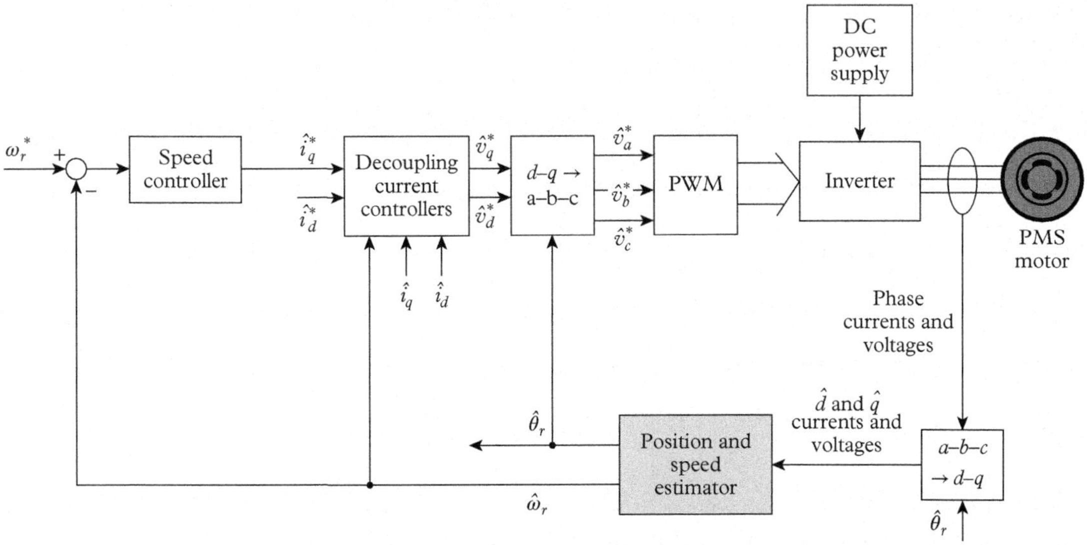

Figure 6.4 *Implementation of rotor position scheme in the VC system of a PMS machine.*

The accuracy is thus improved by using an effective online parameter estimation method to take into account parameter variations.

6.4.2 Back EMF scheme

Another position estimation method can be implemented by considering eqn (6.4.1) with its corresponding hypothetical or estimated reference frame and variables and parameters as defined in the previous sub-section (Kim and Sul 1997).

The estimated rotor position and speed are renewed in every estimation interval. If this interval is sufficiently shorter than the mechanical time constant of the machine, the rotor position can reasonably be assumed constant during the interval. Also, the current commands for the inner current loops in Fig. 6.4 are generated by the outer speed control loop at the beginning of the estimation interval. If the electrical time constant and the sampling time are much less than the position estimation interval, then the motor currents have sufficient time to pass their transient states and approach to their commanded values at the end of the estimation interval. The previously mentioned timing arrangement is easily met in today's motor control systems. Therefore, the deviations of rotor position and stator currents can be ignored in eqn (6.4.1) at the end of each estimation interval, i.e.,

$$p\hat{\theta}_r = 0, \quad p\hat{i}_d = 0, \quad p\hat{i}_q = 0. \tag{6.4.8}$$

The estimated voltage equations are thus reduced to

$$\hat{v}_d = R_s\hat{\imath}_d + \lambda_m p\theta_r \sin \Delta\theta, \tag{6.4.9}$$

$$\hat{v}_q = R_s\hat{\imath}_q + \lambda_m p\theta_r \cos \Delta\theta. \tag{6.4.10}$$

These equations can be rearranged as

$$\hat{e}_d = \hat{v}_d - R_s\hat{\imath}_d = \lambda_m p\theta_r \sin \Delta\theta, \tag{6.4.11}$$

$$\hat{e}_q = \hat{v}_q - R_s\hat{\imath}_q = \lambda_m p\theta_r \cos \Delta\theta, \tag{6.4.12}$$

where $\hat{e}_q$ and $\hat{e}_d$ are the estimated d- and q-axis back EMF quantities. It is assumed that the initial rotor estimation error is sufficiently small. Thus, at the end of each estimating interval the position error can be calculated by

$$\frac{\hat{e}_d}{\hat{e}_q} = \frac{\sin \Delta\theta}{\cos \Delta\theta} = \tan \Delta\theta \approx \Delta\theta. \tag{6.4.13}$$

Therefore, by having the right-hand side of eqn (6.4.13), the estimated rotor position may converge to the actual rotor position in consecutive estimation intervals by the rule

$$\hat{\theta}_r\,(n+1) = \hat{\theta}_r\,(n) + \Delta\theta, \tag{6.4.14}$$

where n and $n + 1$ refer to any two consecutive estimation intervals. Now, the estimated speed can simply be obtained from the time derivative of the estimated rotor position at each interval. The derivation of position signal may lead to performance instability due to the noises in the position signal, however. An alternative speed estimation scheme is to use the equation

$$\hat{e}_d^2 + \hat{e}_q^2 = (\lambda_m\omega_r)^2 \left[\sin^2\Delta\theta + \cos^2\Delta\theta\right] = (\lambda_m\omega_r)^2. \tag{6.4.15}$$

As the direction of rotor speed agrees with the sign of $\hat{e}_q$ in a wide region of position error, i.e., $-\pi/2 < \Delta\theta < \pi/2$, the estimated motor speed can be determined by

$$\hat{\omega}_r = \frac{1}{\lambda_m}\sqrt{\hat{e}_d^2 + \hat{e}_q^2}\ \text{sign}\left(\hat{e}_q\right). \tag{6.4.16}$$

The implementing system is the same as that depicted schematically in Fig. 6.4, except for the d- and q-axes voltage components, which are replaced by the reference ones. Thus, the position and speed estimator block receives the current components and the reference voltage components all in the hypothetical RF to calculate the rotor position according to eqns (6.4.11)–(6.4.14) and the motor speed according to eqn (6.4.16). The estimated speed is used as the feedback signal to the speed controller. The controller generates i_q as in the conventional VC system of PMS motors. This current signal, together with $i_d^* = 0$, is applied to the current controllers to produce the voltage commands both for the estimation block and for being transformed to the phase voltage commands. The estimated rotor position is used for this transformation and for transforming the detected phase currents to the hypothetical RF.

This rotor and speed estimation scheme is dependent on R_s and λ_m. It does not need any voltage measurement as its voltage command signals are given by current controllers. Therefore, it requires fast and accurate current sensing and control.

6.4.3 Feedforward voltage scheme

A very simple speed and position estimation scheme can be implemented by just using a voltage reference signal (Bae *et al.* 2003). In this scheme, feedforward voltage signals are generated by using the commanded current signals and are added to the output of the current controllers. Such signals are used in the decoupling circuit of the VC for IPM motors as presented in Fig. 3.7. In surface-mounted PMS motors the decoupling is mainly maintained by choosing $i_d = 0$. As a result, the q-axis voltage equation is independent from i_d. Although the dependency of the d-axis voltage on the q-axis current still is showing up, no special decoupling circuit or feedforward signals are usually employed. The feedforward decoupling signals can be used for the sake of speed estimation, however. These signals are the same as eqns (3.4.5) and (3.4.6), except for extra resistive voltage drops, as

$$v_{d0} = R_s i_d^* - L_s \omega_r i_q^*, \tag{6.4.17}$$

$$v_{q0} = R_s i_q^* + L_s \omega_r i_d^* + \omega_r \lambda_m, \tag{6.4.18}$$

where $\hat{\omega}_r$ is the estimated motor speed. The voltage equations transferred to the estimated rotor position reference frame can be give as

$$v_d = R_s i_d - L_s \omega_r i_q + \omega_r \lambda_m \sin \Delta\theta, \tag{6.4.19}$$

$$v_q = R_s i_q + L_s \omega_r i_q + \omega_r \lambda_m \cos \Delta\theta, \tag{6.4.20}$$

where the electrical transient is neglected and $\Delta\theta = \hat{\theta}_r - \theta_r$ is the error between the estimated and the actual rotor positions, respectively.

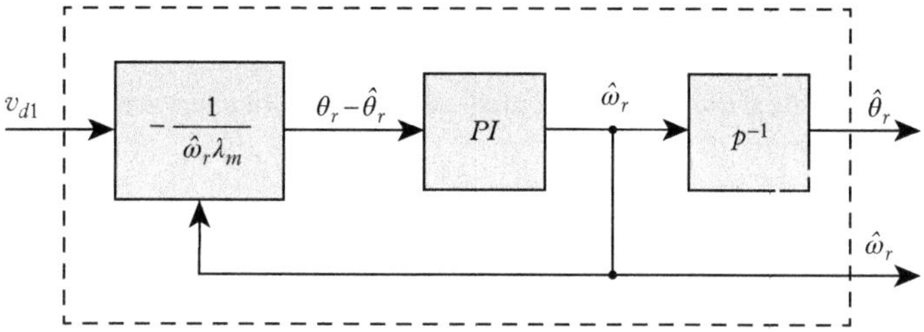

Figure 6.5 *Speed and position estimator by feedforward voltage.*

Referring to Fig. 3.7, it is seen that the d-axis current controller output is given by

$$v_{d1} = v_d^* - v_{d0}. \tag{6.4.21}$$

If the current controllers act accurately, then $v_d^* = v_d$ at the steady-state. Therefore, eqn (6.4.21) can be calculated by subtracting eqn (6.4.17) from eqn (4.6.19). This yields

$$v_{d1} = R_s\left(i_d - i_d^*\right) - L_s\omega_r(i_q - i_q^*) + \omega_r\lambda_m \sin\Delta\theta. \tag{6.4.22}$$

Considering that the current errors are trivial under the PI controllers and also, assuming that $\Delta\theta$ is small, the following relationship is reasonably held:

$$v_{d1} \approx \omega_r\lambda_m \sin\Delta\theta \approx \omega_r\lambda_m\Delta\theta. \tag{6.4.23}$$

Therefore, for estimating the rotor position and speed in VC of surface-mounted PMS motors with decoupling current controller, the output of the d-axis current controller, v_{d1}, is divided by $-\hat{\omega}_r\lambda_m$ to yield $\theta_r - \hat{\theta}_r$. This error is applied to a PI controller to provide the estimated speed as its output. Finally, the estimated speed is integrated to give the estimated rotor position. These are seen in Fig. 6.5.

This method of speed estimation does not need any voltage sensor for detecting either the phase voltages or the DC link voltage. It performs accurately in high-speed range, but does not work properly at standstill and low speeds.

6.5 Saliency-based method

The saliency-based method of rotor position estimation, as mentioned at the opening of this chapter, is traditionally regarded as a main category of rotor position estimation. The method is laid on the inherent dependency of motor phase inductances on rotor position as a structural characteristic of many electrical machines. This is because the inductance of a motor phase winding depends on the magnetic flux linkage of the phase. The flux linkage varies with the rotor position if there is magnetic saliency in the motor as in inset and IPM motors. Therefore, the phase inductance depends on the rotor position. This fact can be used to estimate the rotor position of such a motor during the steady-state and transient-speed operations.

It is possible to estimate the rotor position by this method over the entire speed range, including zero speed. There are several schemes for the saliency-based position estimation method, which can be divided into two main groups, i.e., with and without using high-frequency signals. The position estimation without high-frequency signals is referred to here as offline scheme in which the position is obtained from the measured motor variables. The high-frequency signal schemes use high-frequency signals, which are superimposed on the motor variables. The high-frequency signals are provided by external signals applied to the motor, or they are caused by normal behavior of inverters.

6.5.1 Offline scheme

This scheme calculates offline the phase inductance of a motor from the actual motor voltage and current in different rotor positions and stores the results in a look-up table with the phase inductance values as the input and the rotor positions as the output. Then, as the motor works, the phase inductance is recalculated online at each sampling period and its value is given to the look-up table to obtain the rotor position corresponding to the phase inductance. In this way, the rotor position is estimated in each sampling period (Kulkarni and Ehsani 1992).

The machine voltage equations in stationary *a*–*b*–*c* reference frame presented by eqns (2.3.1) and (2.3.2) show that the voltage of a phase winding is a function of the motor phase currents in addition to the back EMF produced by the PM rotation. If the switching frequency of the inverter that supplies the motor is high (> 10 kHz), however, the variation of inductances with rotor position can be neglected during a switching interval. Considering this assumption, and since the sum of stator phase currents are zero, the instantaneous voltage of phase *a* reduces to

$$v_a = R_s i_a + L_a p i_a + e_{ma}, \tag{6.5.1}$$

where the phase inductance L_a is given by

$$L_a = L_{aa} - M_{ab}, \tag{6.5.2}$$

and L_{aa} and M_{ab} are the self and mutual inductances, respectively. Recall that they are functions of $2\theta_r$ as given by eqns (2.3.4) and (2.3.7). Also, e_{ma} is the magnet-produced back EMF of phase *a*.

L_a is obtained from eqn (6.5.1) as

$$L_a = (v_a - R_s i_a - e_{ma})/pi_a. \tag{6.5.3}$$

If pi_a and e_{ma} are calculated during the switching interval, the phase inductance can be calculated from eqn (6.5.3). Assuming that the switching frequency is high, pi_a can be calculated by using the values of the motor phase current at two previous samplings as

$$pi_a = \frac{\Delta i_a}{\Delta \mathrm{t}} = \frac{i_{a1} - i_{a0}}{t_1 - t_0}, \tag{6.5.4}$$

where subscripts 1 and 0 in eqn (6.5.4) denote the two consecutive most recent samplings. Also, the back EMF can be calculated by referring to the first equation of (2.3.2) as

$$e_{ma} = Kp\theta_r = K\frac{\theta_0 - \theta_1}{t_0 - t_1}, \tag{6.5.5}$$

where $K = \lambda_m \sin\theta_r$ and is assumed constant during the interval. This is in accordance with the assumption of constant back EMF during the interval. Again, subscripts 1 and 0 refer to the two samplings.

It must be noted that L_a is a function of $2\theta_r$ as a result of eqn (6.5.2). This means that each calculated value of this inductance corresponds to four different rotor positions. A solution to find a correct position among the four positions is to use the information from all three phase inductances during a switching interval. Figure 6.6 depicts such inductances during an electrical cycle, where the inductance of each phase is calculated offline for a position interval of 60 electrical degrees and is stored in the look-up table as shown by bold lines in the figure (Kulkarni and Ehsani 1992). It is seen that phase *b* inductance is calculated for position estimation in 0–15 electrical degrees, phase *a* inductance in 15–75 electrical degrees, phase *c* inductance in 75–135 electrical degrees, and phase *b* inductance again in 135–195 electrical degrees, and so on. The look-up table may store values of each phase inductance for every 0.05 electrical degree increment. Because only a segment of 60° wide

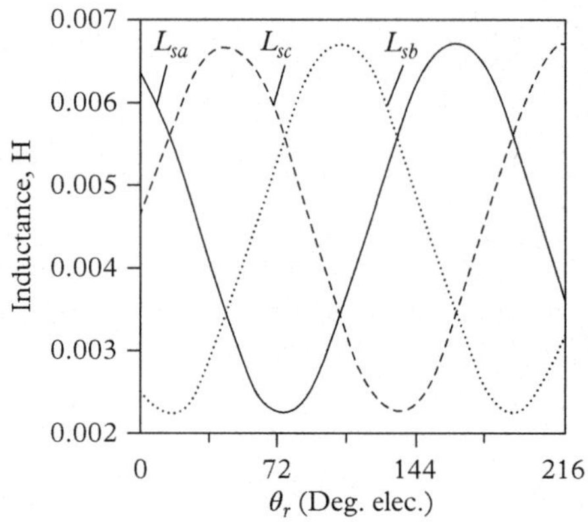

Figure 6.6 *Phase inductances during an electrical cycle (Kulkarni and Ehsani 1992).*

of each inductance profile is utilized for estimating the rotor position, the look-up table needs not much memory.

The method also needs online estimation of phase inductances as inputs to the look-up table to get the rotor position at each instance as the output of the look-up table. The same procedure as presented previously for offline calculations can be used for online calculation of the inductances.

The online part of the estimation process is fast. The offline measurement of inductances is time consuming, however. A dedicated setup can be built to automate the measurement, in the case of high volume production. The estimated position may be subject to high noise due to the calculation of current derivative in eqn (6.5.4). The accuracy of the estimated position depends on both offline and online calculations of inductances. The inductances, in turn, depend on the accuracy of the machine measurements, which is more difficult during the online sampling. The accuracy also depends on the resolution of the look-up table. A higher resolution, of course takes more memory for storing the look-up table. It is possible to determine effective inductance values for position points for which there is no inductance stored in the look-up table by averaging the adjacent values. The operating conditions may adversely affect the accuracy of the estimated position due to motor parameter variations. The variations may occur because of magnetic saturation under heavy loading conditions and high frequency at high speed. Multiple look-up tables may be used to take into account the operating conditions, e.g., for light load and heavy load conditions, at the price of increased memory for storing the look-up tables. Finally, it is reminded that the scheme can only be applied effectively to inset and IPM motors, and not to surface-mounted PMS motors. As a rule of thumb, higher magnetic saliency leads to better position estimation.

6.5.2 High-frequency signal injection scheme for salient motors

This scheme also utilizes the dependency of motor inductances on rotor position to estimate position and speed. This is achieved by injecting high-frequency low-magnitude voltages into the motor stator windings, in addition to the usual voltages. The motor currents are then sensed and processed to produce a signal that is related to the error between the actual rotor position and an estimated rotor position. The error is finally applied to an observer to be compensated, and the position and speed are obtained (Corley and Lorenz 1998). The scheme is elaborated later.

The high-frequency voltage is injected into the stator windings such that it produces a high-frequency current with $i_{di} = 0$, where i_{di} is the d-axis component of the current vector in a two-axis rotating reference frame. The RF rotates with the estimated rotor position, $\hat{\theta}_r$. In this RF thus, an injected flux linkage vector produced only by i_q in steady-state is defined by

$$\lambda_{di} = 0, \tag{6.5.6}$$

$$\lambda_{qi} = \frac{v_{si}}{\omega_i} \sin \omega_i t, \tag{6.5.7}$$

where v_{si}/ω_i is the magnitude of the injected high-frequency flux linkage and ω_i is the frequency of the injected voltage. The flux linkage in the stationary RF is obtained by

$$\lambda_{Di} = \lambda_{si} \sin \hat{\theta}_r, \tag{6.5.8}$$

$$\lambda_{Qi} = \lambda_{si} \cos \hat{\theta}_r. \tag{6.5.9}$$

These flux linkage components in terms of stationary current components i_{Di} and i_{Qi} are obtained as

$$\lambda_{Di} = -L_2 \sin 2\theta_r i_{Qi} + (L_1 - L_2 \cos 2\theta_r) i_{Di}, \tag{6.5.10}$$

$$\lambda_{Qi} = (L_1 + L_2 \cos 2\theta_r) i_{Qi} - L_2 \sin 2\theta_r i_{Di}, \tag{6.5.11}$$

where L_1 and L_2 are given by (2.5.12) and (2.5.13). Now, from eqns (6.5.6)–(6.5.11), the injected current components are obtained as

$$i_{Di} = \left[-I_{i0} \sin \hat{\theta}_r + I_{i1} \sin 2(\theta_r - \hat{\theta}_r)\right] \sin \omega_i t, \tag{6.5.12}$$

$$i_{Qi} = \left[I_{i0} \cos \hat{\theta}_r - I_{i1} \cos 2(\theta_r - \hat{\theta}_r)\right] \sin \omega_i t, \tag{6.5.13}$$

where the injected current magnitudes, I_{i0} and I_{i1}, are

$$I_{i0} = \frac{v_{si}}{\omega_i} \cdot \frac{L_1}{L_1^2 - L_2^2}, \tag{6.5.14}$$

$$I_{i1} = \frac{v_{si}}{\omega_i} \cdot \frac{L_2}{L_1^2 - L_2^2}. \tag{6.5.15}$$

The d-axis component of the injected current vector in estimated rotor position RF is

$$i_{di} = i_{Qi} \sin\hat{\theta}_r + i_{Di} \cos\hat{\theta}_r. \tag{6.5.16}$$

Substituting from eqns (6.5.12)–(6.5.13) into eqn (6.5.16) yields

$$i_{di} = I_{i1} \sin 2(\theta_r - \hat{\theta}_r) \sin \omega_i t. \tag{6.5.17}$$

This is a sinusoidal signal with the same frequency as the frequency of the injected voltage. It also contains a DC magnitude proportional to $\sin 2(\theta_r - \widehat{\theta}_r)$. This is the main signal for estimating the rotor position. The signal is the same as the signal generated in an RTD converter sensing system. Thus, it is processed according to a procedure similar to that of an RTD converter. The signal is first demodulated to obtain a DC signal proportional to the position error, $\theta_r - \hat{\theta}_r$. The DC signal is then applied to an observer to update the estimated values of the motor speed and position to converge the estimated position to the actual one. The implementation of the system is shown in Fig. 6.7.

The actual phase currents are first transformed to the rotor reference frame. The d-axis current then goes through a band-pass filter (BPF) to give eqn (6.5.17). The result passes through a demodulation block to obtain the DC signal to be applied to the observer, which will be explained next.

The system must also manipulate the high-frequency voltage command to ensure $i_{di} = 0$, as mentioned previously. Neglecting the ohmic voltage drop, the d- and q-axis components of the injected voltage are given by

$$v_{di} = p\lambda_{di} - \hat{\omega}_r \lambda_{qi}, \tag{6.5.18}$$

$$v_{qi} = p\lambda_{qi} + \hat{\omega}_r \lambda_{di}. \tag{6.5.19}$$

Substituting eqns (6.5.6) and (6.5.7) into eqns (6.5.18) and (6.5.19) yields

$$v_{di} = -\frac{\hat{\omega}_r}{\omega_r} v_{si} \sin \omega_i t, \tag{6.5.20}$$

$$v_{qi} = v_{si} \cos \omega_i t. \tag{6.5.21}$$

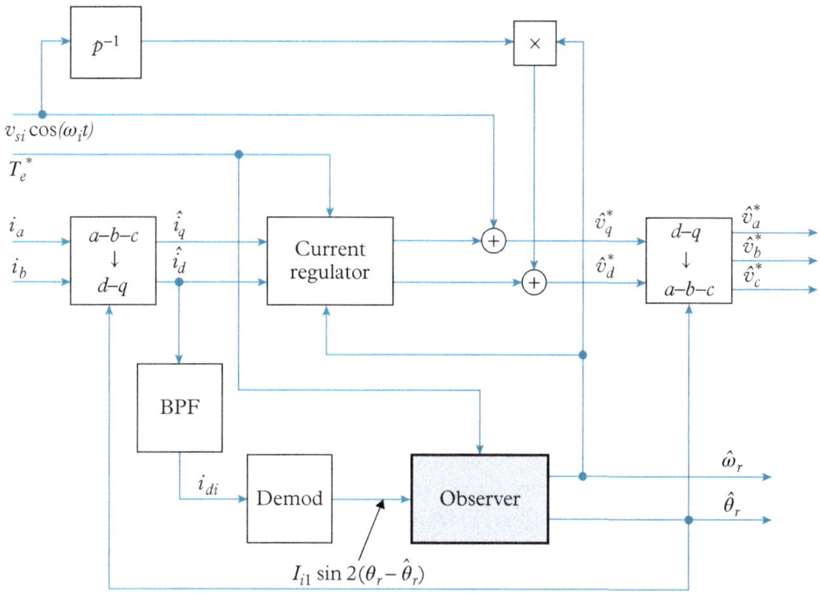

Figure 6.7 *Rotor position and speed estimation system by signal injection.*

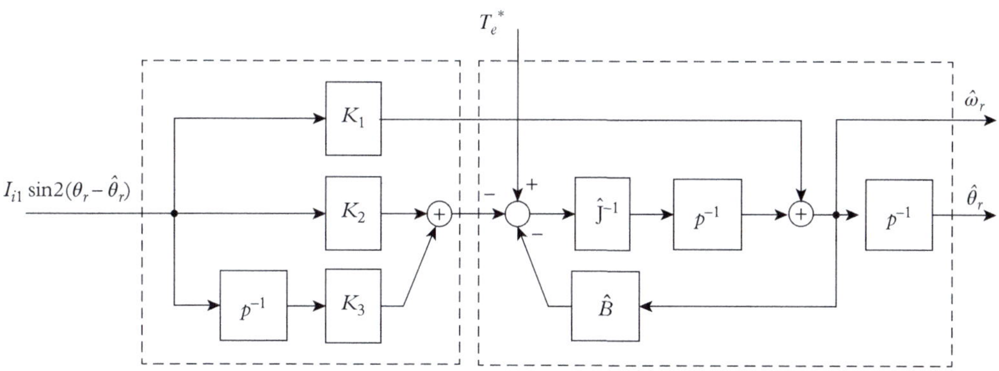

Figure 6.8 *Observer block diagram for estimating rotor position and speed by using the DC magnitude of the injected current signal.*

These equations are used in Fig. 6.7 to generate the injected voltage commands. A current regulator is also included in the system.

A synthesis of the observer of the Luenberger type is given in Fig. 6.8. The observer consists of two parts. In the first part, a *PI* controller reduces the error toward 0 and provides an estimated torque signal. The estimated torque, together with the reference torque, are applied to the mechanical model of the machine as the second part of the estimator to give the estimates of rotor speed and position. A feedforward speed signal, provided by a proportional position controller in the first part of the observer, is added to the speed output of the mechanical model for the sake of compensating the error. The observer works under the load change and speed transient.

The estimation method is independent of the speed value and works over a wide speed range, including zero speed.

The magnitude of the injected voltage has an important effect on the estimation. If it is too small, then the high-frequency current will be small. A high magnitude of the voltage causes high torque ripples due to the high magnitude of the high-frequency current, however. A voltage magnitude of about 10% the rated voltage is suggested for the high-frequency injected voltage. The saliency-based scheme with signal injection has found widespread acceptance in the literature. The schemes different from that mentioned previously are also presented. A variety of such schemes are used in some commercial sensorless products.

6.5.3 Inverter switching harmonics scheme for salient motors

It is possible to utilize the voltage and current harmonics in stator windings of PMS motors caused by the normal switching of the inverter, not by the injected signals, to calculate the motor inductances and thus the motor position. Therefore, this scheme can also be used to estimate the rotor position of IPM motors at different speeds including low and zero speeds (Ogasawara and Akgi 1998b). The scheme is presented next.

Referring to Section 2.4, the voltage equation of IPM motors can be presented in a stationary two-axis reference frame with matrix notation by

$$\boldsymbol{v}_s = \boldsymbol{R}_s \boldsymbol{i}_s + p\boldsymbol{\lambda}_s, \tag{6.5.22}$$

$$p\boldsymbol{\lambda}_s = \boldsymbol{L}_s p \boldsymbol{i}_s + \boldsymbol{e}_0, \tag{6.5.23}$$

where $\boldsymbol{e}_0$ is an induced voltage (Ogasawara and Akgi 1998a) and $\boldsymbol{L}_s$ and $\boldsymbol{R}_s$ are the inductance and resistance matrices as presented by

$$\boldsymbol{L}_s = \begin{bmatrix} L_1 - L_2 \cos 2\theta_r & -L_2 \sin 2\theta_r \\ -L_2 \sin 2\theta_r & L_1 + L_2 \cos 2\theta_r \end{bmatrix}, \tag{6.5.24}$$

$$\boldsymbol{R}_s = \begin{bmatrix} R_s & 0 \\ 0 & R_s \end{bmatrix}. \tag{6.5.25}$$

A bold variable in (6.5.22) and (6.5.23) is a column vector including the D– and Q– axes components. It is seen in eqn (6.5.24) that $\boldsymbol{L}_s$ depends on the rotor position. When the motor is supplied by a voltage-source PWM inverter, the switching harmonics in the voltage and current vectors can be divided into their fundamental and harmonic components as

$$\boldsymbol{v}_s = \boldsymbol{v}_f + \boldsymbol{v}_h, \tag{6.5.26}$$

$$\boldsymbol{i}_s = \boldsymbol{i}_f + \boldsymbol{i}_h. \tag{6.5.27}$$

Assuming negligible resistive voltage drop in $\boldsymbol{v}_h$, the harmonic voltage vector at a certain rotor position is simplified to

$$\boldsymbol{v}_h = \boldsymbol{L}_s p\, \boldsymbol{i}_h. \tag{6.5.28}$$

This equation indicates that the inductance matrix, $\boldsymbol{L}_s$, and thus θ_r can be calculated if the harmonic components of the motor voltage and current vectors are known. The average voltage vector during a sampling period can be defined as

$$\boldsymbol{v}_{av} = \frac{1}{T_s} \sum t_k \boldsymbol{v}_k, \tag{6.5.29}$$

where $\boldsymbol{v}_k$ is an inverter switching voltage vector. Also, T and t_k are the sampling period and the duration of $\boldsymbol{v}_k$, respectively. Therefore, the harmonic voltage vector during a sampling period can be obtained as

$$\boldsymbol{v}'_k = \boldsymbol{v}_f - \boldsymbol{v}_{av}. \tag{6.5.30}$$

Also, it is possible to extract the harmonic current from the total current by observing that eqn (6.5.28) assumes a linear variation of the motor current during a sampling period. This current variation is the sum of all current variations due to all switching vectors applied during a sampling period, as seen in Fig. 6.9. This can be presented by

$$\Delta \boldsymbol{i} = \sum \Delta \boldsymbol{i}_k, \tag{6.5.31}$$

where $\Delta \boldsymbol{i}_k$ devotes the current variation in an interval of t_k. Therefore, the harmonic component of the motor current during t_k can be obtained according to the following equation, as shown in Fig. 6.9:

$$\Delta \boldsymbol{i}'_k = \Delta \boldsymbol{i}_k - \frac{t_k}{T_s} \Delta \boldsymbol{i}. \tag{6.5.32}$$

A linear relationship, governing the voltage and current harmonics in each inverter switching interval, can be obtained from eqn (6.5.28) as

$$\boldsymbol{L}_s \Delta \boldsymbol{i}'_k = \boldsymbol{v}'_k t_k. \tag{6.5.33}$$

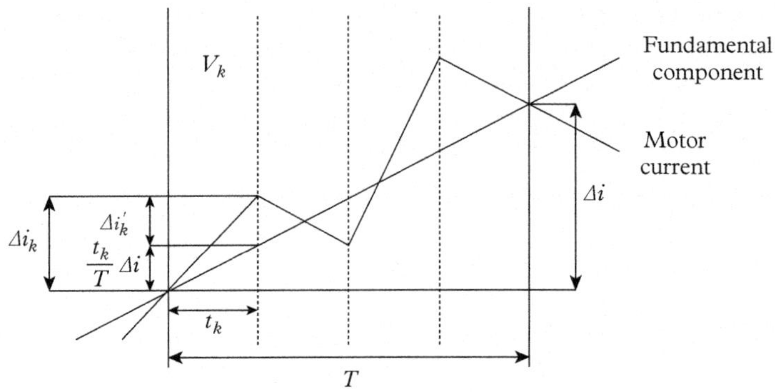

Figure 6.9 *Motor current waveform (Ogasawara and Akgi 1998b).*

By writing eqn (6.5.33) for all inverter switching vectors, the voltage vectors v_0–v_7 can be given collectively as

$$\boldsymbol{L}_s\left[\Delta \boldsymbol{i}_0' \;\; \Delta \boldsymbol{i}_1' \;\; \cdots \;\; \Delta \boldsymbol{i}_7'\right]\Delta \boldsymbol{i}_k' = \left[\boldsymbol{v}_0' t_0 \;\; \boldsymbol{v}_1' t_1 \;\; \cdots \;\; \boldsymbol{v}_7' t_7\right]. \tag{6.5.34}$$

$\boldsymbol{L}_s$ can be found by transposing eqn (6.5.34) to get

$$\begin{bmatrix} \Delta \boldsymbol{i}_0'^{\,T} \\ \Delta \boldsymbol{i}_1'^{\,T} \\ \vdots \\ \Delta \boldsymbol{i}_7'^{\,T} \end{bmatrix} \boldsymbol{L}_s^T = \begin{bmatrix} \boldsymbol{v}_0'^{\,T} \\ \boldsymbol{v}_1'^{\,T} \\ \vdots \\ \boldsymbol{v}_7'^{\,T} \end{bmatrix}, \tag{6.5.35}$$

and then finding its transpose as

$$\boldsymbol{L}_s^T = \begin{bmatrix} \Delta \boldsymbol{i}_0'^{T} \\ \Delta \boldsymbol{i}_1'^{T} \\ \vdots \\ \Delta \boldsymbol{i}_7'^{T} \end{bmatrix}^{LM} \begin{bmatrix} \boldsymbol{v}_0'^{T} t_0 \\ \boldsymbol{v}_1'^{T} t_1 \\ \vdots \\ \boldsymbol{v}_7'^{T} t_7 \end{bmatrix} = \begin{bmatrix} L_{11} & L_{12} \\ L_{21} & L_{22} \end{bmatrix}, \tag{6.5.36}$$

where the notation "*LM*" devotes the left pseudoinverse operator and performs the matrix function

$$A^{LM} = \left(A^T A\right)^{-1} A^T. \tag{6.5.37}$$

Now, from eqns (6.5.36) and (6.5.24) the rotor position is calculated as

$$2\theta_r = \tan^{-1}\frac{L_{12} + L_{21}}{L_{11} - L_{22}}. \tag{6.5.38}$$

The unique solution from eqn (6.5.38) is obtained by distinguishing the polarity of the rotor PM poles. The position estimation scheme also needs a modified PWM pattern (Ogasawara and Akgi 1998b). The scheme provides the rotor position with moderate accuracy of less than 10° in electrical angle.

6.5.4 High-frequency signal injection scheme for non-salient motors

The high-frequency signal injection scheme and the inverter switching harmonics scheme are capable of estimating the rotor position in PMS motors with magnetic saliency at normal operating frequencies where $L_d \neq L_q$. They cannot be used for rotor position estimation in PMS motors with surface-mounted PM poles, however, where $L_d = L_q = L_s$. But the analysis of PMS machines reveals that even the surface-mounted PM machines experience magnetic saliency under injected high-frequency voltages (Jang *et al.* 2004). This is due to the saturation of the stator core around the q-axis at high frequencies due to the PM flux. As a result, appropriate high-frequency signal injections cause larger d-axis high-frequency impendence than q-axis high-frequency impendence. The impedance difference is higher under injected voltages with higher frequencies and lower magnitudes. This high-frequency saliency contains useful information about rotor position. Applying a certain high-frequency voltage to the machine, the rotor position information can be observed and extracted from the high-frequency current and utilized in a rotor position estimation scheme. The principles and the procedure of this scheme are presented here (Jang *et al.* 2004).

If the injected high-frequency components are extracted from the motor voltage and current under signal injection, the following equations in actual rotor reference frame are valid for the high-frequency components,

$$v_{di} = R_{di} i_{di} + L_{di}\, p i_{di}, \tag{6.5.39}$$

$$v_{qi} = R_{qi} i_{qi} + L_{qi}\, p i_{qi}, \tag{6.5.40}$$

where all variables and parameters are the conventional ones in the actual rotor reference frame at the injected frequency. The voltage equations (6.5.39) and (6.5.40) in steady-state are given as

$$v_{di} = (R_{di} + \mathrm{j}\omega_i L_{di}) i_{di} = Z_{di} i_{di}, \tag{6.5.41}$$

$$v_{qi} = (R_{qi} + \mathrm{j}\omega_i L_{qi}) i_{qi} = Z_{qi} i_{qi}, \tag{6.5.42}$$

where Z_{di} and Z_{qi} are the d- and q-axis high-frequency impendences, respectively. Also, ω_i is the frequency of the injected voltage in radians per second. Now, the injected high-frequency current can be transformed from the actual rotor position, θ_r, *RF* to the estimated rotor position, $\hat{\theta}_r$, RF by

$$\begin{bmatrix} \hat{i}_{di} \\ \hat{i}_{qi} \end{bmatrix} = \begin{bmatrix} \cos(\theta_r - \hat{\theta}_r) & -\sin(\theta_r - \hat{\theta}_r) \\ \sin(\theta_r - \hat{\theta}_r) & \cos(\theta_r - \hat{\theta}_r) \end{bmatrix} \begin{bmatrix} i_{di} \\ i_{qi} \end{bmatrix}. \tag{6.5.43}$$

Substituting for i_{di} and i_{qi} in eqn (6.5.43) from eqns (6.5.41) and (6.5.42) yields

$$\begin{bmatrix} \hat{i}_{di} \\ \hat{i}_{qi} \end{bmatrix} = \begin{bmatrix} \cos(\theta_r - \theta_r) & -\sin(\theta_r - \theta_r) \\ \sin(\theta_r - \theta_r) & \cos(\theta_r - \theta_r) \end{bmatrix} \begin{bmatrix} \frac{1}{Z_{di}} & 0 \\ 0 & \frac{1}{Z_{qi}} \end{bmatrix} \begin{bmatrix} v_{di} \\ v_{qi} \end{bmatrix}. \tag{6.5.44}$$

Now, v_{di} and v_{qi} in eqn (6.5.44) are transformed to the estimated rotor position *RF* to yield the current components versus voltage components all in the estimated rotor position *RF* as

$$\begin{bmatrix} \hat{i}_{di} \\ \hat{i}_{qi} \end{bmatrix} = \begin{bmatrix} \cos(\theta_r - \hat{\theta}_r) & -\sin(\theta_r - \hat{\theta}_r) \\ \sin(\theta_r - \hat{\theta}_r) & \cos(\theta_r - \hat{\theta}_r) \end{bmatrix} \begin{bmatrix} \frac{1}{Z_{di}} & 0 \\ 0 & \frac{1}{Z_{qi}} \end{bmatrix}$$

$$\begin{bmatrix} \cos(\theta_r - \hat{\theta}_r) & \sin(\theta_r - \hat{\theta}_r) \\ -\sin(\theta_r - \hat{\theta}_r) & \cos(\theta_r - \hat{\theta}_r) \end{bmatrix} \begin{bmatrix} \hat{v}_{di} \\ \hat{v}_{qi} \end{bmatrix}. \tag{6.5.45}$$

The current components in eqn (6.5.45) can be written as

$$\hat{i}_{di} = \frac{1}{Z_{di}Z_{qi}} \left[Z_{avg} - \frac{Z_{dif}}{2} \cos 2(\theta_r - \hat{\theta}_r)\hat{v}_{di} - \frac{Z_{dif}}{2} \sin 2(\theta_r - \hat{\theta}_r)\hat{v}_{qi} \right], \tag{6.5.46}$$

$$\hat{i}_{qi} = \frac{1}{Z_{di}Z_{qi}} \left[Z_{avg} + \frac{Z_{dif}}{2} \cos 2(\theta_r - \hat{\theta}_r)\hat{v}_{di} - (\frac{Z_{dif}}{2} \sin 2(\theta_r - \hat{\theta}_r)\hat{v}_{qi} \right]. \tag{6.5.47}$$

The last two equations show that the rotor position error depends on one current component and two voltage components. The equations can be simplified if the high-frequency voltage is injected along one axis only. Consequently, the high-frequency current is more influential on the same axis that the voltage is injected. Thus, the injection of high-frequency voltage along the q-axis develops substantial torque ripples when the rotor position error is small. Therefore, it is advantageous to inject the high-frequency voltage along the d-axis. In this case,

$$\hat{v}_{di} = V_i \cos \omega_i t, \tag{6.5.48}$$

$$\hat{v}_{qi} = 0. \tag{6.5.49}$$

The high-frequency currents are then obtained by substituting eqns (6.5.48) and (6.5.49) into eqns (6.5.47) and (6.5.46) as

$$\hat{\imath}_{di} = \frac{V_i \cos \omega_i t}{Z_{di} Z_{qi}} \left[Z_{avg} - \frac{Z_{dif}}{2} \cos 2(\theta_r - \hat{\theta}_r) \right], \tag{6.5.50}$$

$$\hat{\imath}_{qi} = \frac{V_i \cos \omega_i t}{Z_{di} Z_{qi}} \left[-\frac{Z_{dif}}{2} \sin 2(\theta_r - \hat{\theta}_r) \right]. \tag{6.5.51}$$

It is seen that both $\hat{\imath}_{di}$ and $\hat{\imath}_{qi}$ depend on the rotor position error. However, $\hat{\imath}_{di}$ is not zero even if the error is zero, due to the presence of Z_{avg}, whereas i_{qi} is zero when the error is zero. Therefore, $\hat{\imath}_{qi}$ is chosen for the rotor position estimation. The position error can be determined from eqn (6.5.51) if $Z_{dif} \neq 0$.

Assuming that the high-frequency resistance can be neglected in Z_{di} and Z_{qi}, the high-frequency current of eqn (6.5.51) can be written as

$$\hat{\imath}_{qi} = \frac{V_i}{2\omega_i^2 L_{di} L_{qi}} \left[r_{dif} \cos \omega_i t - L_{dif} \sin \omega_i t \right] \sin 2(\theta_r - \hat{\theta}_r), \tag{6.5.52}$$

where r_{dif} and L_{dif} are the differences between the d- and q-axis high-frequency resistances and inductances, respectively. A simple function of rotor position error can be obtained by some signal processing. The processing is done by firstly multiplying $\hat{\imath}_{qi}$ by another produced signal of $\sin \omega_i t$ to achieve

$$\hat{\imath}_{qi} \sin \omega_i t = -\frac{V_i \sin 2(\theta_r - \hat{\theta}_r)}{2\omega_i^2 L_{di} L_{qi}} \left[\frac{\omega_i L_{dif}}{2} - |Z_{dif}| \sin(2\omega_i t - \varphi) \right], \tag{6.5.53}$$

where

$$|Z_{dif}| = \sqrt{r_{dif}^2 + \omega_i^2 L_{dif}^2}, \qquad \varphi = \tan^{-1} \frac{\omega_i L_{dif}}{r_{dif}}. \tag{6.5.54}$$

It is seen that eqn (6.5.53) consists of a DC component and a harmonic component. The DC component can be obtained by passing $\hat{\imath}_{qi} \sin \omega_i t$ to a low-pass filter with appropriate corner frequency. If the rotor position error is sufficiently small, then the DC component can be linearized by an acceptable approximation to have a very simple function of position error as

$$f(\theta_r - \hat{\theta}_r) = \frac{K_{err}}{2} \sin 2(\theta_r - \hat{\theta}_r) \approx K_{err}\left(\theta_r - \hat{\theta}_r\right), \tag{6.5.55}$$

where

$$K_{err} = \frac{V_i L_{dif}}{2\omega_i L_{di} L_{qi}}. \tag{6.5.56}$$

Bearing in mind the discussion presented at the beginning of this subsection, eqn (6.5.56) shows that it is L_{dif} that causes the rotor position to be estimated by this scheme. In low frequencies, L_{dif} approaches $L_d - L_q = 0$, resulting in $K_{err} = 0$. It emphasizes the need for high-frequency signal injection for position estimation.

The linearized position error of eqn (6.5.55) can be used in various types of position estimation procedures to reduce the error to 0 and provide the estimated rotor position. A simple solution is to use a PI controller with appropriate gains to force the position error to 0. An alternative solution is to use a hysteresis controller with a suitable hysteresis band to get to the position signal.

The scheme can be implemented according to Fig. 6.10. The figure shows that the estimated high-frequency q-axis current, $\hat{\imath}_{qi}$, is extracted from the total q-axis estimated current, $\hat{\imath}_q$, by a band-pass filter, provided there is an appropriate high-frequency ω_i. Then, $\hat{\imath}_{qi}$ is multiplied by the high-frequency signal, $\sin \omega_i t$, provided there is by the signal injection (SI) block. The resulting signal, $i_{qi} \sin \omega_i t$, goes through a low-pass filter (LPF) to generate the signal $K_{err}(\theta_r - \hat{\theta}_r)$. This is applied to the PI (or hysteresis) controller to estimate the motor speed. A position estimator, with its simplest form as an integrator, gives the estimated rotor position, $\hat{\theta}_r$. The SI block receiving an appropriate low-voltage magnitude, V_i, generates the high-voltage signal, $V_i \cos \omega_i t$, to be added to the d-axis output of the current controllers to provide the reference d-axis voltage, v_d^*. This voltage command in connection with the q-axis voltage command is transferred to the $\hat{a}$–$\hat{b}$–$\hat{c}$ RF by using the estimated rotor position. The estimated rotor position is also used in the a–b–c to $\hat{d}$–$\hat{q}$ current transformation to provide $\hat{\imath}_d$ and $\hat{\imath}_q$ as the total d- and q-axis current components. The current components are filtered by low-pass filters to produce the feedback currents to the current controllers as seen in Fig. 6.10.

The inverter supplies the motor by the required voltage through the SVM technique. It is important to take care of implementation issues in using this scheme.

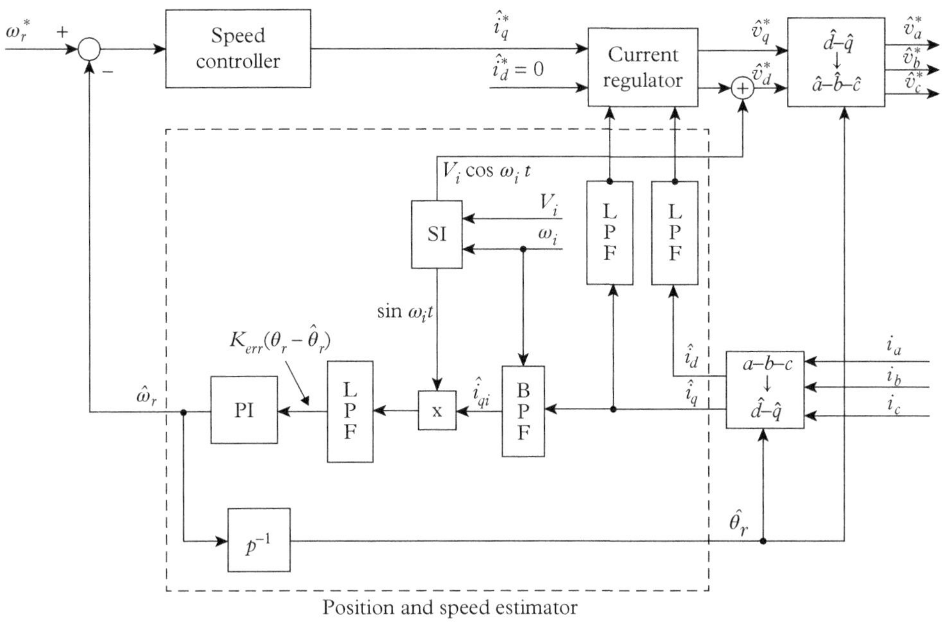

Figure 6.10 *Rotor position and speed estimation scheme for surface-mounted PMS motors by high-frequency signal injection.*

6.6 Closed-loop observer-based method

Closed-loop observers predict the variables and/or parameters of a system by utilizing the system model, and employ the general feedback theory to correct the estimated variables by reducing the error between the actual measured output and the estimated output. The observer is commonly presented in state-space form. Figure 6.11 depicts a schematic view of an observer in which the input of the system and the augmented output error are applied to the observer to provide the estimated states. The observer includes a dynamic model of the system and a state adjustment scheme. Either a linear or a non-linear mathematical model in state-space form, depending on the system dynamics and the type of observer, may present the system. The observer design consists of proper selection of the observer parameters and providing an effective state adjustment scheme. A major criterion for the evaluation of an observer is the convergence of the observer output to the system output and thus the convergence of the estimated stats to the actual states. The theory of observers has been applied to

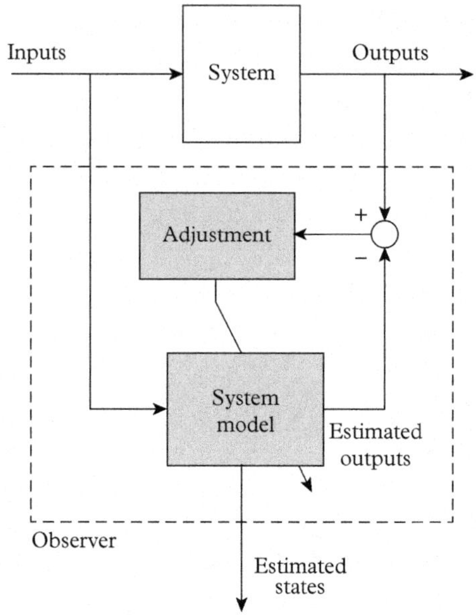

Figure 6.11 *A schematic view of a closed-loop observer.*

the position and speed estimation of PMS machines for more than two decades now and numerous estimation schemes have been presented in the literature; some of the salient ones are presented in this section.

6.6.1 State observer scheme

A state observer is an easy-to-understand, closed-loop observer, which is presented in this sub-section. The fundamental principles of the observer are given first, followed by the state-space model of PMS motors. Finally, the implementation of the observer is presented (Jones and Lang 1989).

6.6.1.1 Principles

The model of a PMS motor in general is non-linear. The electrical dynamics can be linearized if the model is transformed into the rotor reference frame, however. Nevertheless, the whole model is still non-linear. Therefore, considering a non-linear system, it is appropriate to present the general theory of state observers by presenting the dynamic model of the system in state space by the non-linear matrix equations

$$\dot{\boldsymbol{x}}(t) = \boldsymbol{f}[\boldsymbol{x}(t), \boldsymbol{u}(t)], \tag{6.6.1}$$

$$\boldsymbol{y}(t) = \boldsymbol{h}[\boldsymbol{x}(t)], \tag{6.6.2}$$

where $\boldsymbol{x}(t)$ and $\boldsymbol{u}(t)$ are the actual system state vector and the input vector, respectively. Also, $\boldsymbol{y}(t)$ is the measured output vector of the system, where the right-hand side of eqn (6.6.1) represents the state dynamics. Bold symbols are used to present matrices and vectors of appropriate dimensions. Equations (6.6.1) and (6.6.2) are non-linear in general in the sense that the right-hand side of the equations are non-linear functions of the state vector.

The following equation presents the general form of an identity observer, which gives the estimation of entire states,

$$\dot{\hat{\boldsymbol{x}}}(t) = \boldsymbol{f}\left[\hat{\boldsymbol{x}}(t), \boldsymbol{u}(t)\right] + \boldsymbol{G}\left[\boldsymbol{y}(t) - \hat{\boldsymbol{y}}(t)\right], \tag{6.6.3}$$

where $\hat{\boldsymbol{x}}(t)$ and $\hat{\boldsymbol{y}}(t)$ denote the estimated state vector and the estimated output vector, respectively, and G is the observer gain matrix. Equation (6.6.3) denotes that the gain matrix provides the inclusion of innovation, caused by the output error, to the system dynamics. As the error is being vanished, the estimated states approach the actual states. At the same time, the observer equation of (6.6.3) approaches the system equation of (6.6.1).

6.6.1.2 *Machine model*

An appropriate model of the system is utilized in the design of state observer. Therefore, a state-space model of IPM motors in rotor RF is represented as

$$pi_d = -\frac{R_s}{L_d}i_d + \omega_r\frac{L_q}{L_d}i_q + \frac{v_d}{L_d}, \tag{6.6.4}$$

$$pi_q = -\frac{R_s}{L_q}i_q - \omega_r\frac{L_d}{L_q}i_d - \omega_r\frac{\lambda_m}{L_q} + \frac{v_q}{L_q}, \tag{6.6.5}$$

$$p\omega_r = \frac{P}{J}\left(T_e - T_L - B\omega_r - T_f\right), \tag{6.6.6}$$

$$p\theta_r = \omega_r, \tag{6.6.7}$$

where the motor electromagnetic torque, T_e, and drag and friction torque, T_f, respectively, are given as

$$T_e = \frac{3P}{2}i_q\left[\lambda_m + \left(L_d - L_q\right)i_d\right], \tag{6.6.8}$$

$$T_f = C\ \mathrm{sgn}(\omega_r)\,. \tag{6.6.9}$$

In the previous equations, J, B, and C are the rotor inertia, the viscous damping coefficient, and the coulomb friction coefficient, respectively.

The d- and q-axis stator currents, and the electrical rotor speed and position are selected as system states. Also, the d- and q-axis stator voltage components are chosen as the input variables and the d- and q-axis stator current components as the output variables. As a result, the state vector, the input vector, and the output vector are presented, respectively, as

$$\boldsymbol{x} = \begin{bmatrix} i_d & i_q & \omega_r & \theta_r \end{bmatrix}^T, \tag{6.6.10}$$

$$\boldsymbol{u} = \begin{bmatrix} v_d & v_q \end{bmatrix}^T \tag{6.6.11}$$

$$\boldsymbol{y} = \begin{bmatrix} i_d & i_q \end{bmatrix}^T. \tag{6.6.12}$$

6.6.1.3 Observer implementation

The output error, which includes the differences of the measured and the estimated d- and q-axis currents, is multiplied by the gain matrix. The result is then added to the system dynamics as presented in eqn (6.6.3). The gain matrix is a constant matrix of two parts, an electrical part contributing to the dynamics of d- and q-axis currents, and a mechanical part, contributing to the motor speed dynamics as

$$\boldsymbol{G} = \begin{bmatrix} \boldsymbol{G}_i \\ \hline \boldsymbol{G}_\omega \end{bmatrix} = \begin{bmatrix} g & e \\ e & g \\ \hline m & n \end{bmatrix}. \tag{6.6.13}$$

The mechanical part of the gain matrix does not provide any direct contribution to the rotor position. This is because the estimated position is obtained by an integration of the speed signal and such a contribution introduces current measurement noises directly into the estimated position and deteriorates the smoothing effect of integration on the speed signal.

Referring to the model equations and the previous selections, the observer equation of (6.6.3) can be written for IPM motors by the three equations

$$p\begin{bmatrix} \hat{i}_d \\ \hat{i}_q \end{bmatrix} = \begin{bmatrix} -\frac{R_s}{L_d} & \hat{\omega}_r \frac{L_q}{L_d} \\ \hat{\omega}_r \frac{L_d}{L_q} & -\frac{R_s}{L_q} \end{bmatrix} \begin{bmatrix} \hat{i}_d \\ \hat{i}_q \end{bmatrix} - \hat{\omega}_r \lambda_m \begin{bmatrix} 0 \\ \frac{1}{L_q} \end{bmatrix} + \begin{bmatrix} \frac{1}{L_d} & 0 \\ 0 & \frac{1}{L_q} \end{bmatrix} \begin{bmatrix} v_d \\ v_q \end{bmatrix}$$

$$+ \begin{bmatrix} g & e \\ e & g \end{bmatrix} \left(\begin{bmatrix} i_d \\ i_q \end{bmatrix} - \begin{bmatrix} \hat{i}_d \\ \hat{i}_q \end{bmatrix} \right), \tag{6.6.14}$$

$$p\omega_r = \begin{bmatrix}\hat{\imath}_d & \hat{\imath}_q\end{bmatrix} \frac{3P^2}{2} \frac{\lambda_m}{J} \begin{bmatrix} 0 \\ 1 \end{bmatrix} + \frac{3P^2}{2} \frac{L_d - L_q}{J} \begin{bmatrix}\hat{\imath}_d & \hat{\imath}_q\end{bmatrix} \begin{bmatrix} 0 & 1 \\ 1 & 0 \end{bmatrix} \begin{bmatrix} \hat{\imath}_d \\ \hat{\imath}_q \end{bmatrix}$$

$$- \frac{PB\omega_r}{J} - \frac{PC}{J} \text{sgn}\,(\omega_r) - \frac{PT_L}{J}$$

$$+ \frac{3P}{2} \frac{\lambda_m}{J} \begin{bmatrix} m & n \end{bmatrix} \left(\begin{bmatrix} i_d \\ i_q \end{bmatrix} - \begin{bmatrix} \hat{\imath}_d \\ \hat{\imath}_q \end{bmatrix} \right), \qquad (6.6.15)$$

$$p\hat{\theta}_r = \hat{\omega}_r. \qquad (6.6.16)$$

In the previous equations, i_d and i_q are the measured currents transferred to the rotor reference frame. Also, v_d and v_q are the measured voltages transferred to the same RF. It is worth mentioning that these transformations are carried out using the estimated rotor position. A direct voltage measurement can be avoided by calculating the D- and Q-axis voltage components in the stationary RE by using the DC link voltage and the inverter switching functions and then transforming them to the rotor RF. In the case of a short sampling time and high current dynamics, the d- and q-axis voltage references as the outputs of current controllers can be used instead of v_d and v_q in eqn (6.6.14). Also, $\hat{\imath}_d$ and $\hat{\imath}_q$ are the estimated current components as the estimator output. Again, note the lack of correcting feedback on the position dynamics.

The gain matrix can be determined by the pole assignment technique. The selection of proper observer gains depends on the motor speed. Therefore, a constant gain matrix may not result in desirable estimation over the entire speed range. Two solutions to tackle this problem may be followed. One solution is to design different gain matrices at different speeds. Then again, a scheduling scheme is employed to select an appropriate matrix for a specific speed range (Sepe and Lang 1992). Also, the acceptable gain may change under the influence of motor parameter uncertainties. Therefore, it is desirable to adjust the gains online in response to machine conditions. A kind of such adaptive gain adjustment is investigated in connection with the extended Kalman filter in the next sub-section. The solution of eqns (6.6.14)–(6.6.16) needs initial values of the system states. These can all be assumed to be zero. It is true for current components and speed. However, the assumption of a zero value, in general, is not accurate for rotor position. The estimation accuracy of all states is sensitive to the initial rotor position error as it can be seen by eqn (6.6.3). Therefore, the initial rotor position must be determined by some means before the estimation process starts. This is a major limitation of this and some other rotor position estimation schemes. The observer is sluggish at lower speeds as there is less speed voltage in the current measurement to be utilized by the observer. The convergence rate can be improved by increasing the observer gain. This may cause observer instability at high speeds where there is higher speed voltage. A gain scheduling may be used as a solution as mentioned previously. The observer performance is adversely affected by the errors and noises in the measured current and voltage. Therefore, accurate and smooth current and voltage measurements are required (Jones and Lang 1989).

6.6.2 Extended Kalman filter scheme for non-salient motors

A Kalman filter is an optimal algorithm for real-time state and parameter estimation of linear systems. By being optimal, it minimizes the mean square error of the estimated quantities. It takes into accounts modeling inaccuracies, input noises, and disturbances and comes up with an accurate estimation result. This is why it is called "filter." EKF is the modified version of the algorithm for estimating states and parameters of non-linear systems. EKF applies the Kalman filter to linearized versions of non-linear systems. This is done by linearizing the state and measurement equations about the predicted state as an operating point.

When EKF is applied to the position and speed estimation of PMS motors, it does not need information about mechanical parameters. In addition, the accurate initial rotor position and speed, required in many position estimation methods, including the state observer scheme presented previously, are not needed in this method. Instead, EKF uses any available information about the initial conditions of system states, including rotor position, even if they differ considerably from the actual states at start-up. It needs knowledge about the system dynamics and statistical information about system noise, disturbances, uncertainties, and system modeling errors, however. The algorithm is computation intensive due to complex matrix calculations. Thus, it requires fast processing tools. Nevertheless, it possesses rapid convergence to the actual values and avoids swinging around the actual values at start-up. Application-specific DSPs for motor drive control implementation of today's market can handle the computation burden.

6.6.2.1 Principles of EKF algorithm

For the application in hand, it is appropriate to present the non-linear model of the system by the following matrix equations (Bado *et al.* 1992; Bolognani *et al.* 1999),

$$\dot{x}(t) = f[x(t)] + \boldsymbol{B}\boldsymbol{u}(t) + \boldsymbol{\sigma}(t), \tag{6.6.17}$$

$$\boldsymbol{y}(t_k) = \boldsymbol{H}\boldsymbol{x}(t_k) + \boldsymbol{\mu}(t_k), \tag{6.6.18}$$

where $\boldsymbol{x}(t)$ denotes the system state vector with the initial state vector $\boldsymbol{x}(t_o)$ described as a Gaussian random vector with mean x_0 and covariance $\boldsymbol{P}_0$, and $\boldsymbol{u}(t)$ is the deterministic input vector. Also, $\boldsymbol{x}(t_k)$ is the optimal state estimate sequence generated by the EKF as a minimum variance estimate of $\boldsymbol{x}(t)$ with covariance $\boldsymbol{P}(t_k)$, and $\boldsymbol{y}(t_k)$ models the available discrete-time measurements as an output vector. All the vectors are time variable. The model inaccuracy and system disturbances are modeled by $\boldsymbol{\sigma}(t)$ as a zero-mean white Gaussian noise, independent of $\boldsymbol{x}(t_o)$ and with covariance $\boldsymbol{Q}(t)$. The measured noise and inaccuracy are modeled by $\boldsymbol{\mu}(t_k)$ as a zero-mean white Gaussian noise, independent of $\boldsymbol{x}(t_o)$ and $\boldsymbol{u}(t)$, and with covariance $\boldsymbol{R}(t_k)$. Finally, $\boldsymbol{B}$ and $\boldsymbol{H}$ are the input and the output matrices, respectively.

Equation (6.6.17) is non-linear in the sense that the system matrix is combined with the state vector. Also, eqn (6.6.18) models the discrete-time measurements at instance k by a linear equation. Equations (6.6.17) and (6.6.18) do not represent the most general form of a non-linear model that the EKF algorithm can accept, since the system input is independent from the system states and thus from the outputs, and also the output is a linear function of states. Nevertheless, this formulation is appropriate for presenting a PMS motor model, as will be seen later.

In a time interval from t_{k-1} to t_k, the observer works in two stages, i.e., prediction stage and correction stage as follows. The applicant must give the initial values of the state vector and the covariance matrices before the stages start.

1. Prediction stage

In this stage, which is also referred to as the propagation stage, the optimal estimated state vector and the state covariance matrix are calculated online by the mathematical model of the system at the time k by using the most recent estimates of the state vector and input vector at the instant k-1. This is obtained by integrating eqn (6.6.17) from t_{k-1} to t_k, supposing that $\boldsymbol{\sigma}(t)$ is equal to 0 and covariance matrix $\boldsymbol{Q}$ is constant. The formulation in discrete-time notation is presented as

$$\hat{x}_{k/k-1} = \hat{x}_{k-1/k-1} + \left[f\left(\hat{x}_{k-1/k-1}\right) + \langle \boldsymbol{u}_{k-1} \rangle\right] T_s, \tag{6.6.19}$$

$$\boldsymbol{P}_{k/k-1} = \boldsymbol{P}_{k-1/k-1} + \left(\boldsymbol{F}_{k-1}\boldsymbol{P}_{k-1/k-1} + \boldsymbol{P}_{k-1/k-1}\boldsymbol{F}_{k-1}^T + \boldsymbol{Q}\right) T_s, \tag{6.6.20}$$

where

$$F_{k-1} = \left.\frac{\delta f(x)}{\delta x}\right|_{x=\hat{x}_{k-1/k-1}}. \tag{6.6.21}$$

In these calculations, $\hat{x}_{k/k-1}$ refers to the estimate of the state vector at the kth instance based on measurements up to the (k–1)th instance, and it is similar for other variables with a subscript containing two instances.

2. Correction stage

In this stage, which is also referred to as the innovation, propagation, or filtering stage, the Kalman gain is calculated as

$$K_k = P_{k/k-1}H^T\left(HP_{k/k-1}H^T + R\right)^{-1}. \tag{6.6.22}$$

Using the Kalman gain, the estimated state is updated by

$$\hat{x}_{k/k} = \hat{x}_{k/k-1} + K_k\left(y_k - H\hat{x}_{k/k-1}\right), \tag{6.6.23}$$

where a correction term is added to the estimate of the state vector to produce the new one. The correction term is an amplified output error of the system with the help of Kaman gain, K_k. The output error is the difference between the measured and the estimated outputs. The estimated output is calculated by the system output of eqn (6.6.18). In this stage the covariance matrix, P, is also obtained by

$$P_{k/k} = P_{k/k-1} - K_kHP_{k/k-1}. \tag{6.6.24}$$

The observer then repeats the first stage with the updated quantities.

6.6.2.2 State-space model of machine

A state-space model of the machine must be obtained since the whole estimation algorithm is carried out by matrix calculations. A machine model in rotor reference frame provides a more linear model in some aspects. The two-axis stationary RF is preferred due to the lack of reference frame transformation from stationary RF to rotating RF and vice versa, however. This is mainly because the RF transformation from the stationary RF to rotating RF needed for measured currents, itself, needs the estimated rotor position and this can propagate further error to the estimation process. The machine state-space equations in two-axis stationary RF are presented as

$$pi_D = -\frac{R_s}{L_s} i_D + \omega_r \frac{\lambda_m}{L_s} \sin\theta_r + \frac{v_D}{L_s}, \qquad (6.6.25)$$

$$pi_Q = -\frac{R_s}{L_s} i_Q - \omega_r \frac{\lambda_m}{L_s} \sin\theta_r + \frac{v_Q}{L_s}, \qquad (6.6.26)$$

$$p\omega_r = 0, \qquad (6.6.27)$$

$$p\theta_r = \omega_r. \qquad (6.6.28)$$

It is seen that the motor speed is assumed constant during a switching period. The D- and Q-axis stator currents, and the electrical rotor speed and position are selected as system states. Also, the D- and Q-axis stator voltages are chosen as the input variables and the D- and Q-axis stator currents as the output variables. As a result, the model state vector, input vector, and output vector in this RF are presented, respectively, as

$$x = \begin{bmatrix} i_D & i_Q & \omega_r & \theta_r \end{bmatrix}^T, \qquad (6.6.29)$$

$$u = \begin{bmatrix} v_D & v_Q \end{bmatrix}^T, \qquad (6.6.30)$$

$$y = \begin{bmatrix} i_D & i_Q \end{bmatrix}^T. \qquad (6.6.31)$$

Referring to the model equations and the previous selections, the system matrix and the input and output matrices are presented, respectively, as

$$f(x) = \begin{bmatrix} -\frac{R_s}{L_s} i_D & 0 & \omega_r \lambda_m \sin\theta_r & 0 \\ 0 & -\frac{R_s}{L_s} i_Q & \omega_r \lambda_m \cos\theta_r & 0 \\ 0 & 0 & 0 & 0 \\ 0 & 0 & \omega_r & 0 \end{bmatrix}, \qquad (6.6.32)$$

$$B = \begin{bmatrix} \frac{1}{L_s} & 0 & 0 & 0 \\ 0 & \frac{1}{L_s} & 0 & 0 \\ 0 & 0 & 0 & 0 \\ 0 & 0 & 0 & 0 \end{bmatrix}, \qquad (6.6.33)$$

$$H = \begin{bmatrix} 1 & 0 & 0 & 0 \\ 0 & 1 & 0 & 0 \end{bmatrix}. \qquad (6.6.34)$$

It is seen that the motor model is non-linear since the system matrix is a non-linear function of the system states. The model, however, is linear from the input output point of view. The dependency of the system matrix on the states could be overcome if the motor model were presented in the rotor RF. More non-linearity would arise due to the cross coupling by the speed voltages, however. This is why the stationary RF is preferred over a rotating RF for modeling, as mentioned previously.

6.6.2.3 Implementation

The whole process of motor control and rotor position estimation is carried out according to the flowchart of Fig. 6.12 (Bado *et al.* 1992). The process includes initiation, prediction, control, measurement, and innovation. The estimation starts by initiation step in which the initial values of system states are set. It is convenient to choose $\boldsymbol{x}_0 = 0$. Also, the initial state covariance matrix, $\boldsymbol{P}_0$, and the initial covariance matrices, $\boldsymbol{Q}_0$ and $\boldsymbol{R}_0$, are chosen in this step. They can be obtained by monitoring and stochastically analysing the noises over time. If it is difficult to do the monitoring, it is possible to come up with these matrices through trial and error to get desired estimation results out of the observer. To estimate the rotor position in PMS motors, however, diagonal matrices are usually used, due to the lack of sufficient statistical information to evaluate the off-diagonal elements. The diagonal elements are adjusted by trial and error to get the best position and speed estimation in terms of accuracy and swiftness during transient and steady-state motor performances (Dhaouadi *et al.* 1991). The diagonal matrices of the following forms are suggested (Bado *et al.* 1992):

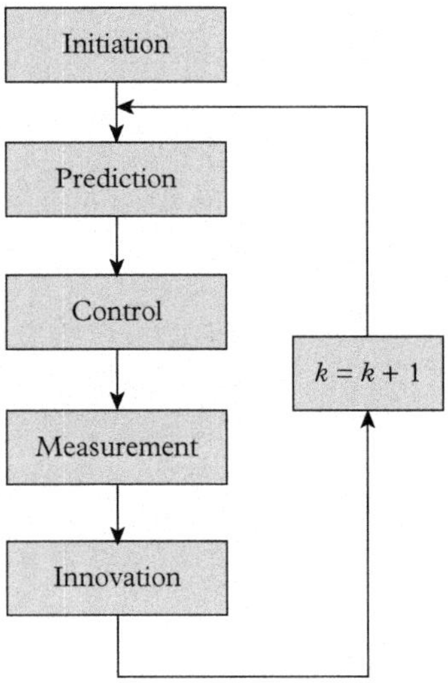

Figure 6.12 *Flowchart of rotor position and speed estimation by EKF algorithm and the motor control.*

$$\boldsymbol{Q} = \boldsymbol{Q}_0 = \begin{bmatrix} a & 0 & 0 & 0 \\ 0 & a & 0 & 0 \\ 0 & 0 & b & 0 \\ 0 & 0 & 0 & c \end{bmatrix}, \boldsymbol{P}_0 = \begin{bmatrix} d & 0 & 0 & 0 \\ 0 & d & 0 & 0 \\ 0 & 0 & e & 0 \\ 0 & 0 & 0 & f \end{bmatrix}, \boldsymbol{R} = \boldsymbol{R}_0 = \begin{bmatrix} g & 0 \\ 0 & g \end{bmatrix}. \tag{6.6.35}$$

The $\boldsymbol{Q}$ and $\boldsymbol{R}$ matrices are fixed to the initial values during the estimation process as presented in eqn (6.6.35). The tuning of the covariance matrices is investigated in the literature (Bolognani *et al.* 2003).

Then, the prediction stage is carried out by executing eqns (6.6.19)–(6.6.21). In the first cycle, the initial values of the state vector and the state covariance matrix, $\boldsymbol{P}$, are used. In addition, in this cycle a null mean voltage vector is employed. In the next cycles, however, the reference voltage vector calculated by the motor control system in the previous cycle replaces the vector of the mean input voltages. Also, an updated $\boldsymbol{P}$ is used in the calculations.

In the control and measurement steps, the usual functions of the motor control system are carried out, including the phase current

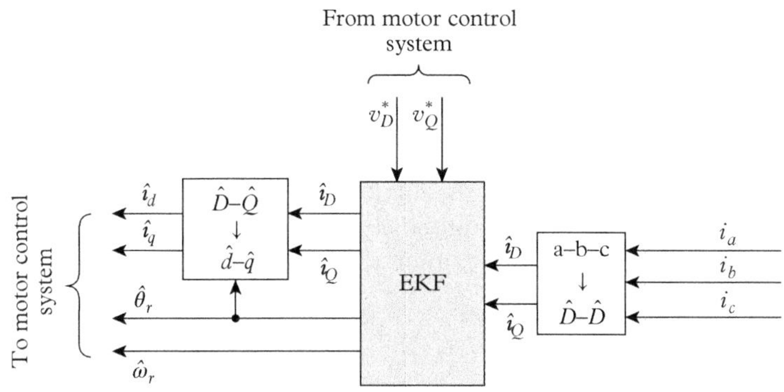

Figure 6.13 *Block diagram of motor control system and EKF.*

measurement, reference frame transformations, and the execution of current and speed control routines, as shown schematically in Fig. 6.13. The speed controller produces the q-axis current command in response to the speed error, which is the difference between the current and the previous estimated speed values. Also, the current controllers produce the voltage commands, v_D^* and v_Q^*.

The correction stage in Fig. 6.12 includes the calculation of eqns (6.6.22)–(6.6.24). In this stage, the latest available measurements and estimated values are used to update the estimated position and speed. The cycle index is then incremented and a new cycle starts, as seen in Fig. 6.12.

6.6.3 Extended Kalman filter scheme for salient motors

Although, a two-axis stationary reference frame is more convenient for rotor and speed estimation of surface-mounted PMS motors as presented previously, the model of IPM motors in this RF is more complex due the differences between the d- and q-axis inductances. As a result, the EKF algorithm would request a long computing time. Therefore, in the case of IPM motors, a model in rotor reference frame is greatly preferred. This needs the modification of the algorithm used in the previous sub-section. In the current sub-section, the state-space model of IPM motors in the rotor reference and a revised version of the EKF algorithm are presented (Bolognani *et al.* 2003).

The model of IPM motors in rotor RF can be presented as

$$pi_d = -\frac{R_s}{L_d} i_d + \omega_r \frac{L_q}{L_d} i_q + \frac{v_d}{L_d}, \tag{6.6.36}$$

$$pi_q = -\frac{R_s}{L_q} i_q - \omega_r \frac{L_d}{L_q} i_d - \omega_r \frac{\lambda_m}{L_q} + \frac{v_q}{L_q}, \tag{6.6.37}$$

$$p\omega_r = 0, \tag{6.6.38}$$

$$p\theta_r = \omega_r. \tag{6.6.39}$$

The d- and q-axis stator currents, and the electrical rotor speed and position are selected as system states. Also, the d- and q-axis stator voltages are chosen as the input variables and the d- and q-axis stator currents as the output variables. As a result, the model state vector, input vector, and output vector in this RF are presented, respectively, as

$$\boldsymbol{x} = \begin{bmatrix} i_d & i_q & \omega_r & \theta_r \end{bmatrix}^T, \tag{6.6.40}$$

$$\boldsymbol{u} = \begin{bmatrix} v_d & v_q \end{bmatrix}^T, \tag{6.6.41}$$

$$\boldsymbol{y} = \begin{bmatrix} i_d & i_q \end{bmatrix}^T. \tag{6.6.42}$$

Referring to the model equations and the previous selections, the system matrix and the input and output matrices are presented, respectively, as

$$f(\boldsymbol{x}) = \begin{bmatrix} -\frac{R_s}{L_d} & \omega_r \frac{L_q}{L_d} & 0 & 0 \\ -\omega_r \frac{L_d}{L_q} & -\frac{R_s}{L_q} & -\lambda_m L_q & 0 \\ 0 & 0 & 0 & 0 \\ 0 & 0 & 1 & 0 \end{bmatrix}, \tag{6.6.43}$$

$$B = \begin{bmatrix} \frac{1}{L_d} & 0 & 0 & 0 \\ 0 & \frac{1}{L_q} & 0 & 0 \\ 0 & 0 & 0 & 0 \\ 0 & 0 & 0 & 0 \end{bmatrix}, \tag{6.6.44}$$

$$\boldsymbol{H} = \begin{bmatrix} 1 & 0 & 0 & 0 \\ 0 & 1 & 0 & 0 \end{bmatrix}. \tag{6.6.45}$$

A non-linear model, consistent with the motor model in rotor reference frame, can be presented by the non-linear matrix equations

$$\dot{\boldsymbol{x}}(t) = f[\boldsymbol{x}(t)]\,\boldsymbol{x}(t) + \boldsymbol{B}\boldsymbol{u}(t) + \boldsymbol{\sigma}(t), \tag{6.6.46}$$

$$\boldsymbol{y}(t) = \boldsymbol{H}\boldsymbol{x}(t) + \boldsymbol{\mu}(t), \tag{6.6.47}$$

where variables are defined as in the previous version of EKF. These non-linear equations can be discretized as

$$\boldsymbol{x}_{k+1} = \boldsymbol{F}_d(\boldsymbol{x}_k)\boldsymbol{x}_k + \boldsymbol{B}\boldsymbol{u}_k + \boldsymbol{\sigma}_k, \tag{6.6.48}$$

$$\boldsymbol{y}_k = \boldsymbol{H}\boldsymbol{x}_k + \boldsymbol{\mu}_k. \tag{6.6.49}$$

The discretized system matrix is obtained from the d–q model as

$$F_d(x_k) = 1 + T_s f(x(kT_s)). \tag{6.6.50}$$

Using the Kalman gain, the estimated state is updated by

$$\hat{x}_{k/k-1} = \hat{x}_{k-1/k-1} + \left[f\left(\hat{x}_{k-1/k-1}\right) + \langle u_{k-1} \rangle\right] T_s, \tag{6.6.51}$$

$$P_{k/k-1} = P_{k-1/k-1} + \left(F_{k-1} P_{k-1/k-1} + P_{k-1/k-1} F_{k-1}^T + Q\right) T_s, \tag{6.6.52}$$

where

$$\hat{x}_{k/k} = \hat{x}_{k/k-1} + K_k \left(y_k - H \hat{x}_{k/k-1}\right), \tag{6.6.53}$$

and

$$F_{k-1} = \left. \frac{\partial f(x(t))\, x(t)}{\partial x} \right|_{x=\hat{x}_{k/k-1}}. \tag{6.6.54}$$

It is seen in eqn (6.6.54) that the correction term is added to the estimate of the state vector to produce the new one. The correction term is an amplified output error of the system with the help of Kalman gain, K_k. The output error is the difference between the measured and the estimated outputs. The estimated output is calculated by the system output of eqn (6.6.49). In this stage the covariance matrix, P, is also obtained by

$$P_{k/k} = P_{k/k-1} - K_k H P_{k/k-1}. \tag{6.6.55}$$

The Kalman gain is calculated as

$$K_k = P_{k/k-1} H^T \left(H P_{k/k-1} H^T + R\right)^{-1}. \tag{6.6.56}$$

The observer then repeats the first stage with the updated quantities.

6.7 Summary

The essential need to rotor position for normal operation of PMS motors, in addition to its usage in closed-loop control of these machines, emphasizes the ultimate importance of rotor position information in these machines. The problems associated with mechanical sensors and their high cost are the reasons for the industry's interest in the rotor and speed sensorless systems as a cost-saving and practical alternative to the motor control with mechanical sensors. However, sensorless control is challenging due to its demanding criteria like accuracy, robustness, swiftness, and capability of working over the entire range of motor operation. There are many sensorless control methods used in the commercial products or presented in the literature responding to these challenges. Therefore, a comprehensive investigation of position and speed estimation methods for PMS machines is cumbersome. Also, the classification of the methods by a single arrangement may not be useful or even possible. Nevertheless, a selection of major position and speed estimation methods together with their corresponding schemes are classified and presented in this chapter in detail. The presentation starts with the traditional back EMF-based method with two of its schemes. The flux linkage estimation method with its four schemes is elaborated. The third category of rotor position estimation presented in the chapter is the hypothetical rotor position method, where a reference frame oriented along a hypothetical rotor position is used in motor modeling. Three rotor estimation schemes of this method, including voltage-based, back EMF-based, and feedforward voltage decoupling schemes, are presented in this chapter. The forth category of estimation methods is the saliency-based method, in which the rotor position is extracted from the position-dependent motor inductances. This method is capable of estimating rotor position over the entire speed range, including zero speed. The method is considered by four different schemes in this chapter, including the offline scheme for salient PMS motors, high-frequency signal injection scheme for salient PMS motors, inverter switching harmonics scheme, and signal injection scheme for non-salient PMS motors. Finally, the observer-based estimation method is presented by three closed-loop schemes, including a simple state observer and two extended Kalman filter-based schemes for non-salient and salient PMS motors. Each scheme is discussed by, firstly, presenting the corresponding fundamental principles, followed by the appropriate motor model and the implementation. The merits and limitations of each scheme are discussed.

PROBLEMS

P.6.1. A method of rotor position estimation is in to inject a high-frequency voltage reference into the output of the d-axis current controller and find out the estimated rotor position from the integration of the estimated high-frequency q-axis current. This is done by taking the FFT of the current component before the integration. The estimated current component is a function of the injected voltage and the rotor position error given by

$$p\hat{i}_q = \frac{L_2}{L_1^2 - L_2^2} V_a \sin \omega_a t \sin 2(\theta_r - \hat{\theta}_r),$$

where the injected voltage is

$$v_i = V_i \sin \omega_i t . \sin 2(\theta_r - \hat{\theta}_r).$$

Prove the first equation. Make any reasonable approximation.

P.6.2. The position estimation schemes are divided into five main methods, each scheme under one method as depicted in Fig. 6.1. However, there are estimation schemes that are related to more than one method. Determine these schemes and all their relations by drawing connections between the schemes with the corresponding methods in Fig. 6.1.

P.6.3. A motor speed signal can be obtained by a derivative of the rotor position. Reversely, the rotor position can be obtained by the integration of the motor speed signal. Discuss the drawbacks of these techniques and consider alternative solutions.

P.6.4. Would it possible to come up with position estimation schemes as a combination of the back EMF method and observer-based method? Derive the governing equation of such an observer.

P.6.5. Design a flux linkage angle estimation system by double-stage low-pass filter. Determine the filter time constant.

P.6.6. Under the hypothesis method, a voltage-based scheme was presented in this chapter. Consider a similar scheme based on the motor current instead of voltage.

Hint: It may be possible to rearrange the equations of the machine voltage components to find the derivatives of the current components.

P.6.7. Categorize the position estimation schemes of Fig. 6.1 in terms of:

1) applicability to non-salient, salient, and both motor types;
2) capability of estimating rotor position at standstill and low speed;
3) dependency on motor parameters;
4) online computation burden;
5) accuracy;
6) required motor signal measurement; and
7) fitting to special motor control methods, e.g., VC.

BIBLIOGRAPHY

Acarnley, P.P. and Watson, J.F. (2006). Review of position-sensorless operation of brushless permanent-magnet machines. *IEEE Trans. Ind. Electron.* 53(2), 352–362.

Aihara, T., Toba, A., Yanase, T., Mashimo, A., and Endo, K. (1999). Sensorless torque control of salient-pole synchronous motor at zero-speed operation. *IEEE Trans. Power Electron.* 14(1), 202–208.

Bado, A., Bolognani, S., and Zigliotto, M. (1992). Effective estimation of speed and rotor position of a PM synchronous motor drive by a Kalman filtering technique. In: *23rd Annual IEEE Power Electronics Specialists Conference*, pp. 951–957. IEEE, Piscataway, NJ.

Bae, B-H., Sul, S-K., Kwon, J-H., and Byeon, J-S. (2003). Implementation of sensorless vector control for super-high-speed PMSM of turbo-compressor. *IEEE Trans. Ind. Appl.* 39(3), 811–818.

Baricuatro, K. (2014). *Sensorless Start-up and Control of Permanent Magnet Synchronous Motor with Long Tieback.* Master of Science in electric power engineering, Norwegian University of Science and Technology.

Bianchi, N., Bolognani, S., Jang, J-H., and Sul, S-K. (2007). Comparison of PM motor structures and sensorless control techniques for zero-speed rotor position detection. *IEEE Trans. Power Electron.* 22(6), 2466–2475.

Bojoi, R., Pastorelli, M., Bottomley, J., Giangrande, P., and Gerada, C. (2013). Sensorless control of PM motor drives—a technology status review. In: *IEEE Workshop on Electrical Machines Design, Control and Diagnosis*, pp. 168–182. IEEE, Piscataway, NJ.

Bolognani, S., Oboe, R., and Zigliotto, M. (1999). Sensorless full-digital PMSM drive with EKF estimation of speed and rotor position. *IEEE Trans. Ind. Electron.* 46(1), 184–191.

Bolognani, S., Tubiana, L., and Zigliotto, M. (2002). EKF-based sensorless IPM synchronous motor drive for flux-weakening applications. In: *37th IAS Annual Meeting, Conference Record of the Industry Applications Conference*, pp. 112–119. IEEE, Piscataway, NJ.

Bolognani, S., Tubiana, L., and Zigliotto, M. (2003). Extended Kalman filter tuning in sensorless PMSM drives. *IEEE Trans. Ind. Appl.* 39(6), 1741–47.

Boussak, M. (2005). Implementation and experimental investigation of sensorless speed control with initial rotor position estimation for interior permanent magnet synchronous motor drive. *IEEE Trans. Power Electron.* 20(6), 141–122.

Briz, F., Degner, M. W., García, P., and Lorenz, R. D. (2004). Comparison of saliency-based sensorless control techniques for AC machines. *IEEE Trans. Ind. Appl.* 40(4), 1107–1115.

Chen, Z., Tomita, M., Doki, S., and Okuma, S. (1998). The sensorless position estimation of salient-pole brushless dc motors and its stability. In: *Proceedings of the Japan Industry Applications Society Conference*, pp. 179–182. IEEE, Tokyo, Japan.

Chen, Z., Tomita, M., Doki, S., and Okuma, S. (2003). An extended electromotive force model for sensorless control of interior permanent-magnet synchronous motors. *IEEE Trans. Ind. Electron.* 50(2), 288–295.

Corley, M.J. and Lorenz, R.D. (1998). Rotor position and velocity estimation for a salient-pole permanent magnet synchronous machine at standstill and high speeds. *IEEE Trans. Ind. Appl.* 34(4), 784–789.

Dhaouadi, R., Mohan, N., and Norum, L. (1991). Design and implementation of an extended Kalman filter for the state estimation of a permanent magnet synchronous motor. *IEEE Trans. Power Electron.* 6(3), 491–497.

Ertugrul, N. and Acarnley, P. (1994). A new algorithm for sensorless operation of permanent magnet motors. *IEEE Trans. Ind. Appl.* 30(1), 126–133.

Fan, Y., Zhang, L., Cheng, M., and Chau, K. T. (2015). Sensorless SVPWM-FADTC of a new flux-modulated permanent- magnet wheel motor based on a wide-speed sliding mode observer. *IEEE Trans. Ind. Electron.* 62(5), 3143–3151.

Foo, G. and Rahman, M. (2010). Sensorless sliding-mode MTPA control of an IPM synchronous motor drive using a sliding-mode observer and HF signal injection. *IEEE Trans. Ind. Electron.* 57(4), 1270–1278.

Foo, G., Sayeef, S., and Rahman, M. (2010). Low-speed and standstill operation of a sensorless direct torque and flux controlled IPM synchronous motor drive. *IEEE Trans. Energy Convers.* 25(1), 25–33.

Fu, M. and Xu, L. (1997). A novel sensorless control technique for permanent magnet synchronous motor (PMSM) using digital signal processor (DSP). In: *Proceedings of the IEEE National Aerospace and Electronics Conference*, pp. 403–408. IEEE, Piscataway, NJ.

Genduso, F., Miceli, R., Rando, C., and Galluzzo, G.R. (2010). Back EMF sensorless-control algorithm for high-dynamic performance PMSM. *IEEE Trans. Ind. Electron.* 57(6), 2092–2100.

Haque, M.E., Zhong, L., and Rahman, M.F. (2003). A sensorless initial rotor position estimation scheme for a direct torque controlled interior permanent magnet synchronous motor drive. *IEEE Trans. Power Electron.* 18(6), 1376–1383.

Harnefors, L. and Nee, H-P. (2000). A general algorithm for speed and position estimation of AC motors. *IEEE Trans. Ind. Electron.* 47(1), 77–83.

Hasegawa, M., Yoshioka, S., and Matsui, K. (2009). Position sensorless control of interior permanent magnet synchronous motors using unknown input observer for high-speed drives. *IEEE Trans. Ind. Appl.* 45(3), 938–946.

Hoque, M.A. and Rahman, M. (1994). Speed and position sensorless permanent magnet synchronous motor drives. In: *Canadian Conference on Electrical and Computer Engineering*, pp. 689–692. IEEE, Piscataway, NJ.

Jang, J-H., Ha, J-I., Ohto, M., Ide, K., and Sul, S-K. (2004). Analysis of permanent-magnet machine for sensorless control based on high-frequency signal injection. *IEEE Trans. Ind. Appl.* 40(6), 1595–1604.

Jang, J-H., Sul, S-K., Ha, J-I., Ide, K., and Sawamura, M. (2003). Sensorless drive of surface-mounted permanent-magnet motor by high-frequency signal injection based on magnetic saliency. *IEEE Trans. Ind. Appl.* 39(4), 1031–1039.

Jeong, Y-S., Lorenz, R.D., Jahns, T.M., and Sul, S-K. (2005). Initial rotor position estimation of an interior permanent-magnet synchronous machine using carrier-frequency injection methods. *IEEE Trans. Ind. Appl.* 41(1), 38–45.

Johnson, J. P., Ehsani, M., and Guzelgunler, Y. (1999). Review of sensorless methods for brushless DC. In: *Conference Record - IAS Annual Meeting (IEEE Industry Applications Society)*, pp. 143–150. IEEE, Piscataway, NJ.

Jones, L.A. and Lang, J.H. (1989). A state observer for the permanent-magnet synchronous motor. *IEEE Trans. Ind. Electron.* 36(3), 374–382.

Kim, H., Huh, K-K., Lorenz, R.D., and Jahns, T.M. (2004). A novel method for initial rotor position estimation for IPM synchronous machine drives. *IEEE Trans. Ind. Appl.* 40(5), 1369–1378.

Kim, J-S. and Sul, S-K. (1997). New approach for high-performance PMSM drives without rotational position sensors. *IEEE Trans. Power Electron.* 1(5), 904–911.

Kim, K-H., Chung, S-K., Moon, G-W., Baik, I-C., and Youn, M-J. (1995). Parameter estimation and control for permanent magnet synchronous motor drive using model reference adaptive technique. In: *21st International Conference on Industrial Electronics, Control, and Instrumentation*, pp. 387–392. IEEE, Piscataway, NJ.

Kulkarni, A.B. and Ehsani, M. (1992). A novel position sensor elimination technique for the interior permanent-magnet synchronous motor drive. *IEEE Trans. Ind. Appl.* 28(1), 144–150.

Lim, K., Low, K., and Rahman, M. (1994). A position observer for permanent magnet synchronous motor drive. In: *20th International Conference on Industrial Electronics, Control and Instrumentation*, pp. 1004–1008. IEEE, Piscataway, NJ.

Linke, M., Kennel, R., and Holtz, J. (2002). Sensorless position control of permanent magnet synchronous machines without limitation at zero speed. In: *IEEE 28th Annual Conference of the Industrial Electronics Society*, pp. 674–679. IEEE, Piscataway, NJ.

Matsui, N. (1996). Sensorless PM brushless DC motor drives. *IEEE Trans. Ind. Electron.* 43(2), 300–308.

Matsui, N., and Shigyo, M. (1992). Brushless dc motor control without position and speed sensors. *IEEE Trans. Ind. Appl.* 28(1), 120–127.

Matsui, N., Takeshita, T., and Yasuda, K. (1992). A new sensorless drive of brushless DC motor. In: *International Conference on Industrial Electronics, Control, Instrumentation, and Automation, Power Electronics and Motion Control*, pp. 430–435. IEEE, Piscataway, NJ.

Moghadam, M.A.G. and Tahami, F. (2013). Sensorless control of PMSMs with tolerance for delays and stator resistance uncertainties. *IEEE Trans. Power Electron.* 28(3), 1391–1399.

Morimoto, S., Kawamoto, K., Sanada, M., and Takeda, Y. (2001). Sensorless control strategy for salient-pole PMSM based on extended EMF in rotating reference frame. In: *Conference Record of the 2001 IEEE Industry Applications Conference, 36th IAS Annual Meeting*, pp. 2637–2644. IEEE, Piscataway, NJ.

Nahid-Mobarakeh, B., Meibody-Tabar, F., and Sargos, F-M. (2004). Mechanical sensorless control of PMSM with online estimation of stator resistance. *IEEE Trans. Ind. Appl.* 40(2), 457–471.

Nahid-Mobarakeh, B., Meibody-Tabar, F., and Sargos, F.-M. (2007). Back EMF estimation-based sensorless control of PMSM: robustness with respect to measurement errors and inverter irregularities. *IEEE Trans. Ind. Appl.* 43(2), 485–494.

Naidu, M. and Bose, B. K. (1992). Rotor position estimation scheme of a permanent magnet synchronous machine for high performance variable speed drive. In: *Industry Applications Society Annual Meeting, Conference Record of the IEEE*, pp. 48–53. IEEE, Piscataway, NJ.

Nakashima, S., Inagaki, Y., and Miki, I. (2000). Sensorless initial rotor position estimation of surface permanent-magnet synchronous motor. *IEEE Trans. Ind. Appl.* 36(6), 1598–1603.

Ogasawara, S. and Akagi, H. (1998a). An approach to real-time position estimation at zero and low speed for a PM motor based on saliency. *IEEE Trans. Ind. Appl.* 34(1), 163–168.

Ogasawara, S. and Akagi, H. (1998b). Implementation and position control performance of a position-sensorless IPM motor drive system based on magnetic saliency. *IEEE Trans. Ind. Appl.* 34(4), 806–812.

Ohnishi, K., Matsui, N., and Hori, Y. (1994). Estimation, identification, and sensorless control in motion control system. *Proc. IEEE,* 82(8), 1253–1265.

Rajashekara, K., Kawamura, A., and Matsuse, K. (1996). *Sensorless Control of AC Motor Drives, Speed and Position Sensorless Operation.* IEEE Press, Piscataway.

Schmidt, P.B., Gasperi, M.L., Ray, G., and Wijenayake, A.H. (1997). Initial rotor angle detection of a nonsalient pole permanent magnet synchronous machine. In: *Conference Record of the IEEE Industry Applications Conference 32nd IAS Annual Meeting*, pp. 459–463. IEEE, Piscataway, NJ.

Schroedl, M. (1996). Sensorless control of AC machines at low speed and standstill based on the "INFORM" method. In: *Conference Record of the IEEE Industry Applications Conference 31st IAS Annual Meeting*, pp. 270–277. IEEE, Piscataway, NJ.

Sepe, R.B. and Lang, J.H. (1992). Real-time observer-based (adaptive) control of a permanent-magnet synchronous motor without mechanical sensors. *IEEE Trans. Ind. Appl.* 28(6), 1345–1352.

Shen, J., Zhu, Z., and Howe, D. (2001). Improved speed estimation in sensorless PM brushless AC drives. In: *IEEE International Electric Machines and Drives Conference (IEMDC)*, pp. 960–966. IEEE, Piscataway, NJ.

Shen, J., Zhu, Z., and Howe, D. (2004). Sensorless flux-weakening control of permanent-magnet brushless machines using third harmonic back EMF. *IEEE Trans. Ind. Appl.* 40(6), 1629–1636.

Shi, J-L., Liu, T-H., and Chang, Y-C. (2007). Position control of an interior permanent-magnet synchronous motor without using a shaft position sensor. *IEEE Trans. Ind. Electron.* 54(4), 1989–2000.

Shi, Y., Sun, K., Huang, L., and Li, Y. (2012). Online identification of permanent magnet flux based on extended Kalman filter for IPMSM drive with position sensorless control. *IEEE Trans. Ind. Electron.* 59(11), 4169–4178.

Vas, P. (1998). *Sensorless Vector and Direct Torque Control.* Oxford University Press, Oxford.

Vas, P. (1999). *Artificial-Intelligence-based Electrical Machines and Drives.* Oxford University Press, Oxford.

Wu, R. and Slemon, G.R. (1991). A permanent magnet motor drive without a shaft sensor. *IEEE Trans. Ind. Appl.* 27(5), 1005–1011.

Xu, P. and Zhu, Z.Q. (2017). Initial rotor position estimation using zero-sequence carrier voltage for permanent-magnet synchronous machines. *IEEE Trans. Ind. Electron.* 64(1), 149–158.

Yang, S.C. (2015). Saliency-based position estimation of permanent-magnet synchronous machines using square-wave voltage injection with a single current sensor. *IEEE Trans. Ind. Appl.* 51(2), 1561–1571.

Yang, S.C., Yang, S.M., and Hu, J.H. (2017). Design consideration on the square-wave voltage injection for sensorless drive of interior permanent-magnet machines. *IEEE Trans. Ind. Electron.* 64(1), 159–168.

Yongdong, L. and Hao, Z. (2008). Sensorless control of permanent magnet synchronous motor—a survey. In: *IEEE Vehicle Power and Propulsion Conference*, pp. 1–8. IEEE, Piscataway, NJ.

Zhao, Y. (2014). *Position/speed sensorless control for permanent-magnet synchronous machines*. PhD, University of Nebraska.

Zhao, Y., Wei, C., Zhang, Z., and Qiao, W. (2013). A review on position/speed sensorless control for permanent-magnet synchronous machine-based wind energy conversion systems. *IEEE J. Emerg. Select. Topics Power Electron.* 1(4), 203–216.

Zhao, Y., Zhang, Z., Qiao, W., and Wu, L. (2015). An extended flux model-based rotor position estimator for sensorless control of salient-pole permanent-magnet synchronous machines. *IEEE Trans. Power Electron.* 30(8), 4412–4422.

Zhu, G., Dessaint, L-A., Akhrif, O., and Kaddouri, A. (2000). Speed tracking control of a permanent-magnet synchronous motor with state and load torque observer. *IEEE Trans. Ind. Electron.* 47(2), 346–355.

Zhu, G., Kaddouri, A., Dessaint, L-A., and Akhrif, O. (2001). A nonlinear state observer for the sensorless control of a permanent-magnet AC machine. *IEEE Trans. Ind. Electron.* 48(6), 1098–1108.

7 Parameter Estimation

Parameters and variables are the basic components of every mathematical model of a system as a set of equations. The parameters differ from the variables by being constant against varying operating conditions. After selecting model variables, it is important to define the model parameters to establish particular relationships between variables. The parameters thus determine how the variables are interconnected in a model. In this sense, a parameter may appear in different ways, e.g., as a coefficient or a power to a single variable or a combination of variables. The numerical values of parameters can simply be determined if the model variables are known under some operating conditions. The variables are usually obtained by measurements of the actual system. The parameter determination is thus reduced to solving the model equations under some operating conditions. As the parameters are not fixed at all operating points, it is not an easy task to obtain accurate parameter values, particularly when the system operating conditions vary over a wide range. In fact, it is usually difficult to model a complex system over wide operating conditions with constant parameters. As the system complexity increases and the operating conditions vary significantly, the parameters themselves may vary. Therefore, the distinguishing property of variables and parameters vanishes. This is why the parameter determination appears as an essential part of many control systems.

In this chapter, the variations of PMS motor parameters with operating conditions are discussed. Then, the effect of these variations on the ability of the motor model to predict the motor behavior is investigated and the necessity of considering the variations in developing accurate models is emphasized. Finally, several offline and online parameter determination schemes are presented as the main body of the chapter.

7.1 Motor parameters and estimation methods

If a motor control system is supplied with accurate parameter values, the motor full potential can be materialized under the appropriate

Control of Permanent Magnet Synchronous Motors. Sadegh Vaez-Zadeh. © Sadegh Vaez-Zadeh 2018.
Published in 2018 by Oxford University Press. DOI 10.1093/oso/9780198742968.001.0001

control system. For example, many loss minimization control systems as described in Section 3.9 can only perform well if accurate information about the motor parameters is available. In general, the motor parameters vary over wide ranges, depending on the operating conditions and the ambient temperature. In many IPM synchronous motors, these variations are dominant due to the motor construction. In particular, R_s changes due to a temperature rise; R_c varies with the motor speed, and also due to the saturation in the iron bridges between rotor magnets, and λ_m varies due to aging, temperature effect, and partial demagnetization. These variations affect the motor characteristics and performance. In this section the effects are studied first. Then a detailed sensitivity analysis is carried out to better quantify the variation effects.

7.1.1 Parameter variations

Parameter variations due to saturation and a method of modeling motor parameters as functions of stator current components are presented in Section 2.9. Here, the effects of parameter variations on PMS motor performance are investigated by focusing on machine loss. Loss minimization control as presented in Section 3.9 is implemented as a part of control system when energy saving is important in motor operation. It is shown here that the parameter variations malfunction loss minimization control if the variations are not taken into account in the control.

The model of IPM synchronous motors including copper and iron losses in a synchronously rotating reference frame is presented in Fig. 2.20. Using this model, the motor electrical loss can be formulated as

$$\begin{aligned} P_L = {} & \frac{3}{2}R_s i_{dT}^2 + \frac{3(R_s+R_c)}{R_c^2}L_d\lambda_m\omega_r^2 i_{dT} + \frac{2}{P}\frac{R_s}{R_c}\omega_r T_e \\ & + \frac{2T_e^2}{3P^2(\lambda_m + (L_d - L_q)i_{dT})^2}\left[\frac{(R_s+R_c)}{R_c^2}L_q^2\omega_r^2 + R_s\right] \\ & + \frac{3(R_s+R_c)}{2R_c^2}(\lambda_m^2 + L_d^2 i_{dT}^2)\omega_r^2. \end{aligned} \tag{7.1.1}$$

The loss is plotted versus i_{dT} in Fig. 7.1, for a motor with specifications as presented in Table 3.1. The figure shows that the motor electrical loss reaches a minimum at point A if i_{dT} is adjusted to an optimal value by an LMC. Also it is shown that the individual variations of R_s, R_c, and λ_m near the nominal speed and torque operation shift the minimum value of P_L to points B, C, and D, respectively, if a corresponding optimal value of i_{dT} is applied to the machine in each case (Vaez-Zadeh and Zamanifar 2006). In an offline or model-based LMC system, however, without a means for inclusion of parameter variations into the control system, a fixed value of i_{dT} corresponding to point A is determined from the machine model with constant parameters and applied to the motor, as an assumed optimal i_{dT} command signal. Of course, this value of i_{dT} is not a true optimal i_{dT} when motor parameters vary and results in suboptimal values of P_L corresponding to points A_1, A_2, and A_3 in Fig. 7.1. Fortunately, at this nominal operation, the suboptimal values are slightly (up to 3%) different from the minimum values of P_L, corresponding to points B, C, and D in Fig. 7.1, respectively.

Figure 7.2 shows the motor electrical loss at a non-nominal motor operation of 50% the nominal torque and 200% the nominal speed as an example. Under these operating conditions the differences between the optimal and the suboptimal values of P_L are much larger (up to about 10%) than those of Fig. 7.1. Simultaneous variations of parameters may result in even larger differences between the optimal and suboptimal values of P_L. These differences clarify the effect of parameter variations on motor performance. They also show the necessity of taking into account true values of motor parameters in control systems. Online parameter estimation is used in model-based LMC of IPM synchronous motors in order for the control system to be able to determine and apply to the motor a true optimal i_{dT} as a command signal under a wide range of operating conditions.

Another aspect of the motor parameter variations can be studied by changing the motor parameters over wide ranges and observing this

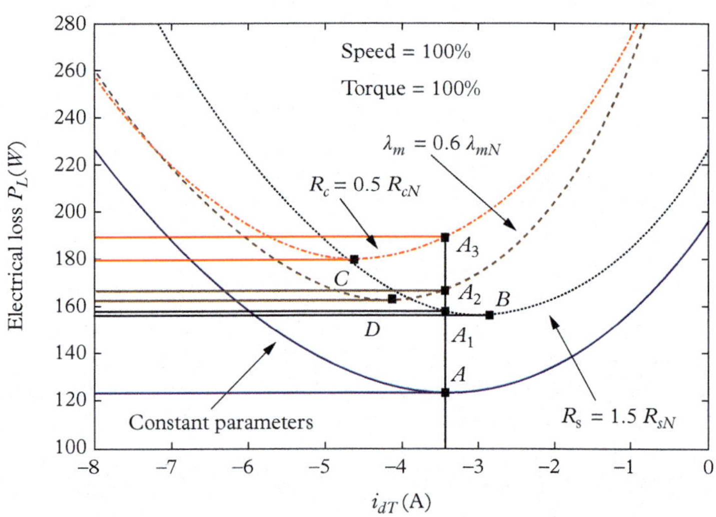

Figure 7.1 *Influence of parameter variations on minimum electrical loss under nominal operating conditions.*

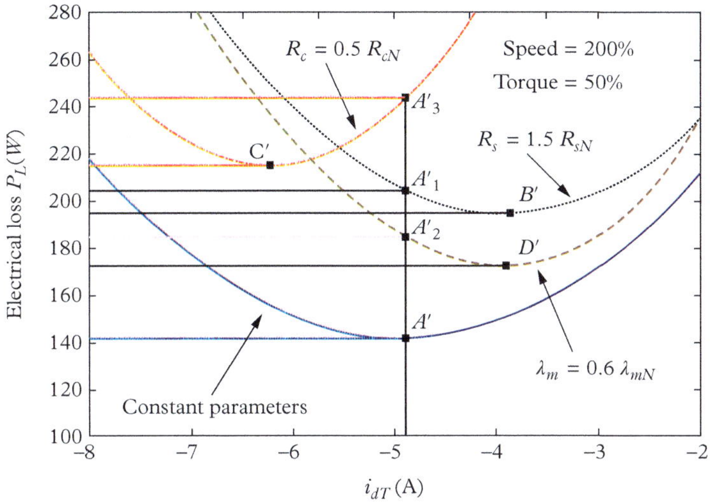

Figure 7.2 *Influence of parameter variations on minimum electrical loss under non-nominal operating conditions.*

effects on the optimal i_{dT} under different operating conditions. This can be carried out by differentiating the right-hand side of (7.1.1) with respect to i_{dT} and equating the result to 0 to obtain the equation

$$a_4 i_{dT}^4 + a_3 i_{dT}^3 + a_2 i_{dT}^2 + a_1 i_{dT} + a_0 = 0, \tag{7.1.2}$$

where $a_0 - a_4$ are functions of motor parameters plus motor speed. An optimal value for i_{dT} is achieved by solving eqn (7.1.2).

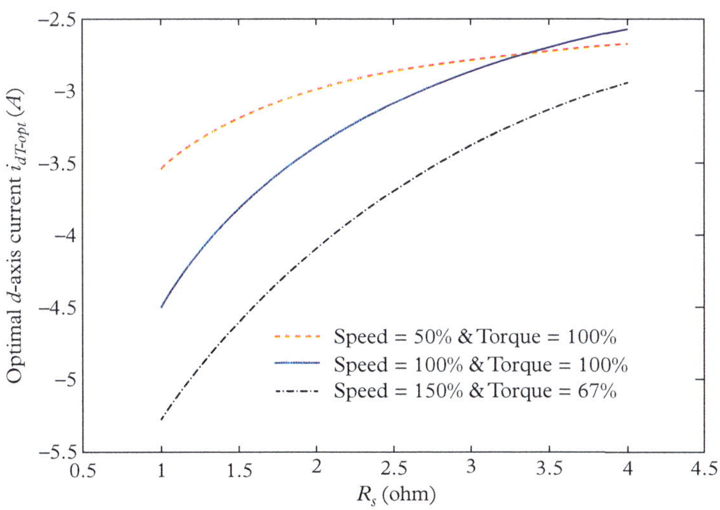

Figure 7.3 *Influence of R_s variations on optimal i_{dT}.*

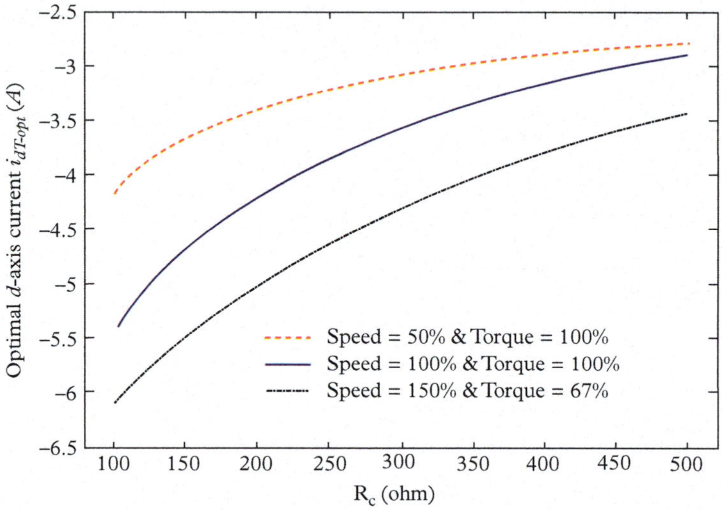

Figure 7.4 *Influence of R_c variations on optimal i_{dT}.*

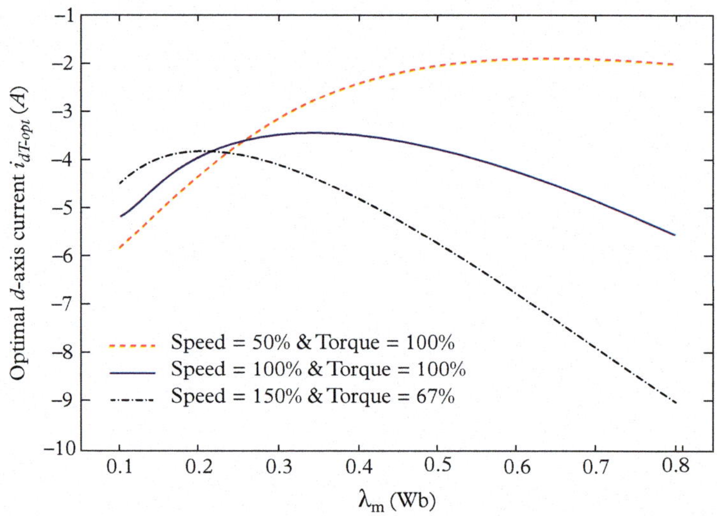

Figure 7.5 *Influence of λ_m variations on optimal i_{dT}.*

Figures 7.3–7.5 show the variations of the optimal value of i_{dT} versus R_s, R_c, and λ_m, respectively, under different operating conditions. It can be seen that except near the nominal operating conditions, gradients of the curves are significant, emphasizing the necessity for machine parameter estimation in offline LMC, as presented in Section 3.9.

A sensitivity analysis is carried out to comprehend the influence of parameter variations on the optimal i_{dT}. The sensitivities of optimal

i_{dT} to R_s, R_c, and λ_m can be found by partial differentiation of eqn (7.1.2) with respect to these parameters and rearranging the results to obtain

$$S_{Rs} = \frac{\partial i_{dT}}{\partial R_s} = -\frac{1}{K}\left(\frac{\partial a_4}{\partial R_s}i_{dT}^4 + \frac{\partial a_3}{\partial R_s}i_{dT}^3 + \frac{\partial a_2}{\partial R_s}i_{dT}^2 + \frac{\partial a_1}{\partial R_s}i_{dT} + \frac{\partial a_0}{\partial R_s}\right), \tag{7.1.3}$$

$$S_{Rc} = \frac{\partial i_{dT}}{\partial R_c} = -\frac{1}{K}\left(\frac{\partial a_4}{\partial R_c}i_{dT}^4 + \frac{\partial a_3}{\partial R_c}i_{dT}^3 + \frac{\partial a_2}{\partial R_c}i_{dT}^2 + \frac{\partial a_1}{\partial R_c}i_{dT} + \frac{\partial a_0}{\partial R_c}\right), \tag{7.1.4}$$

$$S_{\lambda m} = \frac{\partial i_{dT}}{\partial \lambda_m} = -\frac{1}{K}\left(\frac{\partial a_4}{\partial \lambda_m}i_{dT}^4 + \frac{\partial a_3}{\partial \lambda_m}i_{dT}^3 + \frac{\partial a_2}{\partial \lambda_m}i_{dT}^2 + \frac{\partial a_1}{\partial \lambda_m}i_{dT} + \frac{\partial a_0}{\partial \lambda_m}\right), \tag{7.1.5}$$

where $K = 4a_4 i_{dT}^3 + 3a_3 i_{dT}^2 + 2a_2 i_{dT} + a_1$. Figures 7.6–7.8 show these sensitivities under a wide range of motor operating conditions. Figures 7.6 and 7.7 imply that the sensitivities of i_{dT} to R_s and R_c at high speeds are significant. The figures also show that the sensitivities are almost independent of motor torque. Figure 7.8 shows that the sensitivity of i_{dT} to λ_m is particularly significant under high speed and low torque conditions and, vice versa, under low speed and high torque conditions. Parameter estimation methods as presented later take into account the variations as the motor operating conditions change. Thus, incorporating the appropriate parameter estimation methods into the control system prevents control system malfunction and improves motor performance.

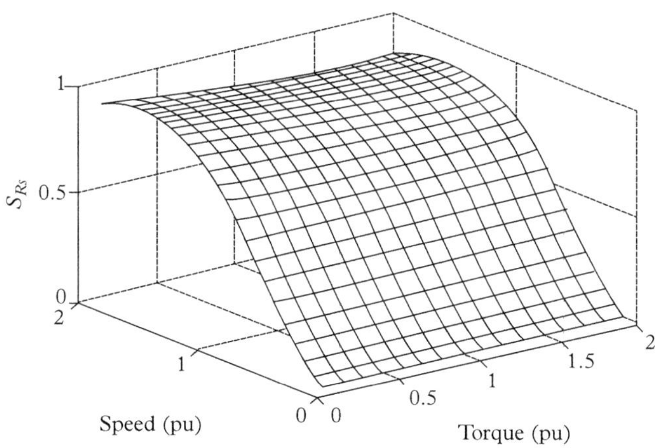

Figure 7.6 *Sensitivity of optimal i_{dT} to R_s.*

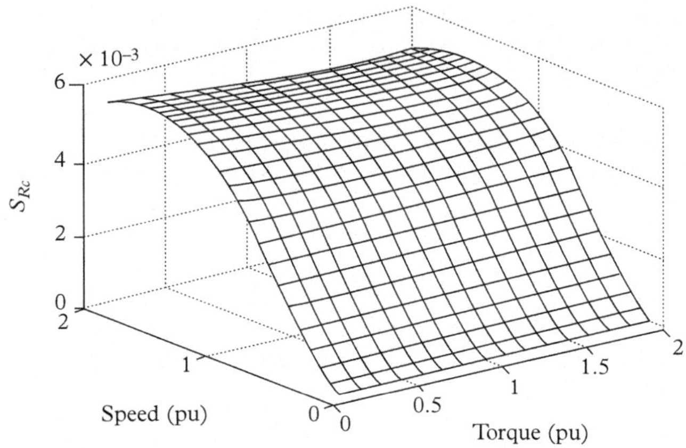

Figure 7.7 *Sensitivity of optimal i_{dT} to R_c.*

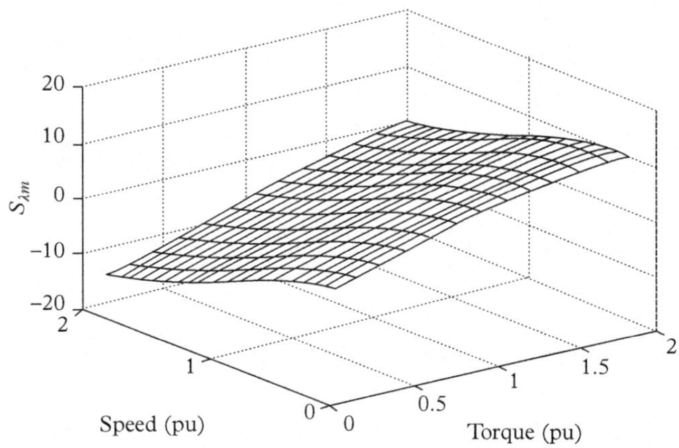

Figure 7.8 *Sensitivity of optimal i_{dT} to λ_m.*

7.1.2 Parameter estimation methods

Motor parameter estimation has appeared as an attractive research interest and a practical task during the past two decades. As a result, many estimation schemes are now available in the literature or in the market for different machines, including PMS motors. Therefore, it is very difficult to come up with a comprehensive account of the progress made in this active field.

Nevertheless, it is necessary to have an overview of some major developments in the field. In this regard, the PMS motor parameter estimation schemes can be divided into two main methods, i.e., offline method and online method. The offline schemes use the equivalent circuit model or vector diagram in connection with the measurement of motor voltage and current to calculate motor parameters at particular operating point or over a range of operating points. They may need special test arrangement for measuring machine voltage and current. Several offline schemes are presented in this chapter for estimating all major motor parameters. These include the DC standstill test for measuring stator resistance; AC standstill test for measuring inductances; no-load test for measuring back EMF; iron loss resistance and d-axis inductance; load test for calculating q-axis inductance and back EMF; determination of inductances and back EMF without information of rotor position; calculation of iron loss resistance without measuring mechanical loss; and finally calculation of motor inductances under VC.

Online parameter estimation schemes use online measurement of motor variables, usually carried out primarily for control systems, to estimate motor parameters with the help of closed-loop observers. The schemes take into account motor parameter variations caused by any sources. They are increasingly being implemented as a part of modern control systems in the market.

A number of online schemes for estimating PMS motor parameters, which are among the more accepted schemes in the field, are presented in this chapter. These include closed-loop observers for estimating motor inductances; recursive least-squares (RLS) scheme for estimating motor inductances, and also for estimating inductances and magnet flux linkage; MRAS-based estimation of λ_m, R_s, and motor inductances; and finally, extended Kalman filter (EKF) scheme for estimating stator resistance and synchronous inductance.

7.2 Offline parameter estimation method

The offline parameter estimation schemes are carried out by special tests, while the motor is not in service. The schemes use machine models and measurements of machine signals to calculate the parameters. The measurements are performed at no load and/or certain machine operating points. It is also possible to do the measurements over a range of operating points to consider the effect of machine operation on the parameters. Nevertheless, the method cannot take into account parameter changes caused by factors other than operating point like aging. All motor parameters including stator resistance, iron loss resistance, magnitude of PM flux linkage, and inductances are measured by offline schemes, as presented in this section.

7.2.1 DC standstill test for measurement of R_s

Stator phase winding of PMS motors, R_s, can be measured by a standard DC resistance test. The winding resistance equals the half of total resistance seen from the machine terminals, assuming the phase windings are star connected. The actual AC resistance is different from the value of R_s from the DC test due to skin effect, however. Nevertheless, the error is negligible for small to medium size motors (Haque and Rahman 1999).

On the other hand, the motor resistance is highly dependent on temperature. Therefore, R_s at any temperature, R_{sT}, is determined from its measured values, R_s, by

$$R_{sT} = R_{s0}\frac{K+T}{K+T_0}, \tag{7.2.1}$$

where T and T_0 are the current and the test temperature and K is a constant, depending on the winding material. $K = 234.5$ for copper. The value of R_s at 25°C is usually used in motor data sheets. The DC value is modified by a factor for large size machines to consider skin effect.

7.2.2 AC standstill test for calculation of inductances

The direct and quadrature inductances, L_d and L_q can be determined by several test methods. They may be classified into two main categories, i.e., running tests and standstill tests. The former tests are suited to fixed-speed line start PM motors. The latter tests are usually used for determining the inductances of inverter-driven PMS motors. This category of tests itself is divided into several schemes, including mainly the AC standstill test, the DC bridge test, the instantaneous flux linkage test, and the standstill torque test. The DC bridge test is also suited to line start PM motors to prevent any induced voltage from the rotor cage. The AC test is suited to inverter-driven PMS motors and is presented here. The test can be carried out with and without neutral connection. The latter is more convenient since the neutral point of windings may not be easily accessible. Figure 7.9 shows the test circuit (Haque and Rahman 1999).

Using the measured line voltage and current, v_L and i_L, the self-inductance of phase a is obtained for a range of rotor position, θ_r, as

$$L_{aa}(\theta_r) = \frac{2}{3}\frac{\sqrt{\left({}^{v_L}/_{i_L}\right)^2 - \left(\frac{3}{2}R_s\right)^2}}{\omega_r}. \tag{7.2.2}$$

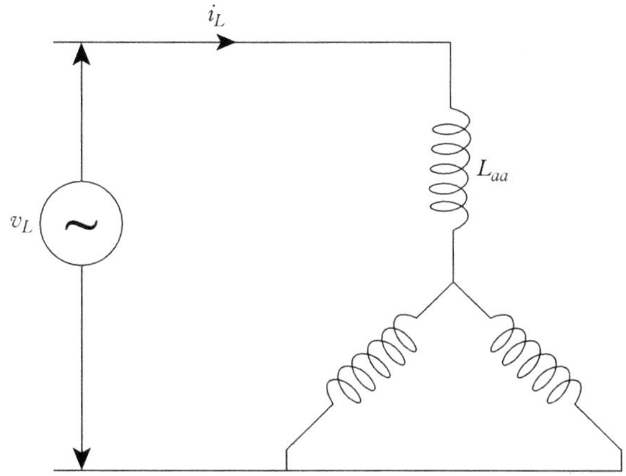

Figure 7.9 *Test circuit for AC standstill test.*

The values of self-inductance are plotted as a function of θ_r. The plot is then fitted to the analytical relationship of phase inductance as

$$L_{aa}(\theta_r) = L_1 + L_2 \cos 2\theta_r. \tag{7.2.3}$$

As a result, the values of L_1 and L_2 are determined. The d- and q-inductances are then obtained from the following equations, which are recalled from eqns (2.5.12) and (2.5.13) as

$$L_d = \frac{3}{2}(L_1 + L_2), \tag{7.2.4}$$

$$L_q = \frac{3}{2}(L_1 - L_2). \tag{7.2.5}$$

The method cannot take into account the saturation.

7.2.3 Measurement of R_c, L_d and λ_m by no-load test

The machine iron loss resistance, R_c, and the magnet flux linkage magnitude (magnet EMF constant), λ_m, can be identified by carrying out a motoring no-load test. A balanced three-phase voltage at the nominal frequency is applied to the motor terminals, while the motor is under no-load conditions. The terminal voltage is adjusted such that the motor runs at unity power factor. This voltage corresponds to a minimum current. Then rms values of the phase voltage and phase current are measured as V_{rms} and I_{rms} and the magnitude of the stator voltage space vector and the stator current space vector are obtained by $v_s = \sqrt{2}V_{rms}$ and $i_{s=}\sqrt{2}I_{rms}$.

Referring to the steady-state equivalent circuit of 2.20, under the previous test conditions, i.e., at no load and unity power factor $i_d = 0$, $i_q = i_s$ and $v_d = 0$, $v_q = v_s$. Thus, the motor equivalent circuit reduces to the q-axis equivalent circuit only, with the voltage equation

$$v_s = R_s i_s + \omega_r \lambda_m, \tag{7.2.6}$$

where i_s contains the iron loss current only, which follows through R_c. From the latter equivalent circuit, R_c is calculated as

$$R_c = \frac{v_s - R_s i_s}{i_s}. \tag{7.2.7}$$

Having v_s and i_s in terms of measured values from the test, R_c is calculated from eqn (7.2.7). The calculated R_c includes windage and friction losses, which are small in comparison with core loss.

The d-axis inductance is also obtained by a no-load test at varying power factor. The test is carried out by changing the terminal voltage to have a range of phase currents from the rated value at lagging power factor to the rated value at leading power factor.

Referring to the torque equation of eqn (2.8.12), the q-axis torque-producing current component, i_{qT}, at no load is zero. Therefore, the speed voltage in the d-axis equivalent circuit of Fig. 2.20 short-circuited. As a result, v_d reduces to a low value such that $v_q \approx v_s$. Therefore, at each current there is

$$e_i = v_s - R_s i_s \cos\varphi, \tag{7.2.8}$$

where e_i is the magnitude of the induced voltage vector. Then

$$i_d = \sqrt{i_s^2 - \frac{e_i^2}{R_c^2}}, \tag{7.2.9}$$

and L_d is calculated as

$$L_d = \frac{e_i - \omega_r \lambda_m}{\omega_r i_d}. \tag{7.2.10}$$

With the magnitudes of voltage and current space vectors obtained from the rms measurements as $v_s = \sqrt{2} V_{rms}$ and $i_{s=}\sqrt{2} I_{rms}$, respectively, R_c and L_d are calculated from eqns (7.2.7) and (7.2.10).

The magnet back EMF constant of PMS motors, λ_m, can also be determined offline by measuring the open-loop terminal voltage, when the machine is working as a generator at no load. It is required to drive the machine by a motor to do the measurement. Assuming a sinusoidal magnet back EMF, it is the same as the measured open circuit voltage:

$$e_{ma} = e_a = \frac{v_{ab}}{\sqrt{3}}, \tag{7.2.11}$$

$$e_{ma} = E_m \cos \omega_r t = \omega_r \lambda_m \cos \omega_r t, \tag{7.2.12}$$

$$\lambda_m = \frac{E_m}{\omega_r}. \tag{7.2.13}$$

7.2.4 Measurement of L_q and λ_m by load test

By referring to voltage equations for a PMS motor at steady-state, which are presented in Chapter 2 as eqns (2.5.30) and (2.5.31), the motor phasor diagram under the conditions of $i_q = i_s$ and $i_d = 0$ can be depicted as in Fig. 7.10. In this figure, v_s and i_s, respectively, show the magnitudes of the stator voltage and current space vectors and φ denotes the power factor angle. The voltage equations under the above-mentioned conditions can then yield the motor parameters as

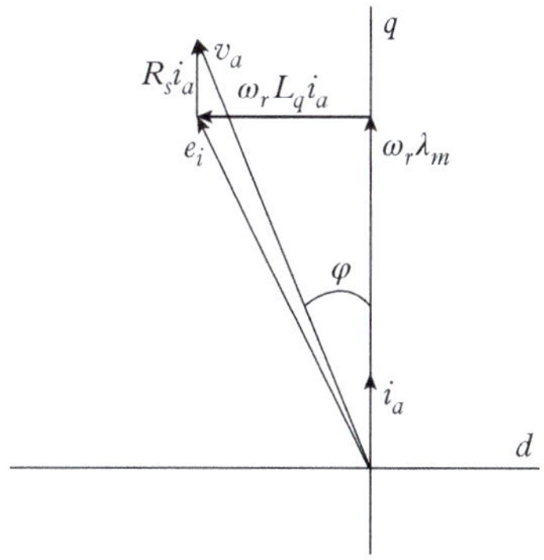

Figure 7.10 *Motor steady-state phasor diagram under load conditions with $i_d = 0$.*

$$L_q = \frac{v_s \sin \varphi}{\omega_r i_s}, \tag{7.2.14}$$

$$\lambda_m = \frac{v_s \cos \varphi - R_s i_s}{\omega_r}. \tag{7.2.15}$$

It is seen that v_s, i_s, and the power factor are needed to calculate L_q and λ_m according to eqns (7.2.14) and (7.2.15). They are obtained from a load test, when the conditions of $i_q = i_s$ and $i_d = 0$ are satisfied by the following procedure. The rotor position sensor is used to index the d-axis alignment of one winding and adjust the motor load by a dynamometer and also the terminal voltage until the zero crossing of sinusoidal current waveform in the winding and the index are coincident (Mellor *et al.* 1991). The motor phase voltage, phase current, and power factor are measured as V_{rms}, I_{rms}, and $\cos\varphi$. Then, $v_s = \sqrt{2}V_{rms}$ and $i_s = \sqrt{2}I_{rms}$.

7.2.5 Determination of L_d, L_q, and λ_m without rotor position

Referring to a typical phasor diagram of a PMS machine as depicted in Fig. 7.11 based on the steady-state voltage equations (2.5.30) and (2.5.31), the d- and q-axis inductances are given as

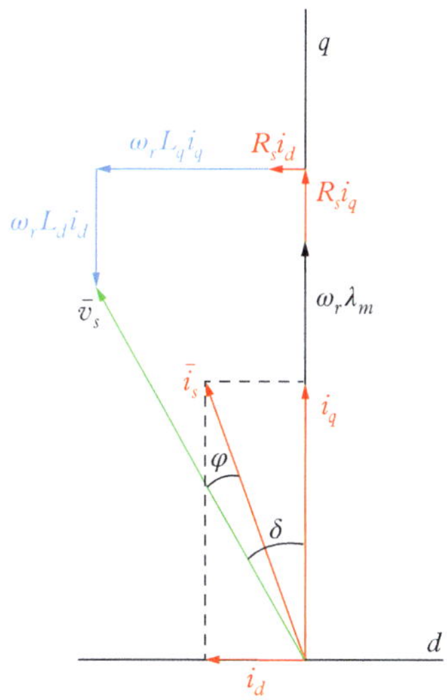

Figure 7.11 *Phasor diagram of a PMS machine when it is supplied directly from the mains.*

$$L_d = \frac{v_s \cos \delta - \omega_r \lambda_m - R_s i_q}{\omega_r i_d}, \tag{7.2.16}$$

$$L_q = -\frac{v_s \sin \delta - R_s i_d}{\omega_r i_q}. \tag{7.2.17}$$

where

$$i_d = i_s \sin(\varphi - \delta), \qquad (7.2.18)$$

$$i_q = i_s \cos(\varphi - \delta), \qquad (7.2.19)$$

and φ and δ are the power factor angle and the load angle, respectively. Also, the input power to the motor is given by

$$P = \frac{3}{2} v_s i_s \cos\varphi. \qquad (7.2.20)$$

It is seen that L_d and L_q in eqns (7.2.16) and (2.7.17) need the value of load angle. Some methods of inductance calculation determine this angle by using hardware like a rotor position sensor. However, if λ_m and R_s are known from the methods presented earlier, it is possible to calculate δ from the steady-state model of the machine as follows (Nee *et al.* 2000):

$$i_d = i_s(\sin\varphi \cos\delta - \cos\varphi \sin\delta), \qquad (7.2.21)$$

$$i_q = i_s(\cos\varphi \cos\delta + \sin\varphi \sin\delta). \qquad (7.2.22)$$

Now, eqns (7.2.21) and (7.2.22) are substituted into eqn (7.2.16) and the result is rearranged as

$$\omega_r \lambda_m = \mathrm{B}\cos\delta + C\sin\delta, \qquad (7.2.23)$$

where

$$B = v_s - \omega_r L_d i_s \sin\varphi - R_s i_s \cos\varphi, \qquad (7.2.24)$$

$$C = \omega_r L_d i_s \cos\varphi - R_s i_s \sin\varphi. \qquad (7.2.25)$$

Then, eqn (7.2.23) is modified by using

$$\sin\delta = \sqrt{1 - \cos^2\delta}, \qquad (7.2.26)$$

and is rearranged and squared on both sides to yield

$$\cos^2\delta - \frac{2\omega_r \lambda_m B}{B^2 + C^2}\cos\delta + \frac{(\omega_r \lambda_m)^2 - C^2}{B^2 + C^2} = 0. \qquad (7.2.27)$$

Finally, eqn (7.2.27) is solved for $\cos\delta$ as

$$\cos\delta = \frac{1}{B^2 + C^2}\left(\omega_r \lambda_m B \pm \sqrt{B^2 C^2 - C^2(\omega_r \lambda_m)^2 + C^4}\right). \qquad (7.2.28)$$

The correct sign in the bracket must be chosen. It is the fact that normally

$$\delta > \varphi, \ \cos\delta < \cos\varphi; \tag{7.2.29}$$

thus, minus sign is usually the correct choice.

Now, having δ, L_d and L_q can be calculated from eqns (7.2.16) and (7.2.17) if i_d and i_q are obtained from eqns (7.2.18) and (7.2.19). It must be noted that if λ_m is calculated by a method leading to eqn (7.2.13), which gives a constant value independent of δ, then the value of L_d from eqn (7.2.16) may not be valid over the entire operating range despite taking into account varying i_d and i_q in eqn (7.2.16) (Rahman *et al.* 1994).

7.2.6 Calculation of R_c without measuring mechanical loss

The method of calculating R_c from the no-load test, presented previously, does not separate the mechanical loss (windage and friction losses) from the iron loss. Referring to the equivalent circuit of Fig. 2.20, the method causes the loss due to R_c to include both iron loss and mechanical loss. However, the mechanical loss is also included in the output power. Therefore, by not separating mechanical loss from R_c, the loss is accounted for twice in the equivalent circuit. The power balance in the machine can further explain the fact as

$$P_{in} - P_{out} = P_{Cu} + P_{Fe} + \Delta P_m, \tag{7.2.30}$$

where P_{in}, P_{out}, P_{Cu}, P_{Fe}, and ΔP_m are input power, output power, copper loss, iron loss, and mechanical loss, respectively. The no-load loss includes iron loss and mechanical loss. If R_c is calculated by the no-load test as in eqn (7.2.7), it will be overestimated. Here, a method of accurately calculating R_c is presented (Urasaki *et al.* 2003).

Expressing the mechanical power as

$$P_m = P_{out} + \Delta P_m, \tag{7.2.31}$$

and substituting eqn (7.2.31) into eqn (7.2.30) yields

$$P_{in} - P_m = P_{Cu} + P_{Fe}. \tag{7.2.32}$$

Rearranging eqn (7.2.32) gives

$$P_{in} - P_{Cu} = P_m + P_{Fe}, \tag{7.2.33}$$

where P_{Cu} and P_{Fe} can be substituted from the equivalent circuit to obtain

$$P_{in} - \frac{3}{2}R_s\left(i_d^2 + i_q^2\right) = P_m + \frac{3}{2}\frac{\omega_r^2\left(\lambda_d^2 + \lambda_q^2\right)}{R_c}. \tag{7.2.34}$$

The left-hand side of eqn (7.2.34) is the air gap power, P_{ag}, and the right-hand side can be written in compact form as

$$P_{ag} = P_m + \frac{3}{2}\frac{e_i^2}{R_c}, \tag{7.2.35}$$

where $e_i = |\bar{v}_s - R_s\,\bar{i}_s|$ is the magnitude of induced voltage. Equation (7.2.34) can also be written in terms of measured phase voltage and phase current, V_{rms} and I_{rms}, as

$$P_{in} - 3R_sI_{rms}^2 = P_m + 3\frac{\left(V_{rms} - R_sI_{rms}\right)^2}{R_c}, \tag{7.2.36}$$

where

$$V_{rms} = \frac{\sqrt{2}}{2}\sqrt{v_d^2 + v_q^2} = \frac{\sqrt{2}}{2}v_s, \quad I_{rms} = \frac{\sqrt{2}}{2}\sqrt{i_d^2 + i_q^2} = \frac{\sqrt{2}}{2}i_s. \tag{7.2.37}$$

R_c can be found by a simple test using eqns (7.2.35) and (7.2.36). The test is carried out by operating the machine under constant speed and load conditions, while changing i_d and recording phase voltage and phase current, V_{rms} and I_{rms}. Changing i_d will change the flux linkage. However, the developed torque remains fixed since the machine is a non-salient PMS motor. Therefore, P_m in eqn (7.2.35) remains constant. Now, depicting P_{ag} versus $1.5\,e_i^2$ according to eqn (7.2.35) provides a linear curve with a slope equal to R_c, as seen in Fig. 7.12.

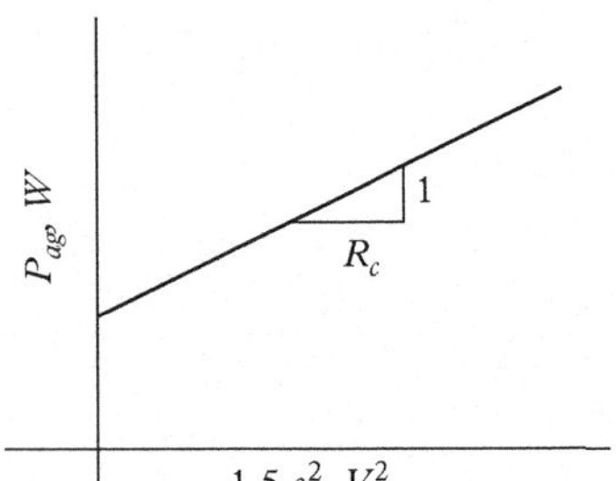

Figure 7.12 *Air gap power versus square of* 1.5 × *(inner voltage) (Urasaki* et al. *2003).*

7.2.7 Calculation of motor inductances under VC

If a PMS motor is controlled under VC method in rotor RF, it is possible to adjust the d- and q-axis current components to desired values. In particular, at $i_d = 0$, the steady-state voltage equations are given as

$$v_d = -\omega_r L_q i_q, \tag{7.2.38}$$

$$v_q = R_s i_q + \omega_r \lambda_m. \tag{7.2.39}$$

Having known ω_r, then L_q is obtained as

$$L_q = \frac{\sqrt{v_s^2 + \left(R_s i_q + \omega_r \lambda_m\right)^2}}{\omega_r i_{q,}}. \tag{7.2.40}$$

Also, at $i_q = 0$, the steady-state voltage equations are given as

$$v_d = R_s i_d, \tag{7.2.41}$$

$$v_q = \omega_r \lambda_m + \omega_r L_d i_{d.} \tag{7.2.42}$$

Having known ω_r, and also obtaining λ_m from a method like the open-loop test as described previously, then L_d can be calculated by

$$L_d = \frac{\sqrt{v_s^2 + (R_s i_d)^2} - \omega_r \lambda_m}{\omega_r i_d}. \tag{7.2.43}$$

The values of v_d and v_q can be obtained from the measured phase voltages to calculate v_s. An easier alternative is to use voltage commands, v_d^* and v_q^*, as the current controllers outputs. The calculation of eqns (7.2.40) and (7.2.43) can be done with different currents to obtain L_d and L_q as functions of i_d and i_q, respectively, thus considering the saturation. The cross-coupling saturation cannot be considered in this method, since a current component is always zero.

It is also possible to calculate the inductances under rated load conditions with different stator current vector angles. The steady-state voltage equation in this scheme yields

$$L_d\,(\alpha) = \frac{v_q - R_s i_q - \omega_r \lambda_m}{\omega_r i_d}, \tag{7.2.44}$$

$$L_q\,(\alpha) = \frac{R_s i_d - v_d}{\omega_r i_q}, \tag{7.2.45}$$

where α is the angle of $\bar{i}_s$ with respect to the d-axis. The cross-coupling saturation effect on L_d and L_q is considered in this case since both i_d and i_q are changing.

7.3 Online parameter estimation method

Online parameter estimation schemes for PMS motors are carried out during the normal operation of the machines. They are done by using motor model and online measurement of motor variables. The main advantage of these schemes is that the parameters are updated during the motor operation, taking into account almost all variations, regardless of their causes. They take into account parameter variations caused by saturation, cross coupling, and temperature. Even the effect of mechanical damages can be considered. Nevertheless, the schemes usually need huge data processing according to complex algorithms. Therefore, they require fast computing processors. They may use some offline calculated results to speed up the online computations.

7.3.1 Closed-loop observers for estimation of motor inductances

A simple closed-loop observer is employed to estimate L_d and L_q, while R_s and rotor flux linkage magnitude are assumed constant (Kim and Lorenz 2002). The assumption is justified to a limited extent since the non-estimated parameters are usually subject to less variations than the inductances. Also, a constant R_s is justified because the effect of R_s is negligible at high speed due to the terminal voltage domination. L_d and L_q are estimated independently by two simple signal processing routines and two control loops that form a closed-loop observer. The observer improves the estimations in the presence of transient noise.

The machine steady-state voltage equations in rotor reference frame are used for the estimation as follows:

$$R_s i_d - v_d = \hat{L}_q \omega_r i_q, \tag{7.3.1}$$

$$v_q - R_s i_q - \omega_r \lambda_m = \hat{L}_d \omega_r i_d. \tag{7.3.2}$$

The model eliminates the deferential terms from the voltage equations, thus simplifying the calculations. It also reduces the estimation sensitivity to noise, since noise is amplified by differentiation. This in turn prevents the risk of instability. The estimation of L_d and L_q from eqns (7.3.1) and (7.3.2) are as follows:

$$\hat{L}_d = \frac{v_q - R_s i_q}{\omega_r i_d} - \frac{\lambda_m}{i_d}, \tag{7.3.3}$$

$$\hat{L}_q = -\frac{v_d - R_s i_d}{\omega_r i_q}. \tag{7.3.4}$$

These estimation equations are synthesized as in Fig. 7.13. The estimated inductances contain undesirable high-frequency harmonics due to current harmonics. These are eliminated by low-pass filters, as seen in Fig. 7.13.

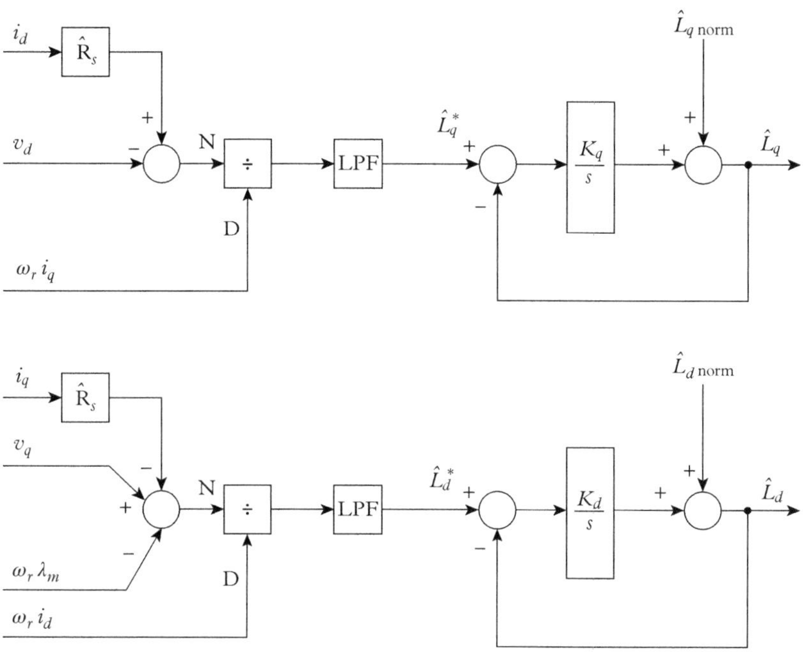

Figure 7.13 *Estimation of machine inductances by closed-loop observer.*

The estimated parameters are used as command signals to simple control loops, which act as a closed-loop observer with the integration gains of K_d and K_q.

The observer uses the nominal values of parameters at start-up. These values are corrected as the estimation goes on. The actual d–q currents and the commanded d–q voltages (current controller outputs) in addition to the rotor speed are used in estimating the parameters. Figure 7.14 shows the inputs and outputs of the parameter estimation system.

The parameter estimation method works at all operating points of PMS machines and during steady-state and transient operations. The method takes into account the effect of saturation on the inductances as it uses the online values of the machine variables in the estimation.

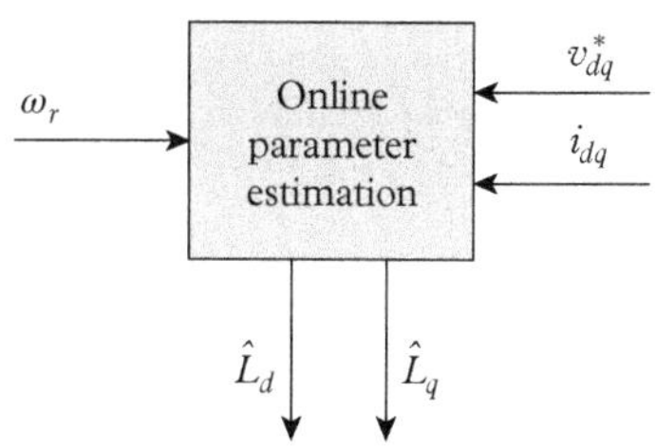

Figure 7.14 *The overall view of the observer input–output.*

7.3.2 MRAS-based estimation of λ_m and R_s

The parameter estimation by MRAS is an online estimation scheme that uses two machine models, as shown in Fig. 7.15

(Kim *et al.* 1995). The first one, which is referred to as the reference model, is a state-space model of the machine and the second one is an adaptive observer with estimated parameters. An error vector, $\boldsymbol{e}$, is produced as the difference of the two model outputs. The error vector is applied to an adaptation mechanism, which works based on a dynamic equation of the error vector to renew the estimated parameters in order to decrease the error. As a result, the estimated parameters approach the actual parameters.

The MRAS can be transformed mathematically to a standard non-linear time varying feedback system based on the error vector, as shown in Fig. 7.16. The mathematical system consists of two subsystems:

1. A feedforward time-invariant linear subsystem providing $\boldsymbol{e}$ as the output, and
2. A feedback non-linear time-varying subsystem, accepting $\boldsymbol{e}$ as the input and providing $\boldsymbol{W}$ as the output.

Then, the following two conditions must be held:

1. The transfer function matrix of the first subsystem is strictly positive real (SPR).

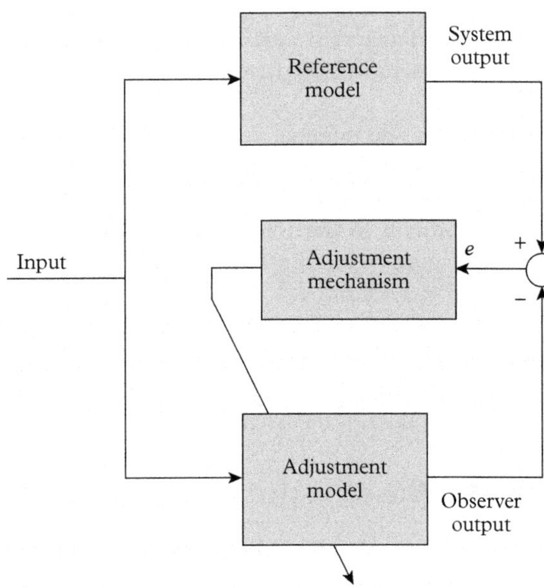

Figure 7.15 *General structure of MRAS (Kim* et al. *1995).*

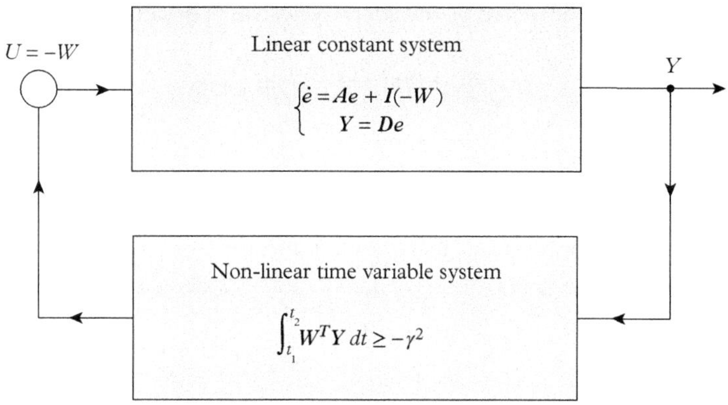

Figure 7.16 *MRAS transferred to a standard non-linear, time-varying feedback system.*

2. Popov's inequality is satisfied for the input and the output of the second subsystem during a time interval of 0 to t_1 for all $t_1 \geq 0$, which is

$$\int e^T W \, dt \geq 0. \tag{7.3.5}$$

Consequently, the adaptive law is developed to fulfill the second condition by estimating suitable values for the model parameters.

The MRAS is applied to a surface-mounted PMS motor to simultaneously estimate stator resistance and magnet flux linkage, considering a constant synchronous inductance (Kim *et al.* 1995).

A mathematical model of the PMS machine in rotor reference frame in connection with the VC in the same RF is used. The model is presented in state-space form by taking the current components in this RF, i_d and i_q, as the state variables,

$$\dot{\boldsymbol{i}}_s = \boldsymbol{A}\boldsymbol{i}_s + \boldsymbol{B}\boldsymbol{v}_s + \boldsymbol{d}, \tag{7.3.6}$$

where

$$\boldsymbol{i}_s = \left[i_q \; i_d\right]^T, \tag{7.3.7}$$

$$\boldsymbol{v}_s = \left[v_q \; v_d\right]^T \tag{7.3.8}$$

$$A = -\frac{R_s}{L_s}\boldsymbol{I} + \omega_r \boldsymbol{J}, \quad \boldsymbol{B} = \frac{1}{L_s}\boldsymbol{I}, \quad \boldsymbol{d} = \begin{bmatrix} -\frac{\lambda_m}{L_s}\omega_r \\ 0 \end{bmatrix}, \tag{7.3.9}$$

$$\boldsymbol{I} = \begin{bmatrix} 1 & 0 \\ 0 & 1 \end{bmatrix}, \quad \boldsymbol{J} = \begin{bmatrix} 0 & -1 \\ 1 & 0 \end{bmatrix}. \tag{7.3.10}$$

Also, the state observer, as the adaptive system is designed as

$$\dot{\hat{\boldsymbol{i}}}_s = \hat{\boldsymbol{A}}\hat{\boldsymbol{i}}_s + \boldsymbol{B}\boldsymbol{v}_s + \hat{\boldsymbol{d}} + \boldsymbol{G}\left(\hat{\boldsymbol{i}}_s - \boldsymbol{i}_s\right), \tag{7.3.11}$$

where the matrices with a ^ are the ones in which the estimated parameters replace the actual parameters and $\boldsymbol{G}$ is the observer gain matrix. The gain matrix is designed such that the poles of the observer are located in appropriate locations in the complex plane as

$$\boldsymbol{G} = -g_1\boldsymbol{I} + g_2\boldsymbol{J} = \begin{bmatrix} -g_1 & -g_2 \\ g_2 & -g_1 \end{bmatrix}. \tag{7.3.12}$$

The gains are determined such that the closed-loop observer poles are k times the poles of the motor model where $k \geq 1$. This is satisfied by

$$g_1 = (k-1)\,\frac{\hat{R}_s}{L_s}, \tag{7.3.13}$$

$$g_2 = (k-1)\,\omega_r. \tag{7.3.14}$$

In this case the first condition, which is required for the stability of the adaptive system, is also satisfied.

Now, the error dynamic equation is formed by subtracting eqn (7.3.11) from eqn (7.3.6) to give

$$\dot{\boldsymbol{e}} = (\boldsymbol{A} + \boldsymbol{G})\,\boldsymbol{e} - \boldsymbol{W}, \tag{7.3.15}$$

where $\boldsymbol{e} = \boldsymbol{i}_s - \boldsymbol{i}_s$ and $\boldsymbol{W}$ is a non-linear, time-varying vector, which is defined as

$$\boldsymbol{W} = -\Delta\boldsymbol{A}.\hat{\boldsymbol{i}}_s - \Delta\boldsymbol{d}, \tag{7.3.16}$$

where $\Delta\boldsymbol{A}$ and $\Delta\boldsymbol{d}$ are the error matrices caused by the parameter variations. They can be presented as

$$\Delta\boldsymbol{A} = \boldsymbol{A} - \hat{\boldsymbol{A}} = -\frac{\Delta R_s}{L_s}\boldsymbol{I}, \tag{7.3.17}$$

$$\Delta\boldsymbol{d} = \boldsymbol{d} - \hat{\boldsymbol{d}} = \begin{bmatrix} -\frac{\omega_r}{L_s} \\ 0 \end{bmatrix} \Delta\lambda_m, \tag{7.3.18}$$

$$\Delta R_s = R_s - \hat{R}_s, \qquad \Delta\lambda_m = \lambda_m - \hat{\lambda}_m. \tag{7.3.19}$$

The general system of Fig. 7.16 in this case is presented as Fig. 7.17, where two non-linear functions of state error are used to provide the

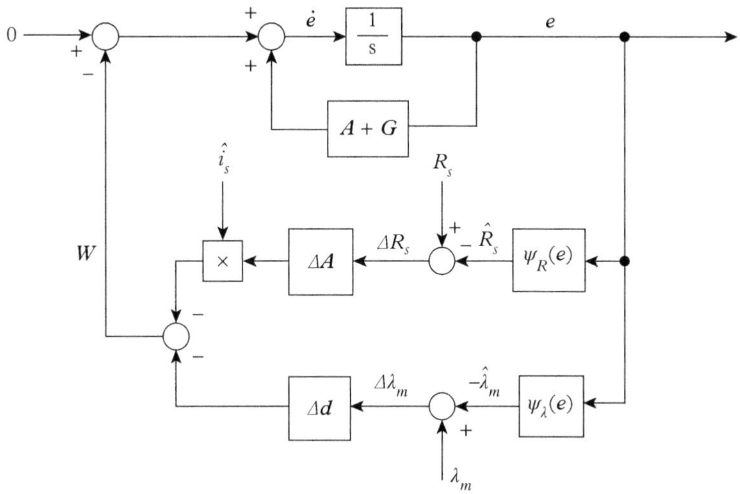

Figure 7.17 *MRAS for estimating magnet flux linkage magnitude and stator resistance of PMS motors (Kim* et al. *1995).*

estimation of parameters (Kim *et al.* 1995). The second condition as presented previously is satisfied in this system as

$$\int_0^{t_1} \boldsymbol{e}^T \boldsymbol{W}\, dt = \int_0^{t_1} \left(\boldsymbol{e}^T \hat{\boldsymbol{i}}_s \frac{\Delta R_s}{L_s} + \boldsymbol{e}^T \begin{pmatrix} \omega_r \\ 0 \end{pmatrix} \frac{\Delta \lambda_m}{L_s} \right) dt \geq -\gamma_0^2, \text{ for all } t_1 \geq 0, \tag{7.3.20}$$

where ${\gamma_0}^2$ is a finite positive constant. It is seen that eqn (7.3.20) has two error components caused by the stator resistance error and the flux linkage error. The error components can be separated as

$$\int_0^{t_1} \boldsymbol{e}^T \hat{\boldsymbol{i}}_s \frac{\Delta R_s}{L_s} dt \geq -\gamma_1^2, \tag{7.3.21}$$

$$\int_0^{t_1} \boldsymbol{e}^T \begin{pmatrix} \omega_r \\ 0 \end{pmatrix} \frac{\Delta \lambda_m}{L_s} dt \geq -\gamma_2^2, \tag{7.3.22}$$

where ${\gamma_1}^2$ and ${\gamma_2}^2$ are finite positive constants.

To improve the transient performance of the estimation, PI adaptation law is usually used (Kubota *et al.* 1993). Therefore,

$$\hat{R}_s = -\left(K_{PR} + \frac{K_{IR}}{s} \right) \cdot \left(e_q \hat{i}_q + e_d \hat{i}_d \right), \tag{7.3.23}$$

where K_{PR} and K_{IR} are the proportional and integral gains, respectively, for the stator resistance estimation. The flux linkage can also be estimated by a similar adaptation law as

$$\hat{\lambda}_m = -\left(K_{P\lambda} + \frac{K_{I\lambda}}{s}\right) . \left(e_q \omega_r\right), \tag{7.3.24}$$

where $K_{P\lambda}$ and $K_{I\lambda}$ are the proportional and integral gains for the flux linkage estimations, respectively. When the estimated resistance and flux linkage converge to the corresponding actual values, the following closed-loop observer governs the error dynamics:

$$\dot{e} = (A + G)\, e = \begin{pmatrix} -k\frac{R_s}{L_s} & -k\omega_r \\ k\omega_r & -k\frac{R_s}{L_s} \end{pmatrix} . e. \tag{7.3.25}$$

The error dynamics is k times faster than the dynamics of PMS motor. The core of the MRAS-based parameter estimation scheme is shown in Fig. 7.18 for use in connection with VC of the machine. It is seen that the measured current components, in addition to the voltage component commands in the two-axis stationary reference frame, are required by the observer to calculate the estimated current components in the rotor reference frame. The latter together with its actual values are used in the parameter estimator to give the estimation of stator resistance and magnet flux linkage magnitude.

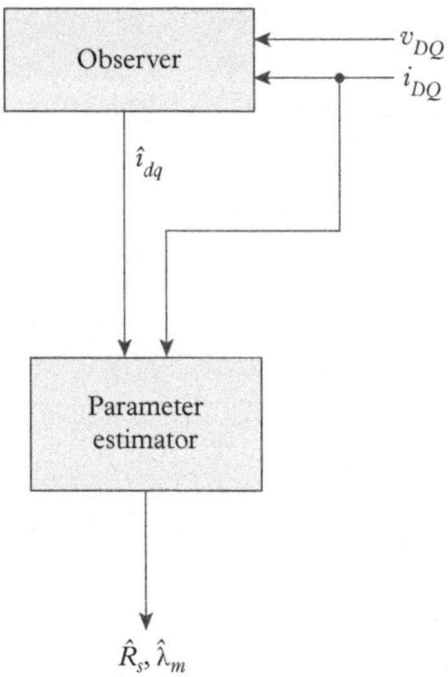

Figure 7.18 *MRAS-based parameter estimation.*

7.3.3 MRAS-based estimation of motor inductances

MRAS can also be used to estimate stator resistance and machine inductances of PMS motors of both salient and non-salient types (Boileau *et al.* 2011). In this scheme, the magnitude of magnet flux linkage is assumed to be known as a constant parameter. A technique based on the output error cancellation is employed. The machine model is presented in the rotor reference frame in connection with a decoupling control technique that improves convergence dynamics and overall system stability. The parameter estimation is carried out under the transient and steady-state conditions. The reference model in state-space form is presented as

$$\hat{L}_d . p\dot{i}_d = -\hat{R}_s i_d + \omega_r \hat{L}_q i_q + G_v . v_{dr}, \tag{7.3.26}$$

$$\hat{L}_q . p\dot{i}_q = -\hat{R}_s i_q - \omega_r \hat{L}_d i_d - \omega_r \lambda_m + G_v . v_{dr}, \tag{7.3.27}$$

where G_v models the inverter as a constant gain and v_{dr} and v_{qr} are the outputs of decoupling current controllers as will be explained. In contrast to the previous scheme, the estimated parameters, instead of the actual ones, are used in the reference model. The model is coupled due to the speed voltages in the d- and q-axis equations. These voltages cause the dependency of v_d on i_q and v_q on i_d. A feedforward decoupling technique was presented in sub-section 3.4.2 to simplify the design of current controllers. Here, an alternative decoupling technique is employed based on decoupling feedback signals, as in Fig. 7.19. The decoupling eliminates the undesirable interaction between d and q current control loops by removing the speed voltages from the system model. The feedback gain matrix, $\boldsymbol{K}$, and the resulting decoupled model is given by

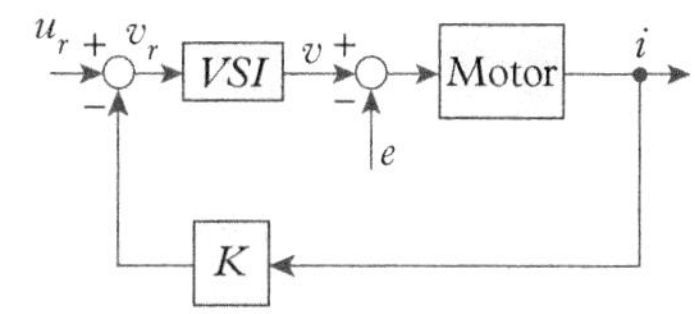

Figure 7.19 *Feedback decoupling technique.*

$$\boldsymbol{K} = \frac{\omega_r}{G_v}\begin{bmatrix} 0 & \hat{L}_q \\ -\hat{L}_d & 0 \end{bmatrix}, \tag{7.3.28}$$

$$\hat{L}_d.pi_d = -\hat{R}_s i_d + G_v.v_{dr}, \tag{7.3.29}$$

$$\hat{L}_q.pi_q = -\hat{R}_s i_q - \omega_r \lambda_m + G_v.v_{dr}. \tag{7.3.30}$$

The decoupling ensures that the estimation of machine parameters to be independent of each other and prevents the influence of error in the initial values of one parameter on the estimation convergence of the other.

The estimator is driven by the current error vector to reduce the error. The vector is defined as

$$\tilde{\boldsymbol{i}}_s = \hat{\boldsymbol{i}}_s - \boldsymbol{i}_s = \begin{bmatrix} \tilde{i}_d & \tilde{i}_q \end{bmatrix}^T. \tag{7.3.31}$$

The adaptation law is chosen as integration of the current error to correct the estimated parameters. For salient machines, R_s and L_q are estimated by the error of i_q and L_d is estimated by the error of i_d. Thus, the estimated values of R_s, L_d, and L_q are given by

$$\hat{R}_s(t) = \hat{R}_s(t_0) - \int_{t_0}^{t} K_R.\tilde{i}_q(\sigma)\,.\,d\sigma, \tag{7.3.32}$$

$$\hat{L}_d(t) = \hat{L}_d(t_0) - \int_{t_0}^{t} K_{Ld}.\tilde{i}_q(\sigma)\,.\,d\sigma, \tag{7.3.33}$$

$$\hat{L}_q(t) = \hat{L}_q(t_0) - \int_{t_0}^{t} K_{Lq}.\tilde{i}_d(\sigma)\,.\,d\sigma, \tag{7.3.34}$$

where K_R, K_{Ld}, and K_{Lq} are the estimator gains determining the convergence rate of the estimation. These are real numbers and must satisfy the following conditions for the sake of estimation convergence:

$$K_{Ld}.\omega_r i_d < 0, \tag{7.3.35}$$

$$K_{Lq}.\omega_r i_q < 0, \tag{7.3.36}$$

$$K_R.i_q < 0. \tag{7.3.37}$$

In the case of non-salient machines, $L_d = L_q = L_s$. As a result, the machine model and the decoupling feedback signals are modified accordingly. Consequently, the parameter estimation laws reduce to

$$\hat{R}_s(t) = \hat{R}_s(t_0) - \int_{t_0}^{t} K_R.\tilde{i}_q(\sigma)\,.\,d\sigma, \tag{7.3.38}$$

$$\hat{L}_s(t) = \hat{L}_s(t_0) - \int_{t_0}^{t} K_L.\tilde{i}_d(\sigma)\,.\,d\sigma. \tag{7.3.39}$$

Again, the following gain conditions must be satisfied for the sake of estimation convergence:

$$\omega_r.K_L.i_q > K_R.i_q, \tag{7.3.40}$$

$$\omega_r.K_L.K_R < 0. \tag{7.3.41}$$

The following gains may be chosen to satisfy eqns (7.3.40) and (7.3.41), respectively,

$$K_R = -K_{R0}.sign\left(i_q\right), \tag{7.3.42}$$

$$K_L = K_{L0}.sign\left(\omega_r i_q\right), \tag{7.3.43}$$

where K_{R0} and K_{L0} are positive real numbers. The overall motor parameter estimation and control system is depicted in Fig. 7.20.

7.3.4 RLS scheme for estimation of motor inductances and R_s

Estimation of R_s, L_d, and L_q can be carried out by the RLS method without the need for speed and position signals (Ichikawa *et al.* 2006). The method is thus insensitive to position and speed errors. The machine magnet flux linkage is assumed to be available in this method. The estimation calculations are carried out by using a machine model in rotor reference frame. An estimated rotor position, instead of an actual rotor position, is used for the reference frame transformation as in

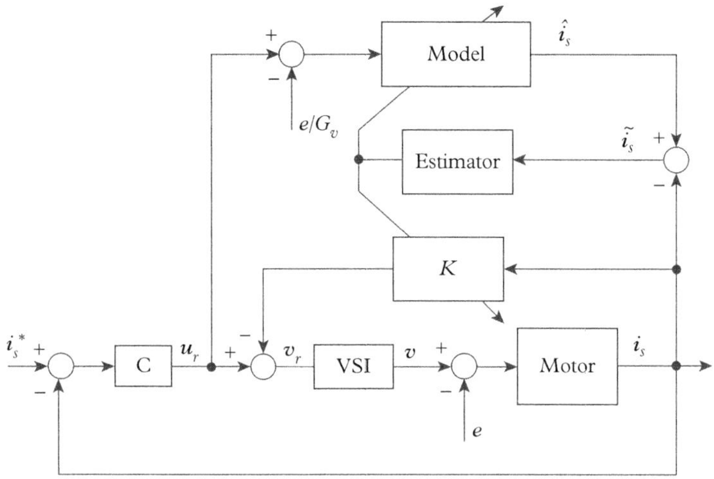

Figure 7.20 *Overall motor parameter estimation and control system based on feedback decoupling.*

Fig. 6.3. A discrete machine model in the aforementioned reference frame can be presented as

$$\begin{bmatrix} \hat{\imath}_d(k+1) \\ \hat{\imath}_q(k+1) \end{bmatrix} = \boldsymbol{A} \begin{bmatrix} \hat{\imath}_d(k) \\ \hat{\imath}_q(k) \end{bmatrix} + \boldsymbol{B} \begin{bmatrix} \hat{v}_d(k) \\ \hat{v}_q(k) \end{bmatrix} + \boldsymbol{C}[1], \qquad (7.3.44)$$

where

$$\boldsymbol{A} = \begin{bmatrix} a_{11} & a_{12} \\ a_{21} & a_{22} \end{bmatrix} = \frac{R_s L_1 \Delta T + L_d L_q}{L_d L_q} \boldsymbol{I} + \frac{R_s L_2 \Delta T}{L_d L_q} \boldsymbol{Q}$$

$$+ \frac{\omega_r (L_d^2 + L_q^2) \Delta T}{2 L_d L_q} \boldsymbol{J} - \frac{\omega_r (L_d^2 - L_q^2) \Delta T}{2 L_d L_q} \boldsymbol{S}, \qquad (7.3.45)$$

$$\boldsymbol{B} = \begin{bmatrix} b_{11} & b_{12} \\ b_{21} & b_{22} \end{bmatrix} = \frac{L_1 \Delta T}{L_d L_q} \boldsymbol{I} - \frac{L_2 \Delta T}{L_d L_q} \boldsymbol{Q}, \qquad (7.3.46)$$

$$\boldsymbol{C} = \begin{bmatrix} c_1 \\ c_2 \end{bmatrix} = \frac{\omega_r \lambda_m \Delta T}{L_q} \begin{bmatrix} \sin \Delta\theta_r \\ -\cos \Delta\theta_r \end{bmatrix}, \qquad (7.3.47)$$

and $\boldsymbol{I}$ and $\boldsymbol{J}$ are defined in eqn (7.3.10). In addition, $\boldsymbol{Q}$ and $\boldsymbol{S}$ are given as

$$\boldsymbol{Q} = \begin{bmatrix} \cos \Delta\theta_r & -\sin \Delta\theta_r \\ -\sin \Delta\theta_r & -\cos \Delta\theta_r \end{bmatrix}, \quad \boldsymbol{S} = \begin{bmatrix} \sin \Delta\theta_r & \cos \Delta\theta_r \\ \cos \Delta\theta_r & -\sin \Delta\theta_r \end{bmatrix}. \qquad (7.3.48)$$

Equation (7.3.44) is transformed into a new form as

$$y = \boldsymbol{\Theta} z, \tag{7.3.49}$$

where y and z are vectors of motor current and voltage components and Θ is a matrix that contains motor parameters as

$$y = \begin{bmatrix} \hat{\imath}_d(k+1) & \hat{\imath}_q(k+1) \end{bmatrix}^T, \tag{7.3.50}$$

$$z = \begin{bmatrix} \hat{\imath}_d(k) & \hat{\imath}_q(k) & \hat{v}_d(k) & \hat{v}_q(k) & 1 \end{bmatrix}^T, \tag{7.3.51}$$

$$\boldsymbol{\Theta} = \begin{bmatrix} A & B & C \end{bmatrix} = \begin{bmatrix} a_{11} & a_{12} & b_{11} & b_{12} & c_1 \\ a_{21} & a_{22} & b_{21} & b_{22} & c_2 \end{bmatrix}. \tag{7.3.52}$$

Now an error matrix is defined as the differences of the estimated system output and the actual output. Referring to the system output of eqn (7.3.49), the square value of the error is formed as

$$\epsilon_i = \left(y - \hat{\boldsymbol{\Theta}} z\right)^2. \tag{7.3.53}$$

Then the parameter matrix is calculated as the error reaches a minimum. This is achieved by an RLS method according to the relations

$$\hat{\boldsymbol{\Theta}}(k) = \hat{\boldsymbol{\Theta}}(k-1) + \left(y - \hat{\boldsymbol{\Theta}}(k-1)\, z\right) z^T \boldsymbol{P}(k), \tag{7.3.54}$$

$$\boldsymbol{P}(k) = \frac{1}{\lambda}\left\{\boldsymbol{P}(k-1) - \boldsymbol{P}(k-1)\, z \left(\lambda + z^T \boldsymbol{P}(k-1)\, z\right)^{-1} z^T \boldsymbol{P}(k-1)\right\}, \tag{7.3.55}$$

where $\boldsymbol{P}$ is a correction gain matrix and λ is the forgetting factor, i.e., a weighting coefficient, which gives higher weights to more recent data and deletes the past data. This means that the method does not assume constant parameters during the estimation procedure.

The parameter matrix depends on the rotor position and speed. It is possible to obtain relationships with no position and speed dependency by the following operations on the matrix elements (Ichikawa *et al.* 2006),

$$M_1 = b_{11} + b_{22} = \frac{2L_1 \Delta T}{L_d L_q}, \tag{7.3.56}$$

$$M_2 = a_{11} + a_{22} - 2 = -\frac{2R_s L_1 \Delta T}{L_d L_q}, \tag{7.3.57}$$

$$M_3 = \sqrt{(b_{11} - b_{22})^2 + (b_{12} + b_{21})^2} = -\frac{2L_2 \Delta T}{L_d L_q}, \tag{7.3.58}$$

where ΔT is the sampling period of the identification system. The estimated parameters can then be calculated from eqns (7.3.56)–(7.3.58) without the need to have any information of position and speed as follows:

$$\hat{R}_s = \frac{-M_2}{M_1}, \tag{7.3.59}$$

$$\hat{L}_d = \frac{2\Delta T}{M_1 + M_3}, \tag{7.3.60}$$

$$\hat{L}_q = \frac{2\Delta T}{M_1 - M_3}. \tag{7.3.61}$$

The RLS parameter estimation can be implemented in connection with a VC system in rotor reference frame, as in Fig. 7.21. The current commands are generated by torque (speed) and flux controllers as is common in VC systems. Then, an injected current signal is added to both current commands. The signal, which is small in comparison with the motor rated current, is required to identify motor parameters. The injected signal consists of M-sequence signals. They are pseudorandom sequence signals used in system identifications.

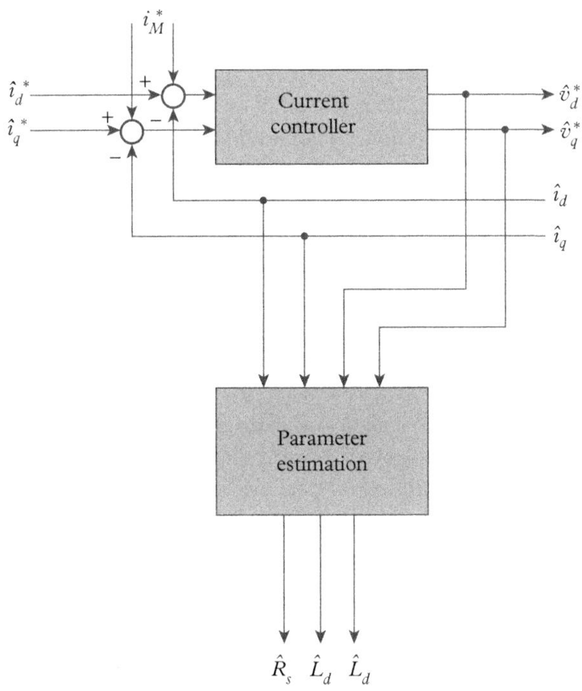

Figure 7.21 *The RLS estimation of motor parameters in connection with a current VC system.*

The RLS parameter estimator receives actual current components in the rotor reference frame in addition to the voltage commands as the outputs of the decoupling current controllers to estimate stator resistance and inductances.

7.3.5 RLS scheme for estimation of motor inductances and λ_m

The RLS parameter estimation scheme presented estimates the stator resistance and machine inductances, but not the magnet flux linkage magnitude. The latter parameter must be identified by offline measurement if the previous scheme is used. Thus, it may vary from its measured value during the machine operation due to changes in the machine operating point and influences the estimation accuracy of the estimated parameters. Here, a double RLS scheme is presented to estimate all four machine parameters, including the magnet flux linkage magnitude (Underwood and Husain 2010). Both IPM and surface-mounted PMS machines are studied. The two RLS algorithms, a slow one and a fast one, are carried out simultaneously. The slow algorithm is used for temperature-dependent parameters, R_s, and magnet flux linkage magnitude, λ_m, which have slow variations, while the fast algorithm is used for current-dependent parameters, L_d and L_q (L_s in the case of non-salient machines) which have rapid variations. In this way, the computation burden of the estimation reduces substantially compared to a potential case where all parameters are estimated by a single fast algorithm. In fact, the RLS method is generally carried out by online matrix computation, including matrix inversion. The single algorithm case leads to oversized matrices and increased computation effort. This is difficult to do during a short sampling time of the motor drive control system. Also, solving a single set of equations for all four parameters is confronted by lack of consistency, since there are only two electrical equations and four unknown parameters. The obstacle may be overcome by using information from a previous sampling to form four equations. However, it increases the risk of estimation instability. The double algorithm scheme estimates two parameters by each algorithm. The fast algorithm runs in every sampling period and the slow algorithm runs every several sampling periods.

Each RLS algorithm uses the relations (Trigeassou *et al.* 2009)

$$\hat{\boldsymbol{\Theta}}(k) = \hat{\boldsymbol{\Theta}}(k-1) + \boldsymbol{K}(k)\,.\,\epsilon(k), \tag{7.3.62}$$

$$\epsilon(k) = y(k) - \boldsymbol{\varphi}^T(k)\,.\,\hat{\boldsymbol{\Theta}}(k-1), \tag{7.3.63}$$

$$K(k) = P(k-1)\,.\,\varphi(k)\,.\left[\lambda\,.\,I + \varphi^T(k)\,.\,P(k-1)\,.\,\varphi(k)\right]^{-1}, \tag{7.3.64}$$

$$P(k) = \left[I - K(k)\,.\,\varphi^T(k)\right]^{-1}.\,P(k-1)/\lambda, \tag{7.3.65}$$

where $\boldsymbol{y}$ is the output matrix, $\hat{\boldsymbol{\Theta}}$ is the estimated parameter vector, $\boldsymbol{\varphi}$ is the feedback matrix, λ is the forgetting factor, $\boldsymbol{I}$ is the identity matrix, and ϵ is the estimation error. Also, $\boldsymbol{K}$ and $\boldsymbol{P}$ are correction gain matrices.

The fast algorithm makes use of the decoupling circuit of the current control system, as presented in sub-section 3.4.2. The decoupling circuit separates the speed voltages from the dynamic part of the machine model. As a result, the dynamic part of the machine voltage equations in the rotor reference frame can be presented by v_{q1} and v_{d1} voltages as

$$v_{q1} = (R_s + L_q p)i_q, \tag{7.3.66}$$

$$v_{d1} = (R_s + L_d p)i_d. \tag{7.3.67}$$

Using eqns (7.3.66) and (7.3.67) to estimate L_d and L_q eliminates speed variations from the estimation. Thus, the motor speed can be regarded as constant during the execution of the fast RLS algorithm. Also, using eqns (7.3.66) and (7.3.67) in the fast algorithm, instead of full voltage equations, speeds up the computations. This is consistent with the short execution interval of the algorithm, which is included in the current sampling period.

Using the linear equations (7.3.66) and (7.3.67), the fast RLS algorithm is presented by the matrices (Underwood and Husain 2010)

$$\boldsymbol{y} = \begin{bmatrix} v_q - R_s i_q - \lambda_m.\omega_r \\ v_d - R_s.i_d \end{bmatrix} \tag{7.3.68}$$

$$\boldsymbol{\varphi}^T = \begin{bmatrix} 0 & \omega_r i_d \\ -\omega_r i_q & 0 \end{bmatrix} \tag{7.3.69}$$

$$\boldsymbol{\Theta} = \begin{bmatrix} L_q \\ L_d \end{bmatrix}. \tag{7.3.70}$$

The following matrices are also used by the slow RLS algorithm (Underwood and Husain 2010):

$$\boldsymbol{y} = \begin{bmatrix} v_q \\ v_d \end{bmatrix}, \tag{7.3.71}$$

$$\boldsymbol{\varphi}^T = \begin{bmatrix} 0 & \omega_r.i_d & i_q & \omega_r \\ -\omega_r.i_q & 0 & i_d & 0 \end{bmatrix}, \tag{7.3.72}$$

$$\boldsymbol{\Theta} = \begin{bmatrix} L_q \\ L_d \\ R_s \\ \lambda_m \end{bmatrix}. \tag{7.3.73}$$

A small signal is added to a current command signal to provide rich data for the slow RLS algorithm. It must be a stepwise signal with small magnitude in order not to disturb the machine torque, but large enough to be felt at the current feedback.

The two RLS algorithms work under a supervision program, as shown in Fig. 7.22. The program manages the data exchange between the algorithms and controls the injected signal. The fast algorithm, running at the sampling rate or the PWM switching frequency, e.g., 10–20 kH, estimates the inductances and sends them to the slow RLS algorithm. The slow algorithm, running at a much lower rate, e.g., 500 Hz, estimates all parameters and sends them to the fast algorithm and updates its inductance estimates with those it receives from the fast algorithm. At the start of the estimation process, there are rough estimates of the parameters, usually from the machine data sheet.

7.3.6 EKF scheme

Extended Kalman filter can be employed to estimate the parameters of PMS machines. An estimation scheme may include the estimated

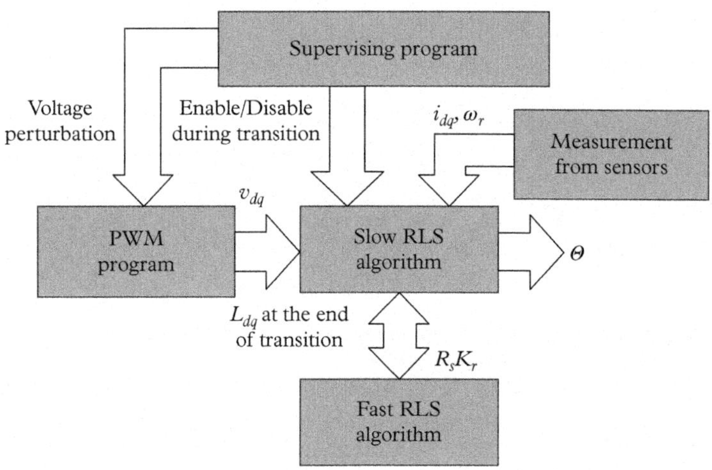

Figure 7.22 *The structure of the double RLS parameter estimation.*

parameters in the state vector together with the motor variables. However, this will increase the size of the state vector and the matrices of EKF, thus causing computational burden dramatically. One solution is to distinguish the variables and parameters in the state vector by their dynamics (Boileau *et al.* 2011). This scheme is employed in parameter estimation of surface-mounted PMS motors. The stator resistance and the synchronous inductance are estimated by including them in the state vector as R_s/L_s and $1/L_s$ together with the stator current components. The motor speed variation, the motor parameter variations, and the current dynamics are assumed very slow, slow, and rapid, respectively. This assumption on timescale of the variable variations leads to the following motor model for the EKF algorithm,

$$p\begin{bmatrix} i_d \\ i_q \\ a \\ b \end{bmatrix} = \begin{bmatrix} -a.i_d + \omega_r.i_q + b.G_v\ v_{dr} \\ -a.i_q - \omega_r.i_d + b.(G_v\ v_{qr} - \omega_r\lambda_m) \\ 0 \\ 0 \end{bmatrix}, \tag{7.3.74}$$

$$y = \begin{bmatrix} 1 & 0 & 0 & 0 \\ 0 & 1 & 0 & 0 \end{bmatrix}\begin{bmatrix} i_d \\ i_q \\ a \\ b \end{bmatrix}, \tag{7.3.75}$$

where $a = R_s/L_s$ and $b = 1/L_s$.

The non-salient PMS motors usually work under $i_d = 0$. Therefore, the conditions for the identifiability of the parameters are that ω_r and i_q are non-zero. A feedback decoupling control technique that improves convergence dynamics and overall system stability, the same as that presented in connection with the MRAS method in this chapter, is used. The feedback decoupling reduces the machine model of eqn (7.3.74) to

$$p\begin{bmatrix} i_d \\ i_q \\ a \\ b \end{bmatrix} = \begin{bmatrix} -a.i_d + b.G_v\ v_{dr} \\ -a.i_q + b.(G_v\ v_{qr} - \omega_r\lambda_m) \\ v0 \\ 0 \end{bmatrix}, \tag{7.3.76}$$

with the same output vector, $\boldsymbol{y}$, as in eqn (7.3.75). The new current controller outputs are v_{dr} and v_{qr}. Having the model, the EKF can be realized by deciding on the initial covariance matrix, $\boldsymbol{P}_0$ and its weighting matrices, $\boldsymbol{Q}$ and $\boldsymbol{R}$. The matrices can be chosen by some means including trial and error. The following matrices have been presented for a certain PMS motor (Boileau *et al.* 2011):

$$P_0 = diag\,(0.1, 0.1, 100, 100)\,, \tag{7.3.77}$$

$$Q = diag\,(1, 1, 500, 500)\,, \tag{7.3.78}$$

$$R = diag\,(1, 1)\,. \tag{7.3.79}$$

The estimator performance shows that the convergence of the estimated parameters are coupled; i.e., the resistance error cannot vanish, unless the inductance error vanishes.

The magnet flux linkage variations also affect the estimation. A method for taking the flux linkage variations into account in the EKF resistance and inductance estimation has been proposed (Boileau *et al.* 2011).

7.4 Summary

In this chapter, the estimation methods of PMS motor parameters are presented. The motor parameters are explained first and their importance in modeling the motors is elaborated upon. The parameters vary with motor operating conditions. These variations affect the motor performances, as presented with regard to minimum loss operation of PMS motors.

The PMS motor parameter estimation schemes are divided into two main methods, i.e., offline method and online method.

Several offline schemes for determining all major motor parameters are presented in these chapters. The offline schemes can be divided into DC standstill and AC standstill tests, no-load test, load test, and VC scheme. These schemes use equivalent circuit equations or vector diagram equations in connection with measurement of voltage and current to calculate motor parameters for a particular operating point or over a range of operating points. They may need special test arrangement for measuring machine voltage and current.

Online parameter estimation schemes use the online values of motor variables usually used in the control system to estimate motor parameters with the help of closed-loop observers. A number of online schemes used to estimate PMS motor parameters are presented in this chapter. These include closed-loop observers for estimation of motor inductances; MRAS-based estimation of λ_m and R_s, and also motor inductances; RLS scheme for estimation of motor inductances, and also, for estimation of inductances and magnet flux linkage; and finally, the EKF scheme. The online schemes take into account the motor parameter variations caused by any source. They are being implemented as a part of modern control systems in the market.

PROBLEMS

P.7.1. Calculate motor parameters in a two-axis stationary reference frame by an offline scheme.

P.7.2. Investigate the effects of inductance variations on MTPA operation of IPM machines.

P.7.3. Investigate the effects of inductance variations on unity power factor conditions of IPM machines.

P.7.4. Investigate the effects of variations in magnet flux linkage magnitude on MTPA operation of IPM machines.

P.7.5. Consider the effects of error in magnet flux linkage magnitude on the estimation of PMS motor inductances under VC.

P.7.6. Investigate parameter estimation of PMS motors under VC in a stator flux linkage reference frame.

BIBLIOGRAPHY

Boileau, T., Leboeuf, N., Nahid-Mobarakeh, B., and Meibody-Tabar, F. (2011). Online identification of PMSM parameters, parameter identifiability and estimator comparative study. *IEEE Trans. Ind. Appl.* 47(4), 1944–1957.

Bolognani, S., Zigliotto, M., and Unterkofler, K. (1997). On-line parameter commissioning in sensorless PMSM drives. In: *Proceedings of the IEEE International Symposium on Industrial Electronics*, pp. 480–484. IEEE, Piscataway, NJ.

Chang, L. (1996). An improved FE inductance calculation for electrical machines. *IEEE Trans. Magnet.* 32(4), 3237–3245.

Chiasson, J. (2005). *Modeling and High Performance Control of Electric Machines*. Wiley-IEEE Press, Hoboken, NJ.

Du, T., Vas, P., and Stronach, F. (1995). Design and application of extended observers for joint state and parameter estimation in high-performance AC drives. *IEEE Proc. Electric Power Appl.* 142(2), 71–78.

Dutta, R. and Rahman, M. (2006). A comparative analysis of two test methods of measuring d-and q-axes inductances of interior permanent-magnet machine. *IEEE Trans. Magnet.* 42(11), 3712–3718.

Feng, Y., Yu, X., and Han, F. (2013). High-order terminal sliding-mode observer for parameter estimation of a permanent-magnet synchronous motor. *IEEE Trans. Ind. Electron.* 60(10), 4272–4280.

Gopalarathnam, T. and McCann, R. (2001). Saturation and armature reaction effects in surface-mount PMAC motors. In: *IEEE International Electric Machines and Drives Conference*, pp. 618–621. IEEE, Piscataway, NJ.

Hamida, M.A., De Leon, J., Glumineau, A., and Boisliveau, R. (2013). An adaptive interconnected observer for sensorless control of PM synchronous motors with online parameter identification. *IEEE Trans. Ind. Electron.* 60(2), 739–748.

Haque, M.E. and Rahman, M.F. (1999). Dynamic model and parameter measurement of interior permanent magnet synchronous motor. In: *Proceedings of the Australasian Universities Power Engineering Conference*, 10, pp. 10–13. Northern Territory University, Darwin, Australia.

Husson, R. (2013). *Control Methods for Electrical Machines.* John Wiley & Sons.

Hwang, C.-C. and Cho, Y. (2001). Effects of leakage flux on magnetic fields of interior permanent magnet synchronous motors. *IEEE Trans. Magnet.* 37(4), 3021–3024.

Ichikawa, S., Tomita, M., Doki, S., and Okuma, S. (2006). Sensorless control of permanent-magnet synchronous motors using online parameter identification based on system identification theory. *IEEE Trans. Ind. Electron.* 53(2), 363–372.

Inoue, Y., Kawaguchi, Y., Morimoto, S., and Sanada, M. (2011). Performance improvement of sensorless IPMSM drives in a low-speed region using online parameter identification. *IEEE Trans. Ind. Appl.* 47(2), 798–804.

Kilthau, A. and Pacas, J. (2001). Parameter-measurement and control of the synchronous reluctance machine including cross saturation. In: *Conference Record of the 2001 Industry Applications Conference, 36th IAS Annual Meeting*, pp. 2302–2309. IEEE, Piscataway, NJ.

Kim, H., Hartwig, J., and Lorenz, R. D. (2002). Using on-line parameter estimation to improve efficiency of IPM machine drives. In: *IEEE 33rd Annual Power Electronics Specialists Conference*, pp. 815–20. IEEE, Piscataway, NJ.

Kim, H. and Lorenz, R. D. (2002). Improved current regulators for IPM machine drives using on-line parameter estimation. In: *Conference Record of the 2002 Industry Applications Conference, 37th IAS Annual Meeting*, pp. 86–91. IEEE, Piscataway, NJ.

Kim, K-H., Chung, S-K., Moon, G-W., Baik, I-C., and Youn, M-J. (1995). Parameter estimation and control for permanent magnet synchronous motor drive using model reference adaptive technique. In: *Proceedings of the IEEE IECON 21st International Conference on Industrial Electronics, Control, and Instrumentation*, pp. 387–392. IEEE, Piscataway, NJ.

Kubota, H., Matsuse, K., and Nakano, T. (1993). DSP-based speed adaptive flux observer of induction motor. *IEEE Trans. Ind. Appl.* 29(2), 344–348.

Lee, K-W., Jung, D-H., and Ha, I-J. (2004). An online identification method for both stator resistance and back-EMF coefficient of PMSMs without rotational transducers. *IEEE Trans. Ind. Electron.* 51(2), 507–510.

Liu, K. (2013). *Novel Techniques for Parameter Estimation of Permanent Magnet Synchronous Machines*. University of Sheffield.

Liu, K. and Zhang, J. (2010). Adaline neural network based online parameter estimation for surface-mounted permanent magnet synchronous machines. *Proc. CSEE* 30(30), 68–73.

Liu, K., Zhang, Q., Chen, J., Zhu, Z., and Zhang, J. (2011). Online multiparameter estimation of nonsalient-pole PM synchronous machines with temperature variation tracking. *IEEE Trans. Ind. Electron.* 58(5), 1776–1788.

Liu, K. and Zhu, Z. (2015a). Mechanical parameter estimation of permanent-magnet synchronous machines with aiding from estimation of rotor pm flux linkage. *IEEE Trans. Ind. Appl.* 51(4), 3115–3125.

Liu, K. and Zhu, Z. (2015b). Position offset-based parameter estimation for permanent magnet synchronous machines under variable speed control. *IEEE Trans. Power Electron.* 30(6), 3438–3446.

Liu, K. and Zhu, Z. (2015c). Position-offset-based parameter estimation using the adaline NN for condition monitoring of permanent-magnet synchronous machines. *IEEE Trans. Ind. Electron.* 62(4), 2372–2383.

Liu, K., Zhu, Z., Zhang, Q., and Zhang, J. (2012). Influence of nonideal voltage measurement on parameter estimation in permanent-magnet synchronous machines. *IEEE Trans. Ind. Electron.* 59(6), 2438–2447.

Liu, T., Husain, I., and Elbuluk, M. (1998). Torque ripple minimization with on-line parameter estimation using neural networks in permanent magnet synchronous motors. In: *Conference Record of 1998 IEEE Industry Applications Conference, 33rd IAS Annual Meeting*, pp. 35–40. IEEE, Piscataway, NJ.

Lovelace, E.C., Jahns, T.M., and Lang, J.H. (2002). A saturating lumped-parameter model for an interior PM synchronous machine. *IEEE Trans. Ind. Appl.* 38(3), 645–650.

Mellor, P., Chaaban, F., and Binns, K. (1991). Estimation of parameters and performance of rare-earth permanent-magnet motors avoiding measurement of load angle. *IEEE Proc. B-Electric Power Appl.* 138(6), 322–330.

Mohammed, O., Liu, S., and Liu, Z. (2004). Phase-variable model of PM synchronous machines for integrated motor drives. *IEEE Proc. Sci. Measur. Technol.* 151(6), 423–429.

Morimoto, S., Hatanaka, K., Tong, Y., Takeda, Y., and Hirasa, T. (1993a). Servo drive system and control characteristics of salient pole permanent magnet synchronous motor. *IEEE Trans. Ind. Appl.* 29(2), 338–343.

Morimoto, S., Ueno, T., Sanada, M., Yamagiwa, A., Takeda, Y., and Hirasa, T. (1993b). Effects and compensation of magnetic saturation in permanent magnet synchronous motor drives. In: *Conference Record of the IEEE Industry Applications Society Annual Meeting*, pp. 59–64. IEEE, Piscataway, NJ.

Morimoto, S., Sanada, M., and Takeda, Y. (2006). Mechanical sensorless drives of IPMSM with online parameter identification. *IEEE Trans. Ind. Appl.* 42(5), 1241–1248.

Nee, H.-P., Lefevre, L., Thelin, P., and Soulard, J. (2000). Determination of d and q reactances of permanent-magnet synchronous motors without measurements of the rotor position. *IEEE Trans. Ind. Appl.* 36(5), 1330–1335.

Obe, E. (2009). Direct computation of ac machine inductances based on winding function theory. *Energy Convers. Manag.* 50(3), 539–542.

Piippo, A., Hinkkanen, M., and Luomi, J. (2009). Adaptation of motor parameters in sensorless PMSM drives. *IEEE Trans. Ind. Appl.* 45(1), 203–212.

Rafaq, M., Mwasilu, F., Kim, J., ho Choi, H., and Jung, J. W. (2016). Online Parameter identification for model-based sensorless control of interior permanent magnet synchronous machine. *IEEE Trans. Power Electron.* 32(6), 4631–4643.

Rahman, K.M. and Hiti, S. (2005). Identification of machine parameters of a synchronous motor. *IEEE Trans. Ind. Appl.* 41(2), 557–565.

Rahman, M. and Zhou, P. (1994). Accurate determination of permanent magnet motor parameters by digital torque angle measurement. *J. Appl. Phys.* 76(10), 6868–6870.

Ruoho, S., Kolehmainen, J., Ikaheimo, J., and Arkkio, A. (2010). Interdependence of demagnetization, loading, and temperature rise in a permanent-magnet synchronous motor. *IEEE Trans. Magnet.* 46(3), 949–953.

Salvatore, L. and Stasi, S. (1992). Application of EKF to parameter and state estimation of PMSM drive. *IEEE Proc. B-Electric Power Appl.* 139(3), 155–164.

Schaible, U. and Szabados, B. (1999). Dynamic motor parameter identification for high speed flux weakening operation of brushless permanent magnet synchronous machines. *IEEE Trans. Energy Convers.* 14(3), 486–492.

Shi, Y., Sun, K., Huang, L., and Li, Y. (2012). Online identification of permanent magnet flux based on extended Kalman filter for IPMSM drive with position sensorless control. *IEEE Trans. Ind. Electron.* 59(11), 4169–4178.

Stumberger, B., Stumberger, G., Dolinar, D., Hamler, A., and Trlep, M. (2003). Evaluation of saturation and cross-magnetization effects in interior permanent-magnet synchronous motor. *IEEE Trans. Ind. Appl.* 39(5), 1264–1271.

Trigeassou, J.-C., Poinot, T., and Bachir, S. (2009). Parameter estimation for knowledge and diagnosis of electrical machines. In: *Control Methods for Electrical Machines*, pp. 207–243. ISTE Ltd and John Wiley & Sons Inc, Hoboken, NJ.

Vaez-Zadeh, S. and Zamanifar, M. (2006). *Efficiency optimization control of IPM synchronous motor drives with online parameter estimation.* Technical Report, University of Tehran, School of ECE.

Vaez-Zadeh, S., Zamanifar, M., and Soltani, J. (2006). Nonlinear efficiency optimization control of IPM synchronous motor drives with online parameter estimation. In: *2006 37th IEEE Power Electronics Specialists Conference*, pp. 1–6. IEEE, Piscataway, NJ.

Uddin, M.N. and Chy, M.M.I. (2008). Online parameter-estimation-based speed control of PM AC motor drive in flux-weakening region. *IEEE Trans. Ind. Appl.* 44(5), 1486–1494.

Underwood, S. J. and Husain, I. (2010). Online parameter estimation and adaptive control of permanent-magnet synchronous machines. *IEEE Trans. Ind. Electron.* 57(7), 2435–2443.

Urasaki, M., Senjyu, T., and Uezato, K. (2003). A novel calculation method for iron loss resistance suitable in modelling permanent magnet synchronous motors. *IEEE Trans. Energy Convers.* 18(1), 41–47.

Zhou, P., Rahman, M.A., and Jabbar, M.A. (1994). Field circuit analysis of permanent magnet synchronous motors. *IEEE Trans. Magnet.* 30(4), 1350–1359.

Index

R

S

T

V

W

Z

The manufacturer's authorised representative in the EU for product safety is Oxford University Press España S.A. of El Parque Empresarial San Fernando de Henares, Avenida de Castilla, 2 - 28830 Madrid (www.oup.es/en or product.safety@oup.com). OUP España S.A. also acts as importer into Spain of products made by the manufacturer.
Printed and bound by CPI Group (UK) Ltd, Croydon, CR0 4YY
14/09/2026
02216284-0015